Lecture Notes in Physics

Springer-Verlag Berlin Heidelberg GmbH

The Editorial Policy for Proceedings

The series Lecture Notes in Physics reports new developments in physical research and teaching – quickly, informally, and at a high level. The proceedings to be considered for publication in this series should be limited to only a few areas of research, and these should be closely related to each other. The contributions should be of a high standard and should avoid lengthy redraftings of papers already published or about to be published elsewhere. As a whole, the proceedings should aim for a balanced presentation of the theme of the conference including a description of the techniques used and enough motivation for a broad readership. It should not be assumed that the published proceedings must reflect the conference in its entirety. (A listing or abstracts of papers presented at the meeting but not included in the proceedings could be added as an appendix.)
When applying for publication in the series Lecture Notes in Physics the volume's editor(s) should submit sufficient material to enable the series editors and their referees to make a fairly accurate evaluation (e.g. a complete list of speakers and titles of papers to be presented and abstracts). If, based on this information, the proceedings are (tentatively) accepted, the volume's editor(s), whose name(s) will appear on the title pages, should select the papers suitable for publication and have them refereed (as for a journal) when appropriate. As a rule discussions will not be accepted. The series editors and Springer-Verlag will normally not interfere with the detailed editing except in fairly obvious cases or on technical matters.
Final acceptance is expressed by the series editor in charge, in consultation with Springer-Verlag only after receiving the complete manuscript. It might help to send a copy of the authors' manuscripts in advance to the editor in charge to discuss possible revisions with him. As a general rule, the series editor will confirm his tentative acceptance if the final manuscript corresponds to the original concept discussed, if the quality of the contribution meets the requirements of the series, and if the final size of the manuscript does not greatly exceed the number of pages originally agreed upon. The manuscript should be forwarded to Springer-Verlag shortly after the meeting. In cases of extreme delay (more than six months after the conference) the series editors will check once more the timeliness of the papers. Therefore, the volume's editor(s) should establish strict deadlines, or collect the articles during the conference and have them revised on the spot. If a delay is unavoidable, one should encourage the authors to update their contributions if appropriate. The editors of proceedings are strongly advised to inform contributors about these points at an early stage .
The final manuscript should contain a table of contents and an informative introduction accessible also to readers not particularly familiar with the topic of the conference. The contributions should be in English. The volume's editor(s) should check the contributions for the correct use of language. At Springer-Verlag only the prefaces will be checked by a copy-editor for language and style. Grave linguistic or technical shortcomings may lead to the rejection of contributions by the series editors. A conference report should not exceed a total of 500 pages. Keeping the size within this bound should be achieved by a stricter selection of articles and not by imposing an upper limit to the length of the individual papers Editors receive jointly 30 complimentary copies of their book. They are entitled to purchase further copies of their book at a reduced rate. As a rule no reprints of individual contributions can be supplied No royalty is paid on Lecture Notes in Physics volumes. Commitment to publish is made by letter of interest rather than by signing a formal contract. Springer Verlag secures the copyright for each volume.

The Production Process

The books are hardbound, and the publisher will select quality paper appropriate to the needs of the author(s). Publication time is about ten weeks. More than twenty years of experience guarantee authors the best possible service. To reach the goal of rapid publication at a low price the technique of photographic reproduction from a camera-ready manuscript was chosen. This process shifts the main responsibilit for the technical quality considerably from the publisher to the authors. We therefore urge all authors and editors o proceedings to observe very carefully the essentials for the preparation of camera-ready manuscripts, which will supply on request. This applies especially to the quality of figures and halftones submitted or publication. In addition, it might be useful to look at some of the volumes already published. As a special service, we offer free of charge LATEX and TEX macro packages to format the text according to Springer-Verlag's qual ty requirements. We strongly recommend that you make use of this offer, since the result will be a book of considerably improved technical quality. To avoid mistakes and time-consuming correspondence durin the pro uction period the conference editors should request special instructions from the publisher well before the beginn ng of the conference. Manuscripts not meeting the technical standard of the series will have to be returned for improvement.

For further information please contact Springer-Verlag, Physics Editorial Department II, Tiergartenstrasse 17, D-69121 Heidelberg, Germany

Jean Claude Vial   Karine Bocchialini
Patrick Boumier (Eds.)

Space Solar Physics

Theoretical and Observational Issues in the Context of the SOHO Mission

Proceedings of a Summer School
Held in Orsay, France, 1–13 September 1997

 Springer

Editors

Jean Claude Vial
Karine Bocchialini
Patrick Boumier
Institut d'Astrophysique Spatiale d'Orsay
Université Paris XI, Bâtiment 121
F-91405 Orsay Cedex, France

Cataloging-in-Publication Data applied for.

Die Deutsche Bibliothek - CIP-Einheitsaufnahme

Space solar physics : theoretical and observational issues in the
context of the SOHO mission ; proceedings of a summer school, held
in Orsay, France, 1- 13 September 1997 / Jean Claude Vial ... (ed.).
(Lecture notes in physics ; 507)
ISBN 978-3-662-14198-4   ISBN 978-3-540-69746-6 (eBook)
DOI 10.1007/978-3-540-69746-6

ISSN 0075-8450
ISBN 978-3-662-14198-4

© Springer-Verlag Berlin Heidelberg 1998
Originally published by Springer-Verlag Berlin Heidelberg New York in 1998
Softcover reprint of the hardcover 1st edition 1998

Typesetting: Camera-ready by the authors/editors
Cover design: *design & production* GmbH, Heidelberg
SPIN: 10644149   55/3144-543210 - Printed on acid-free paper

Preface

Solar physics has accumulated major successes during recent years with space missions such as Ulysses and Yohkoh. Since the beginning of 1996 the SOHO mission has brought a wealth of new results in areas of helioseismology and corona physics.

The major SOHO breakthrough has been accomplished with an important European contribution associated with an equally important European participation in the data analysis and interpretation. This work will certainly extend beyond the duration of the mission. However, it is now that scientists must become active in planning the new solar missions. Such activities surely need fresh ideas and new people to confront new results and new challenges. These new people must be encouraged now in the wake of the unique SOHO success.

Such a rationale was at the root of the decision to create this Summer School, taken in May 1996, by the Board of the Solar Physics Joint Division of the European Physical Society and the European Astronomical Society.

The School was prepared under the responsibility of the Scientific Organization Committee (M. Coradini, C. Chiuderi, V. Domingo, P. Heinzel, M. Huber, J.-L. Piéplu, P.L. Pallé, G. Simnett, J.-C. Vial (chairman), L. Vlahos) by the Local Organization Committee (J. Aboudarham, F. Baudin, K. Bocchialini (co-chair), P. Boumier, B. Cougrand, C. Cougrand, A.H. Gabriel (I.A.S. Director), D. Samain, I. Scholl, J. Solomon, J.-C. Vial (co-chair)).

More than eighty-five people applied for the School, of whom thirty-six were selected. The selection list was approved by the Scientific Organization Committee and the Board Meeting of the Solar Section of the European Physical Society (May 1997). The list included candidates with different levels of training, from pre-graduate to post-doctoral students, from most European countries.

The Summer School took place in Orsay from September 1 to September 13. The programme concentrated on general education in the physics of the sun and the heliosphere. A significant component was devoted to the design of space instrumentation and to space data analysis. Of course, the SOHO mission and its wealth of new results were at the forefront of most lectures and all practical work. The programme was divided into twenty-five lectures (thirty-seven hours overall) and practical work (twenty-six hours). The students could make full use of the facilities offered by the Multi-Experiment Data and Operations Center (MEDOC) installed at I.A.S. Moreover, four sessions of "free discussion" were provided in order to promote exchanges between lecturers and students. A visit to the I.A.S. Calibration Facility was also organized. The lectures were concluded by a presentation of ESA scientific projects by R.M. Bonnet, ESA Director of Scientific Programmes.

A presentation of the results of the thirteen groups involved in practical work was made during the last morning session of Saturday 13. Each presentation was followed by a general discussion.

Students had a unique opportunity to meet the experts in their fields, coming from nine European countries. Some lecturers stayed a full week to supervise practical work, to attend other lectures and to discuss with students. I.A.S. provided half a dozen assistant teachers for practical work.

We thank all participants and supporters of this Summer School in Space Solar Physics.

We acknowledge the help of the Scientific Organizing Committee, especially Dr. P. Heinzel and Prof. L. Vlahos.

We acknowledge the major contribution from Euroconferences, without which the School could not have been organized, along with that from E.S.A., C.N.E.S., I.N.S.U./C.N.R.S. (including the Groupement de Recherches "Magnétisme dans les Etoiles de Type Solaire"), and Université Paris XI (the Président and the Département de Physique). We greatly appreciated the support from I.A.S. and the participation of its personnel.

We sincerely thank the lecturers for their generous efforts in providing clear and complete presentations.

We hope that the present proceedings, containing the essential lectures of this exceptional School, will continue to be useful to numerous readers, and especially the students of the School in the course of their professional lives.

Orsay, January 5 1998

The Editors

Contents

Contents

List of Participants

STUDENTS

Aletti, Valérie   aletti@medoc-ias.u-psud.fr
   Institut d'Astrophysique Spatiale, Bât. 121, Université Paris XI, Orsay
   Cedex 91405, France,

Audard, Nathalie   audard@ast.cam.ac.uk
   University of Cambridge, Institute of Astronomy, Madingley Road, Cambridge CB3 0HA, England,

Aulanier, Guillaume   aulanier@mesopb.obspm.fr
   Observatoire de Meudon, 5 place Jules Janssen, Meudon Cedex 92195,
   France,

Berghmans, David   david@oma.be
   Royal Observatory of Belgium, Ringlaan 3, Brussels 1180, Belgium,

Corbard, Thierry   corbard@obs-nice.fr
   Observatoire de la Côte d'Azur, CNRS-URA 1362, B.P. 4229 boulevard
   de l'Observatoire, Nice Cedex 4, 06304, France,

Corti, Gianni   corti@arcetri.astro.it
   University of Florence, Department of Astronomy and Space Science,
   Largo E. Fermi 5, Firenze 50125, Italy,

Czaykowska, Anja   acz@mpe-garching.mpg.de
   Max Planck Institut für Extraterrestrische Physik, Giessenbachstrasse,
   Garching 85740, Germany,

de Pontieu, Bart   bdp@mpe-garching.mpg.de
   Max-Planck-Institut für Extraterrestrische Physik, Giessenbachstrasse,
   Garching 85740, Germany,

Del Zanna, Giulio   g.del-zanna@uclan.ac.uk
   Centre for Astrophysics, University of Central Lancashire, Maudland
   Building, Preston PR1 2HE, England,

Delannée, Cécile   delannee@iap.fr
   IAP, 98 bd Arago, Paris 75014, France,

Falewicz, Robert   falewicz@astro.uni.wroc.pl
   Astronomical Institute of Wroclaw University, Kopernika 11, Wroclaw
   51-622, Poland,

Fierry-Fraillon, David   fierry@irisalfa.unice.fr
UNSA - IRIS Network, Université de Nice Sophia-Antipolis, Laboratoire
d'Astrophysique, Parc Valrose, Nice 06000, France,

Fischbacher, Gordon   fischbac@phys.strath.ac.uk
University of Strathclyde, Department of Physics and Applied Physics,
John Anderson Building 107, Rottenrow, Glasgow G4 0NG, Scotland,

Gad El-Mawla, Diaa   diaa@ita.uni-heidelberg.de
Heidelberg University, Institut für Theoretische Astrophysik der Universität Heidelberg, Tiergartenstrasse 15, Heidelberg 69121, Germany,

Gallagher, Peter   p.gallagher@qub.ac.uk
The Queen's University of Belfast, Department of Pure and Applied
Physics, Belfast BT7 1NM, Northern Ireland,

Georgalikas, Alexander   alex@hyperion.astro.noa.gr
National Observatory of Athens, Astronomical Institute, I. Metaxa and
Vas. Paulou, Lofos Koufos, Palaia Penteli 15236, Greece,

Georgoulis, Manolis   georgoul@astro.auth.gr
Section of Astrophysics, Astronomy & Mechanics, University of Thessaloniki, Thessaloniki 54006, Greece,

Gonzalez-Hernandez, Irene   iglez@ll.iac.es
Instituto de Astrofisica de Canarias, via Lactea s/n, La Laguna, Tenerife
38200, Spain,

Gorshkov, Alexey B.   gorshkov@lnfm1.sai.msu.su
Universitetsky pr. 13, Sternberg Astronomical Institute, Moscow State
University, Moscow 119899, Russia,

Issautier, Karine   issautier@obspm.fr
Observatoire de Meudon / DESPA, 5 place Jules Janssen, Meudon Cedex
92195, France,

Jimenez Reyes, Sebastian   sjimenez@ll.iac.es
Instituto de Astrofisica de Canarias, via Lactea s/n, La Laguna, Tenerife
38200, Spain,

Landi, Enrico   enrico@arcetri.astro.it
Dipartimento di Astronomia e Scienza dello Spazio, Università degli Studi
di Firenze, Largo E. Fermi 5, Firenze 50125, Italy,

Leblanc, François   fleblanc@megasx.obspm.fr
Observatoire de Paris/DESPA, 5 place Jules Janssen, Meudon Cedex
92195, France,

Madjarska, Maria   madjarsk@phys.acad.bg
Institute of Astronomy, 72 Trakia blvd, Sofia 1784, Bulgaria,

Maia, Dalmiro   dalmiro.maia@obspm.fr
Observatoire de Meudon, 5 place Jules Janssen, Meudon Cedex 92195,
France,

McDonald, Lee   lm@mssl.ucl.ac.uk
Mullard Space Science Laboratory UCL, Holmbury St Mary, Dorking
Surrey, RH5 6NT, England,

Perez Perez, Maria Elena   epp©star.arm.ac.uk
   Armagh Observatory, College Hill, Armagh BT61 9D9, Northern Ireland,

Pierre, Frédéric   pierre©deneb.ias.fr
   Institut d'Astrophysique Spatiale, Bât. 121, Université Paris XI, Orsay
   Cedex 91405, France,

Pohjolainen, Silja   silja.pohjolainen©hut.fi
   Helsinki Univ. of Technology, Metsähovi Radio Research Station, Met-
   sähovintie 114, Kylmälä 02540, Finland,

Portier-Fozzani, Fabrice   fpf©astrsp-mrs.fr
   Laboratoire d'Astronomie Spatiale, LAS-CNRS, B.P. 8, Marseille Cedex 12
   13376, France,

Rybak, Jan   jrybak©solar.stanford.edu - astrryba©ta3.sk
   Astronomical Institute, Slovak Academy of Sciences, Tatranska Lomnica
   059 60, Slovak Republic,

Saadatnejad, Bard   bards©astro.uio.no
   Institute of Theoretical Astrophysics, University of Oslo, P.O. Box 1029,
   Blindern, 0315 Oslo Norway,

Sollum, Espen   espen.sollum©astro.uio.no
   Institute of Theoretical Astrophysics, P.O. Box 1029, University of Oslo,
   Blindern, 0315 Oslo, Norway,

Van Aalst, Maarten   aalst©fys.ruu.nl
   Universiteit Utrecht, Sterrenkundig Instituut, P.O. Box 80.000, Utrecht
   NL-3508 TA, The Netherlands,

Varady, Michal   varady©asu.cas.cz
   Astronomical Institute of the Academy of Sciences of the Czech Republic,
   Ondrejov Observatory, Ondrejov 25165, Czech Republic,

Yurchishin, Vasyl   vayur©crao.crimea.ua
   Crimean Astrophysical Observatory, Nauchny Crimea 334413, Ukraine.

LECTURERS

Bonnet, Roger M.   dbauer©hq.esa.fr
   ESA-ASE, 8-10 rue Mario Nikis, 75015 Paris, France,

Carlsson, Mats   mats.carlsson©astro.uio.no
   Institute of Theoretical Astrophysics, P.O. Box 1029, Blindern, N-0315
   Oslo, Norway,

Delaboudinière, Jean-Pierre   boudine©ias.fr
   Institut d'Astrophysique Spatiale, Bât. 121, Université Paris XI, Orsay
   Cedex 91405, France,

Felici, Fabrizio   ffelici©estec.esa.nl
   ESTEC - European Space Agency, Noordwijk, The Netherlands,

Gabriel, Alan gabriel@ias.fr
Institut d'Astrophysique Spatiale, Bât. 121, Université Paris XI, Orsay Cedex 91405, France,

Koutchmy, Serge koutchmy@iap.fr
Institut d'Astrophysique de Paris, 98 bv Arago, Paris 75014, France,

Lemaire, Philippe lemaire@ias.fr
Institut d'Astrophysique Spatiale, Bât. 121, Université Paris XI, Orsay Cedex 91405, France,

Marsch, Eckart marsch@linax1.dnet.gwdg.de
Max Planck Institut für Aeronomie, Max Planck str. 2, Katlenburg-Lindau 37191, Germany,

Martens, Petrus pmartens@esa.nascom.nasa.gov
SOHO Experiment Operations Facility, Code 682.3, Goddard Space Flight Center, Greenbelt MD 20771, USA,

Mason, Helen H.E.Mason@damtp.cam.ac.uk
Department of Applied Mathematics and Theoretical Physics, Silver Street, Cambridge CB3 9EW, UK,

Proctor, Mike M.R.E.Proctor@damtp.cam.ac.uk
DAMTP, University of Cambridge, CB3 0HA Cambridge, UK,

Roca Cortés, Teodoro trc@ll.iac.es
Instituto de Astrofísica de Canarias, Universidad de La Laguna, 38205 La Laguna, Tenerife, Spain,

Solanki, Sami solanki@astro.phys.ethz.ch
Institute of Astronomy, ETH-Zentrum, Zürich 8092, Switzerland,

Ulmschmeider, Peter ulm@artemis.ita.uni-heidelberg.de
Insitut für Theoretische Astrophysik der Universität Heidelberg, Tiergartenstr. 15, Heidelberg 69121, Germany,

Velli, Marco velli@astr11pi.difi.unipi.it
Dipartimento di Astronomia e Scienza dello Spazio, Università di Firenze, Firenze 50125, Italy,

Walsh, Robert robert@dcs.st-and.ac.uk
School of Mathematical and Computational Sciences, University of St. Andrews, St. Andrews KY16 9SS, Scotland.

Local Organizing Committee and Practical Work Managers

Aboudarham, Jean abou@ias.fr
Institut d'Astrophysique Spatiale, Bât. 121, Université Paris XI, Orsay Cedex 91405, France,

Baudin, Frédéric baudin@ias.fr
Institut d'Astrophysique Spatiale, Bât. 121, Université Paris XI, Orsay Cedex 91405, France,

Bocchialini, Karine   bocchialini@ias.fr
   Institut d'Astrophysique Spatiale, Bât. 121, Université Paris XI, Orsay
   Cedex 91405, France,
Boumier, Patrick   boumier@ias.fr
   Institut d'Astrophysique Spatiale, Bât. 121, Université Paris XI, Orsay
   Cedex 91405, France,
Cougrand, Bernard   cougrand@medoc-ias.u-psud.fr
   Institut d'Astrophysique Spatiale, Bât. 121, Université Paris XI, Orsay
   Cedex 91405, France,
Cougrand, Catherine   catherine.cougrand@medoc-ias.u-psud.fr
   Institut d'Astrophysique Spatiale, Bât. 121, Université Paris XI, Orsay
   Cedex 91405, France,
Lamartinie, Sujit   sujit@medoc-ias.u-psud.fr
   Institut d'Astrophysique Spatiale, Bât. 121, Université Paris XI, Orsay
   Cedex 91405, France,
Lepeltier, Vanessa
   Institut d'Astrophysique Spatiale, Bât. 121, Université Paris XI, Orsay
   Cedex 91405, France,
Patsourakos, Spyros   spiros@medoc-ias.u-psud.fr
   Institut d'Astrophysique Spatiale, Bât. 121, Université Paris XI, Orsay
   Cedex 91405, France,
Pike, Dave   cdp@astro1.bnsc.rl.ac.uk
   University of Cambridge, Institute of Astronomy, Madingley Road, Cam-
   bridge CB3 0HA, UK,
Régnier, Stéphane   regnier@medoc-ias.u-psud.fr
   Institut d'Astrophysique Spatiale, Bât. 121, Université Paris XI, Orsay
   Cedex 91405, France,
Samain, Denys   samain@ias.fr
   Institut d'Astrophysique Spatiale, Bât. 121, Université Paris XI, Orsay
   Cedex 91405, France,
Scholl, Isabelle   scholl@medoc-ias.u-psud.fr
   Institut d'Astrophysique Spatiale, Bât. 121, Université Paris XI, Orsay
   Cedex 91405, France,
Solomon, Jacques   solomon@ias.fr
   Institut d'Astrophysique Spatiale, Bât. 121, Université Paris XI, Orsay
   Cedex 91405, France,
Vial, Jean-Claude   vial@ias.fr
   Institut d'Astrophysique Spatiale, Bât. 121, Université Paris XI, Orsay
   Cedex 91405, France.

Methods and Techniques in Helioseismology

Teodoro Roca Cortés

Instituto de Astrofísica de Canarias, Universidad de La Laguna, 38205 La Laguna, Tenerife, Spain

Abstract. In the last two decades, the study of global solar oscillations has provided the only effective method to probe the structure of stars, and particularly that of our Sun. As will be seen throughout this short course, we now know the Sun much better than before thanks to Helioseismology. However, the detection of such normal modes of vibration of the Sun has only been possible recently due to the combination of two factors: first the signals to measure, although almost periodic, have very small amplitudes as compared to noise, and second, instrumental and observing techniques have not reached the required sensitivity until now.

Here we will look at the observational parameters that can be best measured, in terms of S/N, as well as the ultimate noise that is required in order to observe the solar oscillations. Further, we will review the current instruments and techniques that are able to do so with special emphasis on those onboard the SOHO satellite. However, getting the observations is not the end of the story; it is only after waiting for the acquisition of long time series and after applying sophisticated analysis techniques, that a comprehensive picture can be seen and observational parameters can be compared to theoretical predictions in what is known as the Forward Problem of Helioseismology. We will also see how the differences between the predictions and the actual Sun lead us to calculate the internal structure of the models through what is called the Inverse Problem in Helioseismology.

1 Introduction

When facing the task of giving a short course on Helioseismology, the first problem is to find an adequate textbook that could cover the many items and details that it will not be possible to cover in the course. In trying to find such a bibliographical material, I have reached the conclusion that a book or a review that would solve the problem didn't exist. The problem was to dig into the exponentially growing number of papers in the subject in the last 20 years of existence which, of course, I could not possibly do. Nevertheless, a few monographs have been found that I believe could meet the needs of any postgraduate student that faces Helioseismology for the first time, and also wants to be up-to-date in the subject; this material is collected before the references at the end.

It is well known in physics that equilibrium structures can be studied by measuring their spectrum of normal modes of oscillation. Particularly geophysicists use this technique to study the internal structure of our planet. It is only very recently that it has been applied to the Sun, once it was perceived that the Sun could make cavities in which waves could resonate becoming what we know as global modes of oscillation. The object of the Helioseismology consists in using the temporal and spatial properties of solar oscillations to study the structure and dynamics of solar interior. The quality of such a study is measured by the precision of the deductions and the ability to probe all parts of the Sun's interior, and therefore it is dependent on the acquisition of observations of the widest possible range of solar oscillation modes, particularly the acoustic or pressure modes (also called p-modes) and the buoyancy or gravity modes (also called g-modes).

The p-modes have pressure as the restoring force. They are predominantly vertical waves that propagate in cavities whose outer turning point is near the solar surface and their inner turning point depends on their spatial distribution across the solar surface, those with larger spatial scale penetrating deeper. Their frequencies vary from ~ 0.2 mHz up to the acoustic cut-off frequency of the solar atmosphere at ~ 5.5 mHz. The mode amplitudes favour the observations of the higher harmonics of these modes (around 5 min period), which are the ones that have been definitely identified in the solar oscillation spectrum, whether the observation is disk-integrated or spatially resolved.

The gravity, or g-modes, make use of buoyancy rather than pressure as the restoring force. Here also a range of modes with different order and degree are predicted, but their propagation characteristics and thus their diagnostic properties are somewhat different. The g-modes can propagate freely in the radiative interior, but are reflected at the base of the convection zone at $r \sim 0.72 R_\odot$, since this zone is unstable to buoyancy disturbances. Because of this different radial distribution, the g-modes are a much more sensitive indicator of the core conditions than are the p-modes. The frequency range for low degree g-modes extends downwards from the fundamental, which is expected in the region of 0.4 mHz. Asymptotic calculations show that the *periods* of high order g-modes are approximately uniformly spaced, analogous to the uniform *frequency* spacing of the p-modes which provides unique signatures for both types of modes. Observations show no confirmed identification of g-modes (Pallé 1991), so that we have no indication of their spectrum, their amplitude distribution or their coherence times, although the latter are predicted to be very long.

2 Observable Magnitudes and the Solar Background Noise

The p-modes are acoustic waves which travel back and forth between the surface and some internal turning point situated deep in the Sun. In their travel they both move and compress the plasma, therefore the movements can be observed as a velocity signal, detected at the surface using the Doppler effect on spectral absorption lines and/or as variations in the solar diameter.

On the other hand, the compressions may also be detected because the gas will heat up adiabatically and consequently will radiate more than it would otherwise, increasing its luminosity and being detectable as fluctuations in the continuum and/or in line brightness.

For any single acoustic mode the amplitude of the signal, i.e. fluctuations in velocity and in brightness, is very small. The highest velocity amplitudes (at the strongest part of the 5 min band) are smaller than 20 cm/s (Pallé et al. 1986); the corresponding relative brightness fluctuations would be smaller than 3 parts per million (Jiménez et al. 1987).

These numbers correspond to temperature fluctuations of as much as 4 mK or to Sun's radius fluctuations of $\delta R_\odot \sim 10^{-4}$ arcseconds. These signals are very small indeed, in fact they are minute by usual astronomical standards. Although the former two have already been measured, thanks to the high number of photons received from the Sun, the later two have not been unambiguously measured so far. The solar diameter measurements, a problem which consists, among others, in a careful photometric measurement of the limb darkening at the edge of the Sun, was studied very hard by Hill (1985), and Hill and Gu (1988); but there is no consensus amongst the scientific community that either p- or g-modes have been measured so far in the diameter. In fact these experiments achieved noise levels of more than a factor of three above the required accuracy. The last parameter mentioned, temperature fluctuations by measuring variations in the equivalent width of the lines in the spectrum has been tried by Kjeldsen et al. (1995) for stars by adding up the equivalent widths of the lines of hydrogen.

2.1   The Type of Signals to Be Measured

The signals we are interested in, when doing Helioseismology, consist in the deformations at the solar surface due to the effect of the modes. In fact such helioseismic signals at the surface are seen as *almost* periodic in space and in time.

The solutions of the linear perturbations to the equilibrium equations of a spherically symmetric Sun are the eigenfunctions of the appropriate eigenvalue problem. These have the form of:

$$A(t) \cdot F_n(r) \cdot P_\ell^m(\cos\theta) \cdot e^{im\phi} \cdot e^{i\omega t} \tag{1}$$

where n is called the radial order, ℓ the degree and m the azimuthal degree. Obviously, if the problem has spherical symmetry the solutions will not depend on m; however, any axial perturbation, such as the rotation or the existence of a magnetic field would cause the solution to be m-dependent. Consequently at the surface, the signals are proportional to the spherical harmonics $Y_\ell^m(\theta, \phi)$ which are strictly periodic in a sphere; nevertheless, the real world reminds us that we can only observe a hemisphere of the Sun and, often, with an instrumental response of our experiments which would even limit such an observation further, leaving us with an "almost" space-periodic signal.

As far as the time variable is concerned, although the frequencies of the waves are real and therefore provide a pure oscillatory solution, the excitation and damping mechanisms yield an amplitude of such a wave which is an unknown function of time. The actual belief is that most, if not all, of the excitation and damping comes from turbulent convection (Goldreich and Keeley, 1977; Goldreich and Kumar, 1988). To first approximation, the oscillations amplitude induced by the forcing function can be approximated by something like:

$$A(t) = \sum_{i=1}^{P} A_i \, \delta(t - t_i) \, e^{\frac{-(t-t_i)}{\tau_d}} \qquad (2)$$

where τ_d is the lifetime of the oscillator, t_i are the times at which the excitation is made (a stochastic variable) and A_i is the amplitude in each excitation of a particular mode. On top, there could also be changes in the phase of the oscillator. These are some of the reasons for saying that the observed signals are "almost" periodic in time.

Therefore, the best approximation to the helioseismic signal we can think of is that it is the sum of a bunch of damped oscillators (with typical damping times from days to months, depending on frequency) stochastically excited by the solar convection in a way not completely understood yet. Then the Sun acts in the same way as any linear system, whereby the observed oscillating function is the result of the damped oscillator signal convolved with the driving function. By applying Fourier analysis, this convolution become the product of the Fourier transforms of both functions. It would be interesting to know the transfer function of such a linear system, however we can not use the standard method of sending an impulse function (or a set of them) and observe the response because the excitation is a stochastic process.

2.2 The Observed Power Spectrum.
Its Statistics and Accuracy Limits

Therefore the power spectrum estimate of a single realisation of a stochastically excited oscillator consists of the product of the power spectrum of the oscillator times the spectrum of a random excitation function,

$$P_{\text{obs}}(\nu) = L(\nu) \cdot E(\nu) \qquad (3)$$

where, $P_{\text{obs}}(\nu)$ is the observed spectrum, $L(\nu)$ the spectrum of a damped oscillator signal (i.e. a Lorentzian shape) and $E(\nu)$ the spectrum of the excitation function. Therefore, this last term masks or rather conceals the function $L(\nu)$ which contains the information about the oscillation mode, providing the observed spectrum of such an oscillator with a very peaky shape with a lot of fine structures (see Fig. 1). The power spectrum values at a given frequency, for different realisations, are distributed as χ^2 with two degrees of freedom (Woodard 1984; Anderson et al. 1990).

The probability density distribution for the observed power spectra has therefore the form of $P(s) = e^{-s}$. Such a distribution has an important property which is that the standard deviation associated is equal to its expectation value, therefore for any given power estimate the error values will be as high as 100%; this is particularly important near the resonances, where the power is high and so will be the measurement uncertainties. Moreover, since the tail of this distribution is higher than the Gaussian one, it is expected that a greater number of cases of large power events will arise. This fact provides a great danger of misinterpretation of the power spectrum, as can be noticed in Fig. 1. Nonetheless, the more realisations we are able to observe the better the observed spectrum will resemble the ultimate limit spectrum $P_{\lim}(\nu)$, which would arrive when an average of an infinite number of realisations is done; in this case the $E(\nu)$ would be a smooth function of frequency.

The first term in eq. (3) $L(\nu)$, is the spectral response of a damped oscillator signal. Since the acoustic modes of the Sun have typically a high Q value (~ 3000 for $\Gamma_{n\ell m} \sim 1\mu$Hz and for $\nu_{n\ell m} \sim 3$ mHz) then near the resonance frequency $L(\nu)$ has a Lorentzian profile, and for a multiplet (n,l) we have:

$$M_{n\ell}(\nu) = \sum_{m=-\ell}^{m=\ell} \frac{S_{n\ell m}(\Gamma_{n\ell m}/2)^2}{(\nu - \nu_{n\ell m})^2 + (\Gamma_{n\ell m}/2)^2} \tag{4}$$

where $\nu_{n\ell m}$ is the resonance frequency of the mode, $\Gamma_{n\ell m}$ its linewidth, $S_{n\ell m}$ its power at the resonance frequency.

Once the power spectrum is obtained we can ask ourselves, to what accuracy can we measure the resonance frequency $\nu_{n\ell m}$? If T is the length of the observations and $\tau_{\rm d}$ the damping time of the mode (obviously related to $\Gamma_{n\ell m}$) and they are such that $T \lesssim \tau_{\rm d}$ then $\delta\nu_{n\ell m}^{\rm rms} = 1/T$, a property which reflects the scaling factor in the Fourier transforms. However, the most common case in Helioseismology when analysing p-modes, is when $T \gg \tau_{\rm d}$ in which case the resonance line is fully resolved and then the stochastic nature of the excitation function is manifested. In this case a good approximation to the $\delta\nu_{n\ell m}^{\rm rms}$ has been given by Libbrecht (1992):

$$(\delta\nu_{n\ell m}^{\rm rms})^2 = \frac{\Gamma_{n\ell m}}{4\pi T}(1+\beta)^{1/2}[(1+\beta)^{1/2} + \beta^{1/2}]^3 \tag{5}$$

where $\beta = N/S$ is the inverse of the signal-to-noise ratio. Consequently for a typical acoustic mode with $\nu_{n\ell m} \approx 3.1$ mHz, $\Gamma_{n\ell m} \approx 1$ μHz, observed during $T \approx 100$ days with $\beta \ll 0.01$, one has an uncertainty in the resonance frequency precision of $\delta\nu_{n\ell m}^{\rm rms} \approx 0.1$ μHz. One may think that this is a good precision and that by only waiting for longer this number will even decrease further. However, the resonance frequencies of p-modes change slightly with the solar activity cycle; indeed a total amount of 0.45 μHz has been observed from minimum to maximum of solar activity (Régulo et al. 1994). This limits

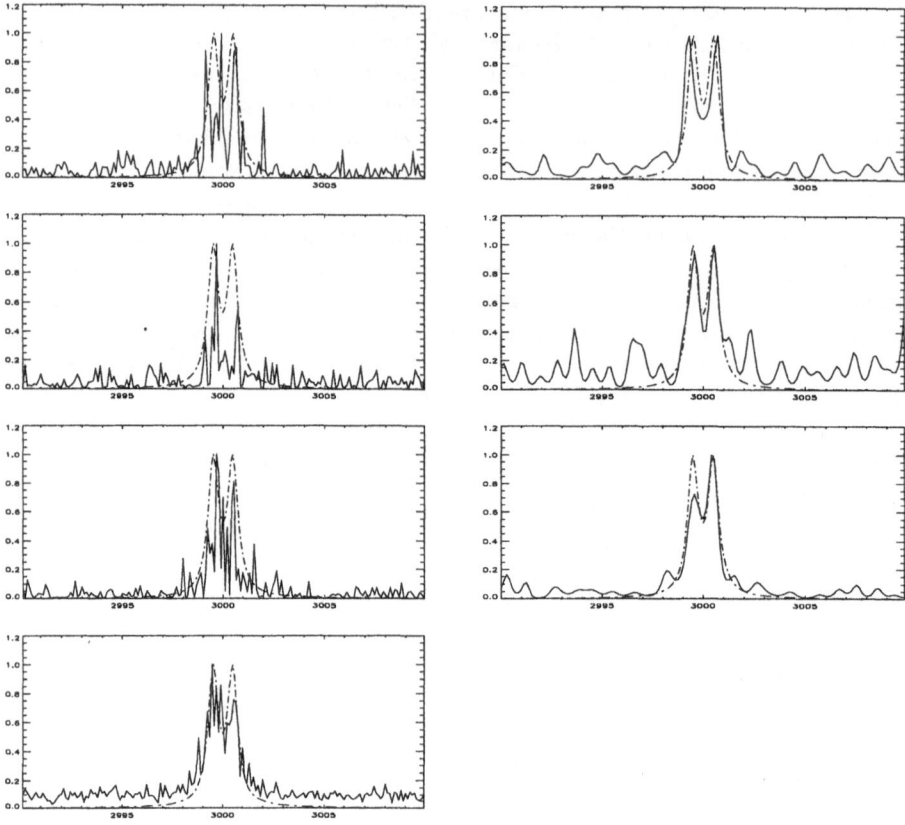

Fig. 1. Power spectra of simulated helioseismic data as a function of frequency (in μHz). Graphs on the left are each the FFT power spectrum of time series $\approx$100 days long in which two oscillators whose frequencies differ in 500 nHz and have lifetimes of 5 days have been simulated. From top to bottom there are 3 different realisations of such series and the fourth one is the average of 10 realisations. The dot-dashed lines are the representations of the oscillators without excitation signal. The right-hand side graphs show the result of the power spectra of the same series (as in their left) but calculated using the Homomorphic Deconvolution technique (see Sect. 4). *Courtesy of F. Baudin.*

the duration of the time interval one can wait. On the other hand, rotational splitting of the multiplet lines, latitudinal differential rotation and magnetic field effects are measured in tens to hundreds of nHz. As time duration is limited, emphasis should be made in the need for measuring as many modes simultaneously as possible, to increase the accuracy with which such subtle effects might be measured.

Depending on the strategy of the observation we can have two types of observed power spectra $P_{\text{obs}}(\nu)$: a) one which contains information on the modes spatially averaged over the solar disk (i.e. the Sun seen as a star), and b) the one which contains spatially resolved oscillations.

When dealing with data of type (a) we are facing spectra of single realizations for modes of the lowest degree (in the range $0 \leq \ell \leq 5$) organised into multiplets with azimuthal degrees ranging from $-\ell \leq m \leq +\ell$; from these, only the modes with $(\ell + m)$ even will be observable since the observations are taken nearly perpendicularly to the rotation axis of the Sun. The separation of these multiplet components is of the same order as their linewidth (specially at $\nu \gtrsim 3$mHz). Thus long observation times are required in order to get reliable estimates of the individual line profile parameters (see Figs. 3 and 4).

If data of type (b) is available, a power spectrum of modes for given values of ℓ and m can, in principle, be obtained. However, due to the fact that only one hemisphere of the Sun is observed the spatial filters employed to isolate such a spectrum are not perfect and this leads to problems like "power leakage" and "overlapping" which also makes difficult the estimation of line profile parameters. However, things can improve if one assumes that, given an ℓ value, the spectra at different m-values can be treated as independent observation of the same process. By averaging these "different realisations" after appropriately shifting the frequencies due to solar rotation, one obtains a much smoother profile. Of course, the statistics of such a "m-averaged spectrum" is different from that for a single realisation. The distribution will still be χ^2 but with a number of degrees of freedom equal to twice the number of averaged spectra. As the number of averaged spectra increases, the distribution becomes close to a normal distribution (Anderson et al. 1990).

Indeed, the observed acoustic spectrum contains many multiplets (n, ℓ) with several peaks, each of which has the Lorentzian shape. Moreover, in addition to the uncertainty due to the excitation function, there is a background noise over which these signals appear, mainly due to the upper part of the convection zone, i.e. granulation. Also an important part of this background is the contribution at a given frequency from the wings of close strong modes (and many unresolved very weak ones) that would add up. In this way, the limit spectrum will be the sum of the limit spectra of each of the modes of oscillation of the Sun. Therefore:

$$P_{\lim}(\nu) = \sum_{\ell}\sum_{n} M_{n\ell}(\nu) + B(\nu) \qquad (6)$$

where $B(\nu)$ is the background noise power spectrum.

2.3   The Solar Background Noise Spectrum

The solar noise background signal can be defined as the contribution from temporal and spatial variations of small scale solar surface structures. From low to high frequencies, surface inhomogeneities such as faculae, sunspots, etc. (coupled with Sun's rotation), convective structures at all scales and their evolution produce the "noise" signal that will hide the smaller oscillatory signals that we are interested in (see Fig. 2).

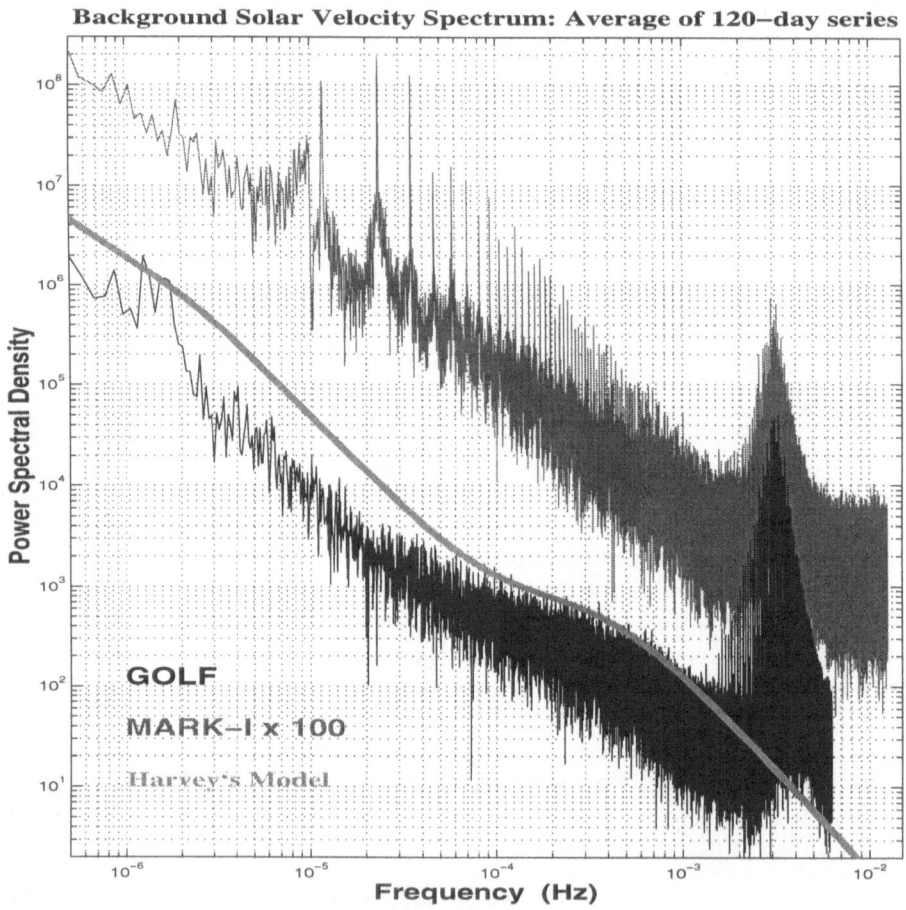

Fig. 2. Disk-integrated velocity background spectra as observed by similar instruments from space (120 days of GOLF) and from ground (5 spectra of 120 days each of Mark I). The later has been multiplied by 100 to clearly separate both plots. Notice the p-modes oscillation spectra clearly seen above background while g-mode spectrum is not. Likewise critical spectral features (1 day^{-1} and its harmonics) are present when observing from a single site (duty cycle $\approx$28%). The first estimate made by Harvey (1985) has also been drawn. *Courtesy of P.L. Pallé.*

The solar noise background in velocity measurements was first estimated by Harvey (1985) who assumed, for integrated sun-disk observations, a power density contribution $B(\nu)$ of the type:

$$B(\nu) = \frac{4\sigma^2\tau}{1 + (2\pi\nu\tau)^2} \qquad (7)$$

where σ is the rms velocity and τ the characteristic time of the process. This background was estimated for the following solar phenomena: granulation, mesogranulation, supergranulation, and active regions. The total background was estimated as the sum of them all, with the appropriate observed (or estimated) values for their σ and τ. Such an estimation was observationally verified soon afterwards by Jiménez et al (1988) and later by Pallé et al (1995) with values not far from the estimated ones. An estimate of this background can be seen in Fig. 2, where the spectra from the GOLF experiment and from the best estimation of the Mark I instrument (of the same kind as GOLF at Obs. del Teide) are shown (to better see some details the later has been shifted by two orders of magnitude upwards). As can be noticed, the p-mode spectrum is clearly seen above the noise while the g-mode spectrum should appear between ~20 and ~200 μHz; notice that the later is not apparent suggesting that such modes should have energies smaller than those found in this frequency range.

An analogous estimate of the solar background irradiance power spectrum has been made recently by Rabello Soares et al. (1997). Using data from different sources and instruments (ground-based and space borne) provides the first estimate at all frequencies (from 1μHz to 30 mHz) of the actual solar background. Notably, the background here is higher than for velocity observations, at both intervals where the p- and g-modes are expected. In both cases it does not seem possible to explain such a signal with a power law with unique exponent for all frequencies, suggesting that they are influenced by specific spatial and temporal scales. No doubt the study of such a spectrum would help to understand these processes, and it is particularly interesting in the case of studying the convection in stars other than the Sun.

The amplitudes of the p-modes are smaller in intensity than in velocity; on the other hand, the background noise is higher due to the granulation, a thermally driven process, which has $\delta T/\delta v$ higher than the same ratio for the p-modes. This means that the solar background noise is higher for continuum intensities than it is for velocity (see Figs. 3 and 4). Therefore it looks as if the S/N ratio would improve if one is measuring higher up in the atmosphere due to: a) the mode amplitudes increase for increasing height in the atmosphere, and b) the granulation contrast decreases with increasing height. This is why most of the current intensity instruments work with Ca II filters, specially those with high spatial resolution.

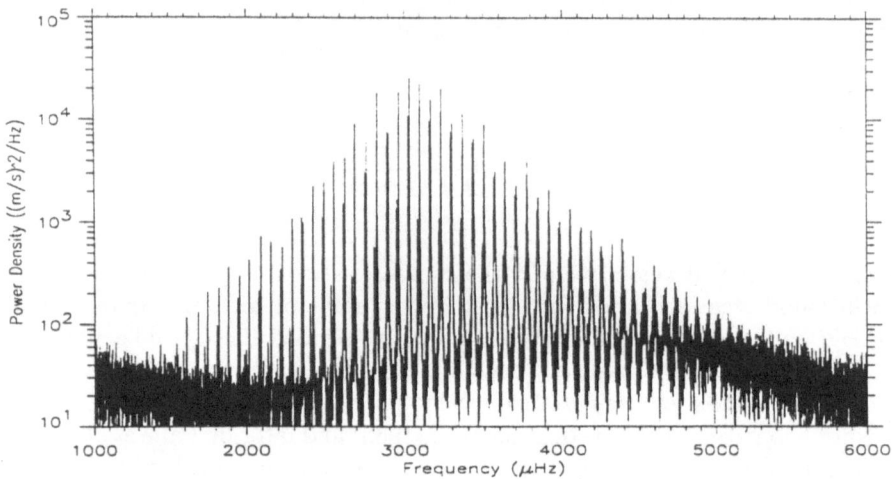

Fig. 3. A slice of the disk-integrated velocity power spectrum obtained using GOLF data showing the p-mode acoustic spectrum.

3   Techniques, Instruments and Observations

The solar background noise spectrum $B(\nu)$ will be further influenced by the instrumentation necessary to obtain the observations, the conditions of such an observation (i.e. the Earth's atmosphere) and the analysis techniques used to obtain the final parameters to be compared with theoretical predictions. As a consequence, these spectra will add to the solar background noise to increase the level of the noise in the observed data. One of the key parameters to consider when building an experiment to observe solar oscillations will be these noise sources which will have to be kept to a minimum and be very stable in time. Special care should be given to preventing the appearence of variations which could introduce power at frequencies in the range of interest.

When trying to observe solar oscillations, the most useful techniques are either photometry or, more generally, spectroscopy. Brightness measurements are the simplest to perform because they only need a photoelectric detector and a filter to isolate a given spectral range. Nevertheless, they do require precise detection techniques because the mode amplitudes are close to the detector intrinsic noise limits; this is particularly true for continuum observations where the amplitudes are very small. Nonetheless as already stated above, the S/N is a factor of 10 higher in velocity measurements and this is why this technique is the preferred one, particularly for ground-based observations where the noise increases due to the presence of the Earth's atmosphere (Jiménez et al. 1987).

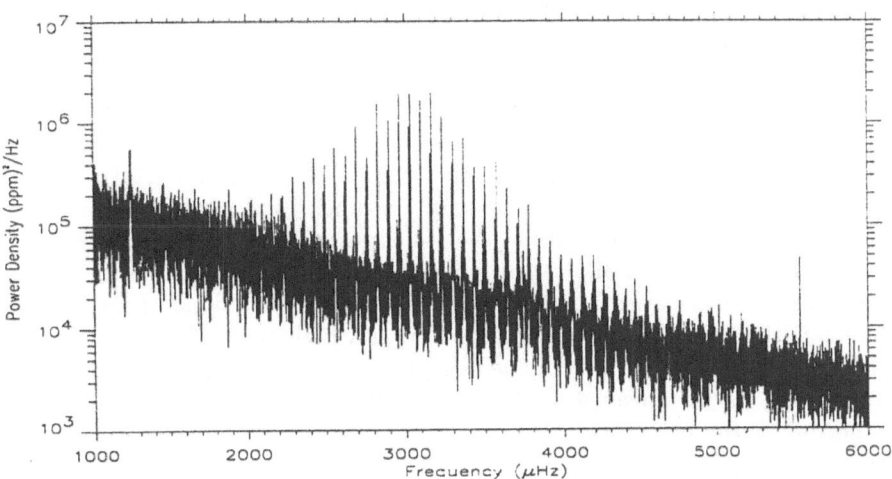

Fig. 4. A slice of the disk-integrated broadband intensity power spectrum obtained using VIRGO/SPM data showing the p-mode acoustic spectrum. Comparison with the velocity spectrum (see Fig.3) shows the differences in background spectra and S/N of the modes (a factor of $\approx$10 lower here), which prevents mode identification below 2 mHz and also those with $\ell >$2.

However, since detector technology is no longer a primary factor in order to choose the instrumentation, the question of what instrument to use is best answered by taking into account the particular observation one wants to perform and then designing the appropriate instrument. In fact, different instruments will perform best in different observing situations, i.e. depending on the frequency range and/or the degree or horizontal wavenumber of the modes. Starting from the very first instruments specifically dedicated to the observation of solar oscillations made in the seventies and ending up to the most modern ones onboard SOHO and on the main ground networks currently working, I will try to condense them, briefly describing the techniques behind each one, neither attempting to make a full description nor giving an historical account. An excellent course reviewing this matter can be found in Brown (1996).

3.1   Instrumental Techniques and Instruments

The easiest way to measure the Doppler shift of an absorption line is the use of a very narrow band filter which can isolate the solar line whose position we want to measure. The observation is made by measuring the intensity at two narrow frequency bands on either wing of the solar line: I_b and I_r. A shift of the line in either direction would decrease the intensity in one wing and increase the one in the opposite wing; therefore, by calculating the ratio: $r = (I_b - I_r)/(I_b + I_r)$, we would get a measure of the shift of the line. Notice that the normalisation to the sum of the intensities reduces the sensitivity to changes in the overall continuum intensity due to the Sun itself, the Earth's atmosphere or even the instrument itself. This ratio is only linearly dependent on the line shift if the shifts are small compared to the linewidth; if these become comparable then corrections for non-linearities should be made if one wants accurate measurements of the velocities. Of course, the critical point here is the stability at short and long term, of the frequency bands used to measure the intensities at the wings. A standard technique to define such bands is the use of a slit spectrograph with a high spectral resolution; with the appropriate set-up the intensities can be measured simultaneously, but a scanning device must be used if the whole Sun is to be measured. On the other hand, the necessary stability to measure such small signals has led to their use only in the case of very high-ℓ acoustic mode measurements.

Observing Very Low-ℓ Modes.

The Resonant Scattering Filters. A way to obtain a very narrow and stable filter (and also the measuring bands mentioned above) is with the use of vapour cells and/or atomic beams which provide an absolute wavelength standard (Isaak 1961; Blamont and Roddier 1961; Fossat and Ricort 1971). This is the kind of technique the GOLF experiment (onboard SOHO) is employing (Boumier, 1991). It uses the optical resonance technique to isolate a narrow region of the solar sodium resonance line. The solar spectrum traverses a sodium vapour cell which has an intrinsic (thermal) absorption linewidth of ~ 25 mÅ. Thus a 25 mÅ slice of the solar line is absorbed and re-emitted in all directions. Part of the signal re-emitted at 90 degrees is suitably recorded by photomultiplier tubes. By placing the cell in a longitudinal magnetic field of ≈ 5000 gauss, the absorption line is Zeeman split into two components displaced from the original by $\pm \approx 108$ mÅ. If the incoming solar flux is now circularly analysed using a linear polariser followed by a quarter-wave plate, it is possible to select one or the other absorption component, respectively left and right circularly polarised, and thus measure the intensity of a point on each of the two wings of the solar profile at: I_b and I_r. This technique is similar, in principle, to that carried out in ground-based measurements of solar oscillations, as for example in the IRIS or BiSON networks (Grec et al., 1991; Brookes et al., 1978); however, some innovations have been introduced which are fully described in Gabriel et al. (1995) as well as the whole GOLF instrument. One of them has been crucial for the experiment, that is a small modulation of the magnetic field in which the cell is placed. If the cell magnetic field is increased/decreased sequentially (modulated) by a small amount, we can measure two more points, further separated in wavelength: I_b^-, I_b^+ and similarly in the red wing of the line. In this way with a four-point measurement technique, a local measurement of the slope at both wings of the line is obtained which provides further information for an accurate calibration of the instrument (Boumier et al. 1994). Further, it is interesting to note here the existence of a 5-point measurement instrument, called MR5 (Robillot et al. 1993) which also provides a measurement of the intensity at line center. The main advantage is the extreme long term stability that this technique provides, and a disadvantage is that only the vapour of alkali atoms can be used in practice, leaving only the Na D and the K resonance lines as the solar lines to be used, thus restricting the measurement to higher in the solar atmosphere.

Broad-band Intensity Measurements. As already stated, broad-band photometry is the technique that requires simpler instrumentation and, indeed these have been the first Helioseismometers in space. Although the ACRIM radiometer onboard SMM satellite was the first space instrument to measure solar oscillations, the IPHIR photometer (Fröhlich et al. 1988) was the first space experiment (onboard PHOBOS) designed to measure p- and g-modes; in fact, this photometer is of the same kind as those which are part of the VIRGO experiment onboard SOHO spacecraft. VIRGO is a multi-instrument experiment which contains two different active-cavity radiometers (DIARAD and PMO6-V) for monitoring the solar 'constant', two three-channel sunphotometers (SPM, one is a back-up) for the measurement of the spectral irradiance at 402, 500 and 862 nm with a bandwidth of 5 nm, and a low-resolution imager (LOI) with 12 pixels, for measuring the radiance distribution over the solar disk at 500 nm. LOI also has an additional 4 detectors, situated in a circular annulus that selects the rim of the Sun, designed to drive two piezoelectric stacks which in turn control the movement of the secondary mirror of the Ritchey-Chrétien telescope at 10 Hz. However their signal can also be used to perform diameter measurements of the Sun and possibly also measure some higher-ℓ modes. While the radiometers and photometers can only measure modes with $\ell \leq 2$, LOI is specifically designed to measure modes of up to $\ell = 8$ and it is the first time that it is flying. The instrumentation and the observed in-flight performance and operational aspects of the irradiance observations are extensively described in Fröhlich et al. (1995).

Observing Intermediate and High-ℓ Modes.

The Magneto-Optical Filters. This is a variation of the resonant scattering technique in which, while maintaining high stability, they also provide high spatial resolution by working in transmission rather than in the scattering mode; in this way the whole solar disk can be passed through the filter simultaneously performing finally an image of the Sun at both frequency bands used to measure the Doppler shift of the line. Pionnering work in this instrumentation has taken place in Rome (Cacciani and Fofi 1978). In brief, the working principle is that solar light passes through two crossed linear polarisers which would transmit nothing if a vapour cell (in a magnetic field) were not placed between them. The cell converts the light in the wings of the solar line into circularly polarised light which traverses the second polariser; at the output of this, the transmitted light is linearly polarised in the same way. Further, by using a second vapour cell the light is returned to the state of opposite circular polarisation; further, the light passes through a quarter-wave plate and a linear polariser which separate the two passbands in linear polarization 90 degrees apart. Different implementations of this filter are possible and have led to several instruments operating currently: LOWL working at Mauna Loa (Tomczyk et al. 1995) whose description is the above, another one at Mt. Wilson (based on only one cell, Rhodes et al. 1986, Korzennik 1990), and at Rome (Cacciani et al. 1988)

Fourier Tachometers. These are special purpose versions of Fourier transform spectrographs. Briefly, incoming light passes through a prefilter (transmission function $T_P(\lambda)$) which isolates the desired solar absorption line. Then it enters a single tunable Michelson interferometer, whose transmission function can be approximated by $T_M(\lambda) = [1+\cos(R\lambda+\varPhi)]/2$ near the filter bandpass, where R is proportional to the mean optical path difference, and $\varPhi$ is a phase determined by the tuning of the MI. After an optical system images the Sun onto a detector, each pixel sees an intensity given by $I(\varPhi) = \int I_S T_P(\lambda)T_M(\lambda)d\lambda$; this integral is the sum of three terms: a constant term related to the mean intensity, and one component each for the sine and cosine transforms of the filtered solar intensity. By measuring $I(\varPhi)$ for three different angles one can obtain estimates of the three terms. These three combinations of the observed intensities called velocity, modulation and intensity (V,M,I) are the outputs of the instrument. This instrument is the one used by the GONG network (Harvey et al., 1988).

Interferometric Based Filters. The use of interferometers as filters with extremely narrow bandpasses has also been a means of measuring Doppler shifts. Such filters have the advantage that they can be tuned to use any desired absorption line in the spectrum; however the difficulty consists in their stability, as the control of the optical length in between the interferometric arms and the exclusion of other interferometric orders than the required wavelength, is quite a demanding task. Another problem is the acceptance angle, which in this case should be a nearby parallel beam (specially for Fabry-Perot type); these demands have prevented the wide use of Fabry-Perot interferometers although some useful observations have been achieved (Rust et al. 1986). The use of MI has more advantages as they have a greater acceptance angle and a lower temperature sensitivity. The idea is to replace one or two of the path difference elements in a Lyot filter with equivalent elements involving polarizing Michelsons; the problems here are that more than one Michelson element is required to achieve the spectral resolution wanted, and that the actual size of them has to be big. The building of such big Michelson blocks of the required quality is not an easy job. This is the approach followed by the MDI/SOI instrument onboard SOHO (Scherrer et al. 1995), in which a 7.5 cm aperture telescope images the entire Sun on a 1024×1024 CCD camera with a projected pixel size of around 2 arcsec. This allows observation of p-modes with $\ell \lesssim 700$. The optional use of an extra lens provides an even higher spatial resolution (nearly a factor of 3) which also allows MDI to measure modes with very high-ℓ indeed. Further, the use of a Doppler measurement sequence of 4 measurements across the line profile provides a way to calibrate the non-linearities of the 2-point technique and it allows the estimation of other quantities such as continuum intensity and line centre depth.

Observing Very High-ℓ Modes.

Ca II line core Photometric Techniques In order to get a S/N ratio better than in the continuum, many observers chose to measure photometrically the flux as emitted in the core of the Ca II K line. Such simple instruments consist of a small telescope, an appropriately controled filter (bandwidth 0.5-1 nm) and a CCD detector, plus a data acquisition system. They are also very robust, and at least three teams have used such a set-up to better measure the very high-ℓ modes: TON (Chou et al 1995), HLH (Bachmann et al. 1995) and POI (LaBonte et al. 1995). These can be better measured in intensity rather than velocity because the velocities associated with acoustic oscillations are predominantly vertical and therefore, at the limb they are almost perpendicular to our line of sight while the intensities are not angle dependent. This means that a larger fraction of the solar surface can be better observed in intensity than in velocity leading to a better discrimination between modes with similar ℓ and m, when these are large.

3.2   Observational Strategies

Major degradation of the solar oscillations spectrum occurs when there are gaps in the data string. In fact, this causes data to be multiplied by a function of value 0 or 1 when data doesn't exist (or it is not useful) or when it does; such a function is called the observing window function. Consequently the observed power spectrum is the convolution of the "true" one (that would have been obtained with 100% duty cycle) with the power spectrum of the window function. The effect of such a convolution is that power that should be at one frequency leaks to another one; the more regular the observing window the worse the effect is, since ghosts frequencies appear near the real ones. This is particularly important when single site observations are made; as a consequence of the 24 hour period in the observing window the degradation of the spectrum results in: a) the appearence of an enormous peak at 1 day^{-1} and its corresponding harmonics (see Fig. 2), and b) the appearence of sidelobes (ghosts peaks) situated at both sides of any peak present in the spectrum ($\pm 11.574 \cdot h$ μHz, with $h = 1,2,..$) leading to an overcrowded spectrum (see also Fig. 5).

Therefore, it became apparent in the early 80's that the observing window function should be improved. Actions were taken in three directions:

1. Observe from very high geographical latitudes, so that in local summer days are close to 24 hours long and therefore the gaps are reduced to the minimum. The first serious attempts were made by Grec et al. (1980) in velocity measurements from the South Pole and by Duvall et al. (1986) from the same site in photometric mode. However, it has almost been abandoned because the gaps created by bad weather prevented obtaining long runs ($\gtrsim$ 2 months) with $\gtrsim 60\%$ duty cycle.

2. Observe from different sites appropriately separated in geographic longitude and at about the same latitude. The first serious attempts in this direction were made by the Birmingham-Tenerife group in 1981 observing spectrometrically from Maui (Hawaii islands) and Tenerife (Canary Islands) and by the Tenerife-ESTEC/SPD group observing photometrically from S. Pedro Mártir (México) and Tenerife (Jiménez, 1988; Jiménez et al. 1990). These experiences led to the later deployement of several networks which are now being working regularly with different duty cycles, from $\sim 87\%$ for GONG down to more modest 35% for TON, passing by intermediate values from BiSON (Chaplin et al. 1996) and IRIS (Fossat 1995).

3. Use space to obtain the best observing conditions. Although the first results on Helioseismology were serendipitiously obtained (Woodard and Hudson, 1983) from the radiometer ACRIM onboard SMM (even having eclipses), the first photo-Helioseismometer sent to space was IPHIR which worked during some 160 days onboard the PHOBOS spacecraft. The final planning of the Helioseismology instruments onboard SOHO (Domingo et al. 1995) has brought the possibility of obtaining long continuous observations, from its position at the L1 point of the Sun-Earth system, with actual measured duty cycle above 99% (GOLF experiment).

The current state of the art in these strategies and instrumentation involved can be found in Pallé (1996) and Toutain (1996).

4   Helioseismic Analysis Tools

Once we get the raw data provided by any of the excellent instrumentation available for Helioseismology nowadays, one realises that this is at most half of the story. Unlike many other astronomical observations where one single good image, or maybe a few of them, is enough to observe an interesting event or describe an interesting phenomenon, here a great amount of data must be

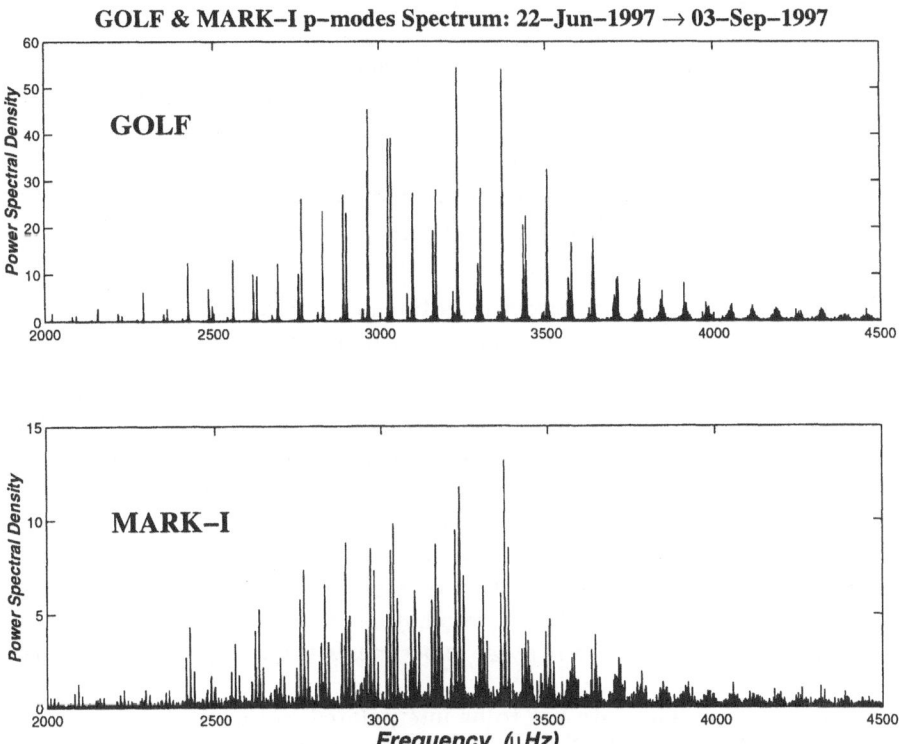

Fig. 5. Spectrum of the radial velocity of disk-integrated sunlight obtained by two resonant scattering instruments, one from ground (Mark I at O. del Teide, Tenerife) and another one from space (GOLF), during the same interval of time. Notice the effects of the observing window function (duty cycles of 44% and 99.9% for Mark I and GOLF respectively). *Courtesy of P.L. Pallé.*

acquired before the data reduction system can yield the measurement of a single parameter suitable to be compared with theoretical predictions. The process will start with "cleaning" the raw data and then Fourier analysing them; the resulting spectra must be appropriately fitted and the characteristics of individual modes obtained; finally, these will give information on the mode physics and the physics of the solar layers where they resonate.

As we have already said in Sect. 2 the p-modes are defined by their eigenfunctions (identified by n, ℓ and m) and their eigenvalues which are their resonant frequencies $\nu_{n\ell m}$. The radial structure, parameterised by n, is not directly accessible to observations because at the solar surface only the angular structure can be used to isolate and identify the different modes. Also by fitting the appropriate model, other parameters of the mode like their linewidths (i.e. lifetimes) and powers can be obtained.

4.1   Global Methods of Extracting Mode Parameters

The calibrated output of our instruments is either observed radial velocity $v_{obs}(x, y, t)$ or intensity $i_{obs}(x, y, t)$ as a function of position on the detector and time. When working with disk-integrated sunlight or similar very low resolution detectors (LOI) the appropriate integration over space coordinates is already provided by the instrumentation; as a consequence the steps to obtaining the power spectra are a bit simpler.

Obtaining the Power Spectrum. Starting from a time series, the data analysis and reduction package should provide the following steps:

1. Cleaning, merging, gap filling... First, data have to be appropriately cleaned rejecting the bad data due to weather, instrument malfunctioning, etc. or at least they should be labeled so that care can be taken later on. Depending on the desired data length to be analysed, data gaps (both in space and time) should be filled out appropriately; if data from another station exists (as in the case of the networks) both series should be merged. Of course, in doing both processes: gap filling and merging, many decisions (as to what data, or combination of them, should be used) have to be taken. Notice that in space experiments these steps are enormously reduced.
2. Interpolation. The data has to be interpolated to the new surface spherical coordinates (θ, ϕ). In doing that several problems will also arise: a) defining the limb and measuring its radius is one of the most critical problems because they will also define the number ℓ of the modes; b) re-orientate appropriately the solar image using the P and $B_{\rm o}$ angles for the solar rotation axis; this process might also take place before the gap filling and merging. At this point we have a time series, say $v(\theta, \phi, t)$.
3. Longitudinal projection. Here we apply an FFT over the ϕ angle which provides the projection of the $v(\theta, \phi, t)$ onto the azimuthal angular part of the spherical harmonics. Doing it in this way will save time because the FFT requires only $N \ln N$ rather than N^2 operations, where N is the dimension of one of our axis. The output of this step will be a set of time series $V_m(\theta, t)$.
4. Latitudinal projection. Here the second projection onto the latitudinal part of the spherical harmonics is made. In short we should perform a Legendre transformation, for each m:

$$B_\ell^m(t) = \int_{S_{\rm obs}} V_m(\theta, t) \, P_\ell^m(\theta) \, \sin\theta \; d\theta \; , \qquad (8)$$

here one of the major problems appears because of the fact that the spherical harmonics are not completely orthonormal on just the observed surface of the Sun (not even a complete solar hemisphere). This leads to the uncomplete separation of modes with similar ℓ and m values (when these are high) therefore yielding a "leakage" of power from one mode to another. In the end, we are left with the coefficients $B_\ell^m(t)$ (complex, where the real and imaginary parts correspond to the amplitudes of the $P_\ell^m(\theta)$ times $\cos\phi$ and $\sin\phi$ respectively).

5. Fourier transformation in time. Now the Fourier transformation of the $B_l^m(t)$ in time is done. However, before doing this a transposition of such coefficients should be made; this is a process that can take a lot of computing time, given the amount of data that one has to treat. The gap filling (in time) could also be made at this point. From here we end up with $P_{\ell m}(\nu)$, the power spectrum of the mode (ℓ, m); this is what we have called in Sect.2, P_{obs}.

An Alternative Method for Step 5: The Homomorphic Deconvolution Technique. As already mentioned some factors determine the shape and width of the observed p-modes such as the limited coherence time of the oscillations and variations in amplitude and phase with time, consequence of its excitation. In some respects, these properties can be considered as limitations in the Fourier analysis, which is only strictly applicable to signals that are stationary with time. In our case Fourier techniques may be limited in their capacity to separate the various effects of amplitude and frequency variations with time.

An alternative analysis method is used which takes into account the particular behaviour of the p-modes. "Homomorphic deconvolution" aims to eliminate the effects of non stationarities of the signal. It has been already used in acoustics (Oppenheim et al. 1968) and geophysics (Ulrych, 1971) for its ability to separate the "echoes" from the signal (due to the reflection of the signal on geological layers). It has been shown to be promising in the case of Helioseismology (Baudin 1993, Baudin et al. 1993) and it is being thoroughly tested on simulated as well as observed data.

This technique tries to separate the two components of the helioseismic signal (the damped oscillator from the excitation) by filtering the unwanted one. This is done in three stages. The first is a complex transformation (involving Fourier transform and logarithm) of the signal. Because of its analogy with the Fourier transform, a similar terminology was invented: the result of this transformation was called the "cepstrum" of the signal, giving the variation of "maplitude" versus "quefrency" (Bogert, Healey & Tuckey 1963). The main property of this cepstrum is that the two above mentioned components are well separated: the excitation is mostly concentrated at high quefrencies while the oscillatory is distributed in the whole quefrency range.

$$\hat{S}(\nu) = \hat{O}(\nu) \cdot \hat{E}(\nu) \rightarrow \log \hat{S}(\nu) = \log \hat{O}(\nu) + \log \hat{E}(\nu) \rightarrow$$
$$\rightarrow Cepstrum = TF^{-1}(\log \hat{S}(\nu)) = Cepstrum(O) + Cepstrum(E)$$

The second step consists in the filtering of the cepstrum in order to eliminate the excitation component of the signal; it can be achieved, either with simple

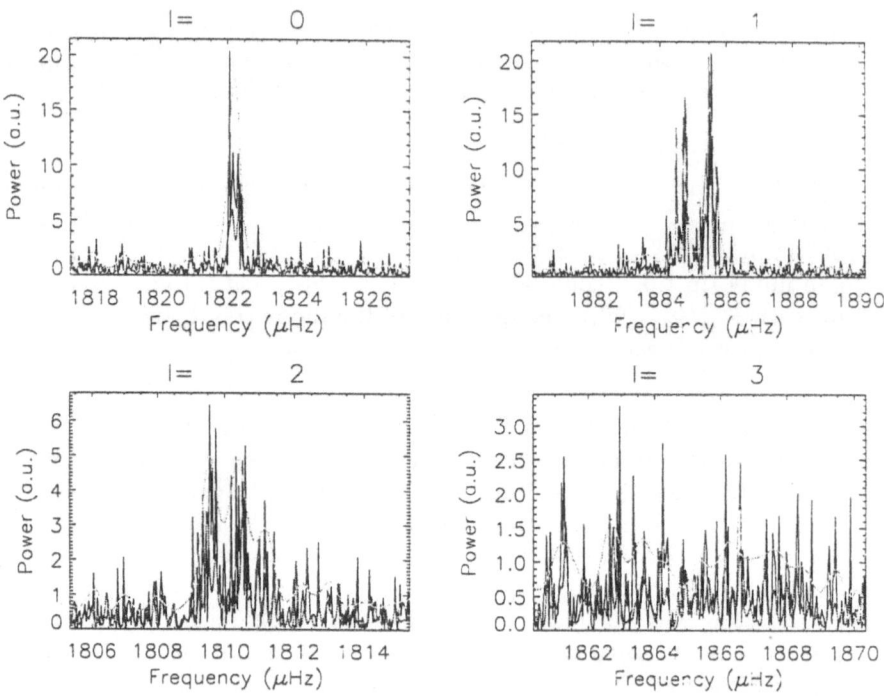

Fig. 6. Slice of the low-ℓ acoustic spectrum of the Sun (based on one year of GOLF data), showing neighbour modes (n, ℓ): (12,0), (12,1), (11,2) and (11,3). In black is the spectrum calculated using standard FFT technique and superimposed in lighter grey colours are the result of the HD technique applied to the same data using two different filters, a soft and a strong one.

filters such as low-pass filters and/or with Cepstra averaging and/or with a more sophisticated quefrency dependent filters,

$$Cepstrum \cdot Filter = C_F \tag{9}$$

Needless to say, this is the critical step in the method because it modifies the contents of the excitation spectrum (and hopefully only this) therefore changing the spectral distribution of the noise in the Lorentzian shape of the peaks. The third step is merely a return from the cepstrum to the deconvolved signal or to its power spectrum.

$$TF(C_F) \to e^{TF(C_F)} = \hat{S_F}(\nu) \to P_{\text{obs}} \tag{10}$$

The different results obtained by the standard FFT technique and this one can be seen in Figs. 1 and 6 in a power spectrum calculated from observed GOLF data and on simulated data respectively.

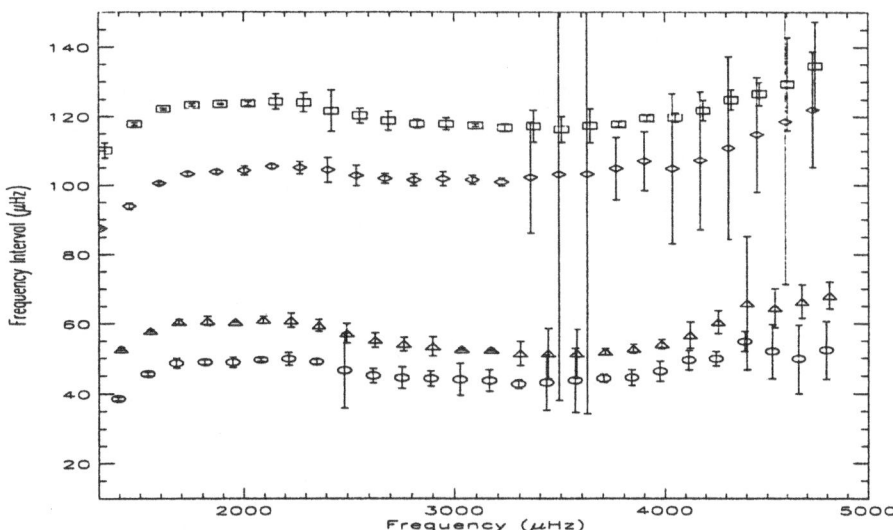

Fig. 7. "Echelle" diagram (with a fixed frequency interval of 135.5 μHz) of the frequencies found with the GOLF experiment. Also drawn, for clarity, are the error bars as $\pm 100\sigma$ for modes with $\nu < 3.7$ mHz and as $\pm 10\sigma$ for modes with higher frequency. Triangles, squares, circles and diamonds are for $\ell = 0$, 1, 2 and 3 respectively.

Peak Fitting in the Observed Spectra. We can fit this power spectrum with a function such as eq. (4) to obtain the values of $\nu_{n\ell m}$, $\Gamma_{n\ell m}$ and $S_{n\ell m}$. There are new problems derived from the non-linear fitting that one has to make and the non-stationarity of the oscillating signals due to the excitation function. This step in the analysis is common to power spectra that come either from a high-ℓ experiment or a low-ℓ one; the fitting strategy may not be exactly the same but the idea behind the process holds for both.

The technique to be used is a Maximum Likelihood one. This procedure consists in looking for a maximum likelihood function able to estimate the parameters **a** of the model $M_i(\mathbf{a}, \nu)$ which best fits the properties of the observed spectra P_{obs}. As we have already explained in Sect.2 the probability density function scaled to any expected value $M_i(\mathbf{a}, \nu)$ is:

$$P_i = \frac{1}{M_i(\mathbf{a}, \nu)} \cdot e^{\frac{-P_{\mathrm{obs},i}}{M_i(\mathbf{a}, \nu)}} \tag{11}$$

where i stands for a given frequency channel in the spectrum. The likelihood function for the whole spectrum is $L = \prod_i P_i$ where L is the likelihood function.

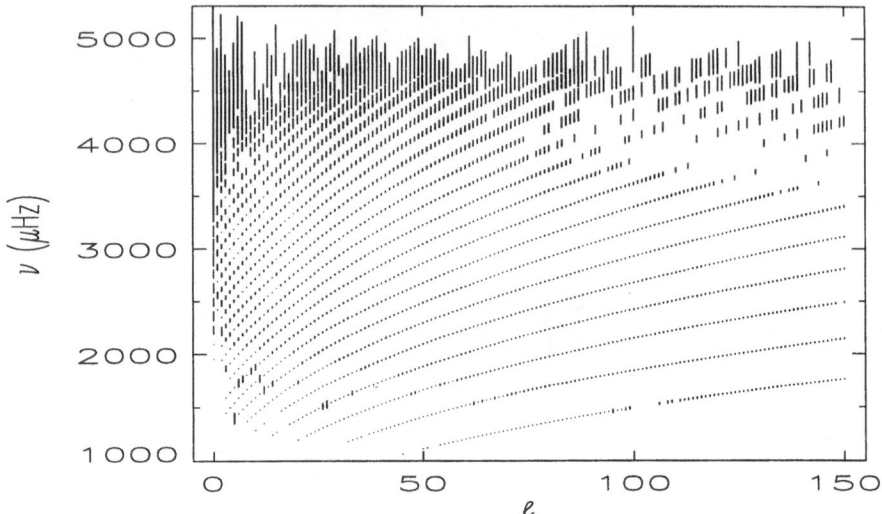

Fig. 8. Central frequencies, concentrated in ridges, obtained by GONG ground-based network in a 2 months run. The errors bars are $\pm100\sigma$ where σ is the formal error on the fitted frequency. Obtained from GONG web home page (see the references).

Therefore we get:

$$-\log L = \sum_i \left(\log M_i(\mathbf{a}, \nu) + \frac{P_{\mathrm{obs,i}}}{M_i(\mathbf{a}, \nu)} \right) \equiv S \qquad (12)$$

Consequently, maximizing L is equivalent to minimizing S with the model:

$$M_i(\mathbf{a}, \nu) = \sum_{j=1}^{N} \frac{S_j(\Gamma_j/2)^2}{(\nu_i - \nu_{o_j})^2 + (\Gamma_j/2)^2} + \sum_{k=1}^{n} c_k \nu_i^{k-1} \qquad (13)$$

with $\mathbf{a} \equiv (S_j, \Gamma_j, \nu_{o_j}, c_k)$, c_k being the coefficients describing the solar background power in the fitting frequency interval. The method has been extensively described in Anderson et al. (1990) and Schou (1992) where also some results can be seen. If the m-averaged spectrum is made, or an average of enough realisations is taken, or the spectra coming from the HD technique is calculated, then the peak fitting can be equally well fitted by using a least squares minimisation technique because the measurement errors approach a normally distributed function. However, it should be stressed that the knowledge of the precise distribution of errors is needed in order to actually use the χ^2 distribution with the appropriate number of degrees of freedom. In this way the errors in the parameters that are the results of the fit will be adequately obtained, otherwise these errors may be underestimated.

The procedure has some problems derived from the assumptions made in the non-linear fitting to the data, such as: a) the selection of the method to solve the non-linear equations; b) the actual number of free parameters in the fit; c) how to fit the background (how many c_k's); d) the initial guesses for such parameters; e) the frequency interval over which the fit is made; f) how many lines are fitted simultaneously; etc.. All of these assumptions require decisions that will influence the output of the fit, in particular the values of a which will be the ones to be compared with theory. All this may lead someone to the conclusion that the actual parameters obtained have large uncertainties and depend on the many assumptions one has to make until arriving at the end of the data analysis. That would be the wrong conclusion because the frequencies of the p-modes are the most precise parameters that have ever been measured for the Sun. Indeed, the achieved precisions are smaller than one part in 10^5 (see Figs. 7 and 8); for linewidths and powers results see e.g. Libbrecht (1988). Nonetheless, these can still, and need to, be improved in order to obtain some clues, e.g. on the solar core, thus the influence of all these "details" in the final measurement need to be well evaluated.

4.2   Local Methods for Very High-ℓ Modes

If our interest is to probe the structure of small-scale features just below the solar surface: spots, local velocity flows, etc... then it is more convenient to use a slightly different approach than the one followed up to now. Indeed in those cases acoustic waves can be regarded as local phenomena; then very high-ℓ modes ($\ell \gtrsim 200$) or very high frequency modes do not survive for long enough to propagate all around the Sun and a local treatment is possible and desirable. Up to now two interesting approaches have been explored:

Rings and Trumpets Analysis. This analysis uses the plane-wave decomposition technique to calculate a 3-dimensional power spectrum of the solar oscillations of a small portion of the solar surface (roughly a few degrees squared); the three dimensions are: k_x, k_y (the two horizontal components of the wavenumber vector $\mathbf{k}$), and $2\pi\nu$ (the temporal frequency). This three dimensional spectrum shows significant power on 2-dimensional surfaces in such a space much like the end of a trumpet; this is the product of the rotation of the ridges (shown in Fig.8), in a (k, ν) space, about the frequency axis since sound waves do not care about their horizontal direction of propagation. Nice circular rings are obtained by taking slices at constant ν centered on the k_x, k_y origin. However, in the presence of horizontal velocity flows in the Sun's surface, the frequencies of the modes are Doppler shifted due to advection of the wavefront by such flows: $2\pi\Delta\nu = \mathbf{k}\cdot\mathbf{U} = k_x U_x + k_y U_y$, where $\mathbf{U}$ is the horizontal velocity flow. This frequency shift displaces the rings (and its center) and by measuring such displacements the velocity of the flow can be also calculated. Moreover, if such flows change their velocity with depth, different modes of given **k** and ν will be displaced differently. Therefore, this information can be inverted to calculate the depth dependence of such sub-surface velocity flows; further, by studying different portions of the Sun one can gain information on their latitude and longitude variation. Details of the method can be found in Hill (1988), Patrón (1994), Haber et al. (1995), Patrón et al. (1995).

Time-Distance Analysis. This method is the most similar one to traditional terrestrial seismology which uses a ray formulation to get a picture of the p-mode phenomenon at the solar surface. It was first described by Duvall et al. (1993). Very simply the method considers an arbitrary point on the solar surface to be the source of acoustic waves. A wave leaving this point, at time t_0 with a particular frequency, in a given direction follows a trajectory and eventually returns to the surface at a distance d away where it causes local displacement of the surface; in the absence of other sources of acoustic waves, the wave signal at d will look identical to the signal at its origin but delayed by a time $T(d) = \int_l \mathrm{d}s/c$, where c is the local sound speed and the integration is taken along the ray-path. The waves arriving at distances close to d travel along slightly different paths, sampling different depths of the Sun, and hence carrying different information. If one could map the dependence $T(d)$ for all possible distances, one is able to sample to a fair depth in the Sun and could estimate the sound speed (and the temperature) there (Kosovichev et al. 1997).

On the other hand, Braun et al. (1988) noticed that p-mode propagation was not adiabatic or could even couple each other in the presence of strong magnetic fields. Recent work using this method has also used phase or reflection information in the time-distance relation to infer things about sunspots and other sources of absorption and scattering of sound waves (Bogdan et al. 1993).

4.3 What About g-Modes Analysis?

As you are already aware, most of what we have said up to now is related to p-modes physics. There are obvious reasons, the most important is that there is a general consensus amongst the scientific community that g-modes have not already been detected. However, this does not mean that it has not been tried; on the contrary, in the 80's it was tried very hard indeed and you can find full account of these trials in Pallé (1991) and Hill et al. (1991). Specifically the main *leit motiv* of the Helioseismology of SOHO is precisely to get rid of the influence of the Earth's atmosphere to improve the signal-to-noise ratio in the frequency range of interest for g-modes. Although the data from SOHO is still flowing it does look as if the problem of detecting g-modes still is similar to the one we had to face in the past; it is a problem of looking for oscillatory signals hidden in random noise with a very poor S/N ratio. Techniques already used for this search in the past, plus new ones that had already not been tried, plus the ones that we will have to invent for this particular problem will be the difficult field in which you may choose to labor and make productive in the near future. I hope that You, young colleagues will have the words and the answers.

5 Comparison Between Model's Frequency Predictions and Reality

Among other objectives Helioseismology includes the derivation of the interior structure from resonant frequency values, the state of the Sun's internal rotation from multiplet's frequency splittings and the excitation and damping of oscillation modes from coherence times and resonant line shapes. In the following, some of the methodologies to tackle the first two problems will be briefly outlined whereas details of the third (probably the less known so far) can be found in Goldreich and Kumar (1988).

The problem of comparing the observations to the predictions of the theory of solar oscillations is what has been called the Forward Problem. In short, it consists in building an equilibrium model of the present Sun (as accurately as one can), perturbing such an equilibrium model and solving the oscillation equations to calculate the frequencies of such a model; then these predictions are compared to the observations and if they don't match, the equilibrium model for the Sun has to be re-calculated again by changing some of its parameters. The process continues by iteration until a good match (within observational errors) is achieved, if possible. Therefore the procedure includes computation of several models and the one that predicts the frequencies that best match the observed ones is selected; indeed, care must be taken in the combination of measured frequencies that allows the isolation of specific parameters or hypothesis in the solar interior calculation. A problem of this procedure is that it does not provide indications on what is deficient in the models. In other words, we should know what parameter and/or assumption has to be changed and at what depth; otherwise, in the best case, we could be changing parameters whose effects may cancel out in the calculated frequencies.

In fact, what we test in doing this is the hypothesis and physical properties that hold the calculation of the theoretical model, which drives us to the problem of calibrating solar models. The best output of this procedure is probably an hypersurface in the model's parameter space in which the Sun is very likely to lie. However, as models are not perfect and have uncertainties in their physics then the borders of such a region become somewhat fuzzy and even the extent of such a region is questionable.

An interesting possibility is to use the Inverse Problem of Helioseismology which consists in deriving, from the observations of the oscillations frequencies, the structural parameters and/or the dynamics of the interior layers of the Sun. The volume edited by Roca Cortés and Sánchez (1996) contains the courses on The Forward and the Inverse problems given by J. Christensen-Dalsgaard and D.O. Gough in the VIth Canary Islands Winter School from which some of the methodology expressed here is taken. Also the volume edited by Cox, Livingston and Matthews (1991) contains several useful and comprehensive reviews.

5.1   The Forward Problem in Helioseismology

The solar oscillations obey the standard equations of motion of the continuity, momentum and energy conservation, along with Poisson's equation which defines the time dependent gravitational potential. This general hydrodynamic description of the star is far too complicated to handle, even numerically with existing computers. However, in solving the solar oscillations problem considerable simplifications are possible that yield the problem tractable: a) the oscillations can be treated as small perturbations to the equilibrium state (the observed amplitudes are very small, i.e. $\delta R_\odot / R_\odot \approx 10^{-7}$, or $\delta L_\odot / L_\odot \approx 2 \times 10^{-6}$); b) the periods of oscillation are very small compared to the Kelvin-Helmholtz characteristic time scale (i.e., $\tau_{osc}/\tau_{KH} \approx 10^{-12}$). These allow the simplifying assumptions of linearity and adiabaticity. With these assumptions, the above mentioned equations are greatly simplified and when supplemented by the appropriate boundary conditions which will select the physically realisable solutions, form a complete fourth-order system of ordinary differential equations. It has non-trivial solutions only for discrete values of the frequency of the oscillations which become eigenvalues of the problem. However as the equations are homogeneous, the solution is determined within a scale factor which leads to an indetermination of the amplitude of the oscillation.

This system can also be solved by using some simplifying assumptions in what is called the asymptotic approximation. Gough (1986) has shown that in such a case the oscillations obey a second order differential wave equation in $c^2 \rho^{1/2} \cdot \mathrm{div} \boldsymbol{\delta r}$. The dispersion relation gives the dependence of a wavenumber k with the oscillation frequencies ω; it also depends on a generalized atmospheric cut-off frequency ω_c, a local acoustic frequency or Lamb frequency S_ℓ and a gravity wave or a buoyancy frequency N_{BV}. The atmospheric cut-off frequency is relevant only in the surface layers. Propagating waves are possible only where the wave frequency is either greater than both the Lamb and buoyancy frequency (p-modes) or smaller than both the Lamb and buoyancy frequency (g-modes). The p-modes are most sensitive to the Lamb frequency which is proportional to the *value* of the sound speed while the g-modes are most sensitive to the buoyancy frequency which is proportional to the difference between the actual and adiabatic density *gradients*. Another interesting fact is that their eigenfunctions are concentrated towards the solar centre in contrast to the eigenfunctions for the p-modes which are more concentrated towards the solar surface. The sensitivity of the mode frequencies to solar structure depends on the amplitude of the mode as a function of position through the Sun's interior. For this reason, g-mode analysis contributes more sensitively to the determination of the solar core structure than p-modes.

Small Changes in the Asymptotic Relation. One of the most interesting results of the asymptotic analysis is the so-called *Duvall's law*. Duvall (1982) found, from an analysis of the observed frequencies that by plotting the observed frequencies in the form of $(n + \alpha)\pi/\omega$ against the parameter ω/L, where $L = (\ell + 1/2)$, all observations fall into a single curve F, for a value of $\alpha{=}1.58$. Although this can be rigorously obtained from theory, it can be roughly justified from the dispersion relation for plane sound waves: $\omega^2 = c^2|\mathbf{k}|^2$, where $\mathbf{k}$ is the wavenumber and c the sound speed, using the following argument. In order to obtain a standing wave we must require that an integral number of oscillations is to be found in the radial direction between the inner turning point and the outer one (photosphere), taking into account that some phase shift may arise at those points; thus,

$$\int_{r_I}^{R} k_r dr = n(\pi + \alpha) \ \rightarrow \ \int_{r_I}^{R} \left(1 - \frac{L^2 c^2}{\omega^2 r^2}\right)^{1/2} \frac{dr}{c} = \frac{[n + \alpha(\omega)]\pi}{\omega} \qquad (14)$$

notice that this is of the form $F(\omega/L) = (n+\alpha)\pi/\omega$ and that the right hand side is known from observation. Therefore the integral can be inverted to obtain the unknown $c(r)$. The results of the first inversion of this kind is to be found in Christensen-Dalsgaard et al. (1986) and later on we will come back to this.

A further interesting thing here is that this relation can be used to perform tests between models that differ solely by a small quantity somewhere deep in the Sun. To introduce this, let's consider $\delta f(r)$ to represent a small change (e.g., of the opacity κ or the sound speed c, or etc.) in the dispersion relation already used above and let's see in what way Duvall's law is modified. Ultimately, we are interested in the effect that such a change will have in the oscillation frequencies, and we obtain (to first order):

$$S\frac{\delta\omega}{\omega} \simeq \frac{1}{2\omega^2} \int_{r_I}^{R} \left(1 - \frac{L^2 c^2}{\omega^2 r^2}\right)^{-1/2} \delta f \frac{dr}{c} + \pi \frac{\delta\alpha}{\omega} \ , \qquad (15)$$

where

$$S = \int_{r_I}^{R} \left(1 - \frac{L^2 c^2}{\omega^2 r^2}\right)^{-1/2} \frac{dr}{c} - \pi \frac{d\alpha}{d\omega} \ , \qquad (16)$$

which is what we wanted. Notice that equation (15) has a simple interpretation as a weighted average of $\delta f/\omega^2$ with a weight that is just the sound

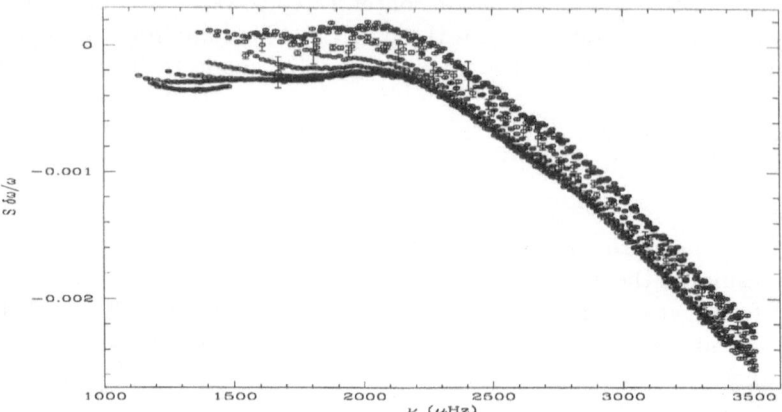

Fig. 9. Scaled frequency differences between observations (from GOLF and LOWL) and model S (see text) in the sense (observations)-(model), drawn against cyclic frequency, ν. Here S is defined in equation (16); τ_0 is the acoustical radius which corresponds approximately to $S(L/\nu = 0)$ (see Fig.10). At this range of frequencies, model and observations already agree to within 3×10^{-4}. However, most of the differences are only a function of ν, which according to equation (17) indicates that most of the model's discrepancies are due to an incorrect modelling of the uppermost layers of the Sun. Nevertheless, since the scatter in the figure is much larger than the frequency errors there must be differences between the Sun and the model in deeper layers too. *Courtesy of F. Pérez Hernández*

travel time along the ray describing the mode. Moreover, notice that (15) can also be interpreted as:

$$S\frac{\delta\omega}{\omega} \simeq \mathcal{H}_1\left(\frac{\omega}{L}\right) + \mathcal{H}_2(\omega) \tag{17}$$

This is an interesting result because both functions can be obtained separately, to within an additive constant, by means of a double-spline fit to the p-mode frequency differences. Moreover, the dependence of $\mathcal{H}_1$ on (ω/L) is determined by the change introduced by δf throughout the whole star, while $\mathcal{H}_2$ dependence on ω depends only on differences in the outer layers. Pérez Hernández (1989) and Pérez Hernández and Christensen-Dalsgaard (1994) have developed this way of testing the assumptions and parameters of the solar models (see Figs. 9 and 10).

Moreover, this approach is also valid to compare the observed frequencies with those that are predicted by a given model, provided the model is close enough to reality. Then, by inverting such an integral one can obtain, e.g. the sound speed variation, along the radius of the Sun with an increased precision (Christensen-Dalsgaard et al. 1989).

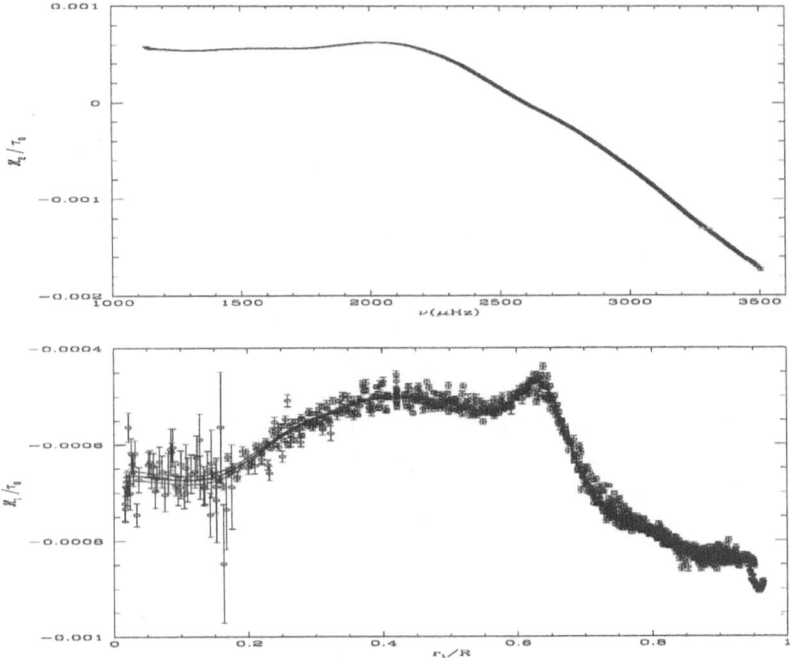

Fig. 10. Top graph: the function $\mathcal{H}_2$ obtained by fitting the scaled frequency differences shown in Fig. 9 to two functions as given by equation (17). Notice that at $\nu < 2$ mHz, $\mathcal{H}_2$ is flat, because at these low frequencies modes have low amplitudes near the surface, hence they are not much affected by uncertainties in these layers. Moreover, from the small oscillatory pattern clearly visible, information about the equation of state and helium abundance can be extracted (Pérez Hernández and Christensen-Dalsgaard, 1994). Modes with higher frequencies are definitely affected by the poorly known physics at the upper layers which have not been properly included in the model yet. In the graph below, points are the scaled frequency differences after subtracting the function $\mathcal{H}_2$, whereas the continuous heavy line is the fitted function $\mathcal{H}_1$ plotted against the inner turning point r_{I}/R, directly related to ν/L; thin continuous lines are 1σ errors estimated from a Monte Carlo simulation. Notice that in equation (15) replacing δf by δc, the function $\mathcal{H}_1$ carries information about sound speed differences between the model and the observations. *Courtesy of F. Pérez Hernández.*

5.2   What Do the Eigenfrequencies Tell Us?

Inspection of the equations of oscillation reveals that the coefficients are determined by the following parameters of the equilibrium model: ρ, p, Γ_1 and g, which in turn are related through the equilibrium equations. In this way only the two functions $\rho(r)$ and $\Gamma_1(r)$ are strictly independent and thus the frequencies of oscillation are determined solely by these two parameters.

Consequently strictly speaking, if no other constraints apply, the frequencies can only give information on such parameters. Obviously, any other pair of variables can also be used that are related to those mentioned. Given that p-modes are standing acoustic waves, their frequencies will be largely determined by the sound speed c; it will be natural to use this variable with any other of the two above mentioned. As we will see, the sound speed can be obtained with high accuracy throughout the Sun and therefore a measurement of T/μ is also posible; but it is important to note that the temperature itself can not be obtained from the knowledge of the frequencies of oscillation alone.

The existence of rotation and/or a magnetic field implies that the equilibrium state is not spherically symmetric. The larger effects to the eigenvalues come from rotation whose effect on a given acoustic mode (n, ℓ) is to split into $(2\ell+1)$ modes, each with a different m value (azimuthal degree) which goes from $+\ell$ to $-\ell$, each separated by an amount proportional to the rotational velocity value in the cavities where the modes propagate.

$$\delta\nu_{n\ell m} = \nu_{n\ell m} - <\nu_{n\ell}> \simeq m\,\frac{\bar{\Omega}}{2\pi} \qquad (18)$$

where $<\nu_{n l}>$ is the frequency averaged over m (or unperturbed energy) and $\bar{\Omega}$ is an appropriate average of $\Omega(r, \theta)$ over the cavity sampled by the modes. Once this is found, one can then calculate higher order terms, quadratic in Ω, like a perturbation in the Coriolis force. An even function of m is also the effect caused by a non-rotating velocity field, a magnetic field and some sort of asphericity contribution to the sound speed. Evidently, the influence of these three perturbations cannot be separated with the help of the observed frequencies alone.

5.3   The Physics in the Sub-Photospheric Layers

Fortunately the knowledge of the eigenfunctions deep into the Sun is quasi-independent of the conditions at the upper boundary. However, uncertainties in the physics of the sub-surface layers of the Sun, particularly in the upper superadiabatic boundary layer of the convection zone, suggest the use of approximate boundary conditions at the surface, which modify in an unknown way the values of the eigenfrequencies. Such effects can be analyzed using the above described method (Pérez Hernández and Christensen-Dalsgaard 1994), where the perturbation can be the incorrectly treated features either in the model (e.g. errors in the hydrostatic structure introduced by the neglect of the turbulent pressure) or in the physics of the oscillations (e.g. the failure of the adiabaticity). At such a shallow region only modes of high ℓ propagate; for other modes the propagation is almost vertical and thus at a given frequency the eigenfunctions are essentially independent of ℓ, thus leaving to the functional $\mathcal{H}_2$ the information on such layers. Overall, this situation is presently highly unsatisfactory and reality must be considered soon;

the availability of increasingly realistic hydrodynamical calculations of these layers and the increasingly accurate inversions from local Helioseismology should help to understand the physical processes there. However, magnetic fields still are an important ingredient in these layers and as yet they are not taken into account in such calculations. A lot of work is waiting to be done.

6   The Inverse Problem and Seismic Models of the Sun

Models of the Sun can be built by making a small number of plausible hypothesis and involving the best known physical processes (both macro- and micro-physics). The models built in this way are called "standard solar models" and a complete account on how they can be built can be found in Bahcall and Ulrich (1988). Also a complete description of the theory of the solar interior with a discussion of the related problems of solar neutrino fluxes has been given by Turck-Chièze et al. (1993). Neutrino flux measurements and Helioseismology have appeared as complementary probes of the solar internal structure. The present detection of thousands of acoustic modes has led to determine precisely the sound speed and the density inside the Sun down to 0.3 $R_\odot$ as it will be commented later on.

However, the observed solar oscillation frequencies provide a wealth of information, in quantity and quality, that can be used to deduce the solar structure. Indeed they provide stronger constraints for the solar models going much further than the knowledge of the global properties of stars like luminosity, mass, radius and neutrino flux. In fact, none of the solar models built so far are able to reproduce the observed frequency measurements available at the level of accuracy of the observational errors. Therefore, one can call any theoretical description of the Sun built by helioseismic inversions of the available frequency measurements a Seismic Model of the Sun.

6.1   The Idea of Inversion: a Simple Case for $\Omega(r)$

The idea of an inversion is to find a representation of the exact solutions of the oscillation equations chosen in such a way that they fit the observed data adequately given their uncertainties. Given this definition and since if such a representation exists, there exists an infinity of alternative possibilities, the problem is how to choose the "best one". One of the criteria can be smoothness and, depending on the method used to perform the inversion, the definition of such a term may vary. Another important issue in the inversions is to get a fair idea of the uncertainty in the solution. Smoothness in the solution and precision will be competing parameters for which a trade off will have to be made.

To illustrate how to proceed in the inversion let's take a practical and simple example. For that we will take the inversion of the rotational splittings of the observed frequencies. Indeed, in building a solar model one assumes that the Sun is spherically symmetric and therefore the eigenfunctions solution of the oscillation equations are degenerate in m. However, this symmetry is lost by solar rotation in a way already described above. For this example let's take only the sectoral modes with $m = \pm\ell$ which are concentrated towards the equator, then the difference between the frequencies of such modes $\delta\omega_{n\ell}$ will be:

$$\delta\omega_{n\ell} = 2\ell\beta_{n\ell}\bar{\Omega}_{n\ell} \tag{19}$$

where $\beta_{n\ell}$ is a constant very close to 1 and $\bar{\Omega}_{n\ell}$ is an appropriately weighted average of the equatorial velocity $\Omega(r)$ over depth

$$\bar{\Omega}_{n\ell} = \int_0^R K_{n\ell}(r) \; \Omega(r)\mathrm{d}r \; , \tag{20}$$

where $K_{n\ell}(r)$ are functions called kernels that are calculated from the eigenfunctions. The above equation can also be written in a similar way as (15). Therefore we can now invert this equation to calculate $\Omega(r)$ given the $\delta\omega_{n\ell}$. Let's re-write expression (19) in the following way,

$$\frac{\delta\omega_{n\ell}}{2\ell\beta_{n\ell}} \equiv \Delta_i = \int_0^R K_i(r) \; \Omega(r)\mathrm{d}r \tag{21}$$

where i stands for all multiplets (n, ℓ) considered in the data set to be inverted. Now, we do a "linear inversion" to find an aproximation $\hat{\Omega}(r)$ to the true angular velocity at a given r_0 in the Sun, therefore it will be linearly dependent on the data $\hat{\Omega}(r) = \sum_i d_i(r_0)\Delta_i$. Therefore, we have:

$$\hat{\Omega}(r_0) = \int_0^R \tilde{K}_i(r_0, r) \; \Omega(r)\mathrm{d}r, \text{where } \tilde{K}_i(r_0, r) = \sum_i d_i(r_0)K_i(r) \tag{22}$$

$\tilde{K}_i$ are the rotationally averaging kernels; notice that if we are inverting another parameter its inversion may require different averaging kernels. These averaging kernels provide the resolution of the inversion through a measure of its width, whereas the inversion coefficients $d_i(r)$ provide the error propagation of the observational errors to the solution $\hat{\Omega}(r)$. Obviously the resolution and the error propagation are competing parameters; therefore there is a trade off between the width of the averaging kernels and the error magnification in order to obtain the optimum solution.

Methods of Inversion. There are different methods of finding the optimal solution which are commonly used:

1. The optimally localized averages (OLA) developped by Backus and Gilbert (1970) which is based on the calculation of these inversion coefficients for each r_0, whose procedure has just been outlined above. It is very slow in finding a solution for the whole Sun.

2. The regularized least squares fit (RLS) (e.g. see Craig and Brown, 1986) which consists in finding the solution that best fits, in the least squares sense, the data at given points in the Sun. Faster than the above but it needs regularization in order to find a "smooth" solution.
3. A procedure that takes the best part of the two above and includes some other refinements; it has been designed by Eff-Darwich (1996).
4. The linearized asymptotic inversions; see Christensen-Dalsgaard et al. (1990) for details.

Of course this particular problem just posed here can be generalized further by performing a two-dimensional inversion of frequencies of all modes measured to find the dependence of the rotational velocity of the Sun with radius and latitude (Korzennik et al. 1988; Thompson et al. 1996; Kosovichev et al. 1997).

6.2   Inverting Relation for Structure Parameters

The dependence of the adiabatic oscillation frequencies on solar structure may be expressed as:

$$\omega_{n\ell}^{\mathrm{ad}} = \mathcal{F}_{n\ell}^{\mathrm{ad}}[\rho(r), c(r)] \tag{23}$$

where $\mathcal{F}_{n\ell}^{\mathrm{ad}}$ is determined through the solution of the equations of adiabatic oscillations. A difficulty in the inverse problem posed in this equation is that the frequencies are non-linearly related to the parameters defining the structure of the Sun. This situation can be solved by assuming that the real Sun can be obtained from a reference model by applying small corrections to it, the differences between the observed frequencies and those predicted by such a reference model can be obtained from a linear perturbation analysis of the oscillations equations, thus resulting in a linear relationship between the frequency differences and the small corrections. Let ρ_0 and c_0 be the density and sound speed of the reference model which predicts a set of frequencies $\{\omega_{(0)n\ell}\}$. Therefore the differences $\delta\rho = \rho - \rho_0$ and $\delta c = c - c_0$ will be the corrections to perform in the reference model to match the differences $\omega_{n\ell}^{\mathrm{obs}} - \omega_{(0)n\ell}$ between the observed frequencies and the predicted ones.

By linearizing equation (23) (assuming $\delta\rho$ and δc to be small), we obtain the following inverting relation:

$$\frac{\omega_{n\ell}^{\mathrm{obs}} - \omega_{(0)n\ell}}{\omega_{(0)n\ell}} = \int_0^R \left(K_{n\ell}^{(\rho)}(r)\frac{\delta\rho}{\rho}(r) + K_{n\ell}^{(c)}(r)\frac{\delta c}{c}(r) \right) dr + \varepsilon_{n\ell}^{-1}\mathcal{G}(\omega) + \epsilon(\omega_{n\ell}^{\mathrm{obs}}) \tag{24}$$

where the kernels $K_{n\ell}^{(\rho)}$ and $K_{n\ell}^{(c)}$ are calculated using the eigenfrequencies of the reference model, the $\varepsilon_{n\ell}^{-1}\mathcal{G}(\omega)$ term is to account for the "unknown" physics at the outer boundary layers and $\epsilon(\omega_{n\ell}^{\mathrm{obs}})$ are the errors of observation.

Such a relation may be solved by any of the methods mentioned above to yield the corrections to the reference model searched for. By incorporating the corrections to the model one ends up with a final seismic model for the Sun.

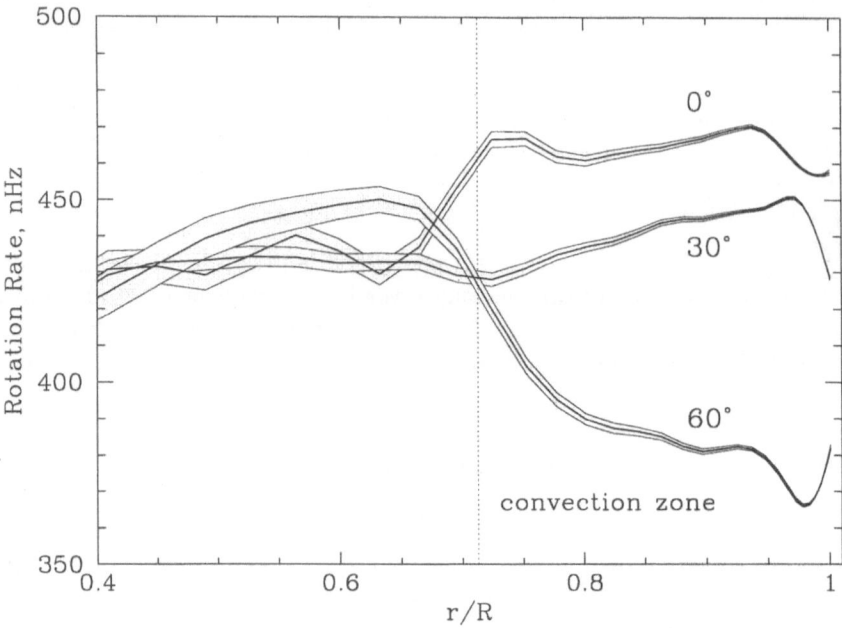

Fig. 11. Inverted solar rotation rate at three latitudes inferred from MDI medium-ℓ data; shaded areas are uncertainties generated in the solution by observational errors and inversion method's accuracy. Notice that the inversion of such modes gives information down to a depth of 60% of the solar radius; to penetrate deeper the splitting of the frequencies of modes of low-ℓ (e.g. those provided by GOLF and VIRGO) must also enter the inversion. Obtained from the SOI/MDI web page.

6.3   Seismic Equivalence and Indistinguishability

Before going into details of the best current seismic model of the Sun let's introduce here the concept of *seismic equivalence*: two solar models are seismically equivalent (with respect to frequency) if they support identical observed eigenfrequencies $\{\omega_{n\ell m}\}$. This concept can be further extended to the concept of *seismic indistinguishability* if one realises that observations have errors. As a consequence one can say that two solar models which differ only in a property such that it can not be determined by constructing data combinations of a given set of measured mode frequencies, because in practice all significant information cancels out leaving only the errors, are called seismically indistinguishable. Obviously better data can remove the indistinguishability between two solar models. These important concepts were introduced by D.O. Gough (1996a) and I draw your attention to his writing and extensive discussion therein, in which he explains the power and danger of the inversion of helioseismic data.

6.4   Seismic Models of the Sun

As already seen, seismic models can be built starting from a "close enough" model of the Sun and by inversion of helioseismic data improving the model incorporating small changes into its physics. Basu (1996) and Basu et al. (1996) have already discussed this procedure and calculated and compiled the possible most accurate seismic models of the Sun. The reference model used nowadays by many inverters is model S of Christensen-Dalsgaard et al. (1996). This model is a standard solar model improved already by taking into account some input from helisoeismology: it is constructed with OPAL equation of state (Rogers et al. 1996), OPAL opacities (improved after input from Helioseismology, Iglesias et al. 1992), observed ratio Z/X at the surface (Grevesse and Noels, 1993) and incorporating diffusion of helium and heavy elements. Quite independently of the methods used for performing the inversions and even of the set of observational data used it is found that:

- The sound speed profile seems to be larger (0.4%) in the Sun than in the model just below the base of the convection zone. This could be due to the accumulation of excess helium in this region and is the signature of some mixing there (Gough et al. 1996b).

- The other difference in sound speed is in the core. Here the differences (±4%) are dependent on the observed data that one uses and even in the method of inversion employed. The latest results obtained using GOLF and LOWL data show a lower sound speed in the Sun respect to the model at $\simeq 0.2R_\odot$ (Turck-Chièze et al. 1997; see Fig. 12).

- The solar Helium abundance, obtained from the variation of the adiabatic index of the solar material in the second helium ionization zone yields results that, depending on inversion methods used, vary from 0.24 to 0.25. Notice that this is compatible with solar evolution theories only if helium settles out from the envelope into the radiative zone.

- The depth of the convection zone can be accurately determined because it leaves its signature in the temperature gradient when it changes from radiative to adiabatic at its base. This can be determined very precisely to be 0.713±0.001 (Basu, 1996).

- The discontinuity of the derivatives of the sound speed at the base of the overshoot layer below the solar convection zone introduces an oscillatory component in the observed p-mode frequencies as a function of frequency (Gough 1990), whose amplitude depends on the severity of the discontinuity whereas the period of the variation gives its position. There is a consensus between the groups that have done this work that any overshoot below the convection zone is small, less than 3000 km.

- The OPAL equation of state seem to be the closest to describe the Sun; however the MHD one shows no significant distinction from it (Mihalas et al. 1988).

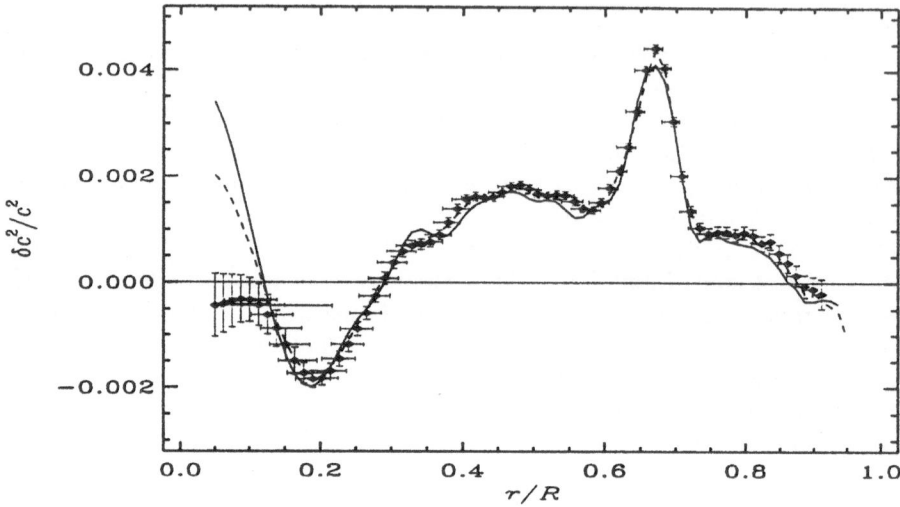

Fig. 12. Difference between the square of the sound speed in the Sun (as from 8 months of GOLF data) and in Model S (see text), in the sense (Sun) - (model). The vertical bars indicate 1σ errors in the inferred differences, while horizontal bars provide a measure of the resolution of the inversion. For comparison, the solid line shows the results obtained by Basu et al. (1997) from analysis of a combination of BiSON and LOWL data, while the dashed line shows the results obtained with just 4 months of GOLF data, combined with LOWL data (Turck-Chièze et al. 1997).

To obtain parameters other than the mechanical properties of the Sun one needs to make use of what is known as secondary inversions. They are called this way because they need some extra assumptions, like equation of state, or fully ionization or any other that allows the determination of e.g. the temperature profile. This is important for the neutrino problem and some researchers are already tackling it. To cite one of them, Takata and Shibahashi (1997) have determined the sound speed profile by inversion of the observed frequencies and solved the basic equations governing stellar structure with the imposition of the sound speed profile found; they finally calculate the neutrino fluxes taking into account the uncertainties on the various input parameters in the model to find that the estimated capture rates are still significantly larger than present solar neutrino observations. Current results seem to confirm that no solar model exists up to now that satisfies both neutrino and helioseismic observational constraints.

Acknowledgements

I would like to thank the organisers for inviting me to such a nice school and to the students who made me think much more than they could imagine. Also I thank Drs. P. Boumier and C. Régulo who carefully read this paper. Collaboration from Drs. P.L. Pallé, F. Pérez Hernández, F. Baudin and SOI and GONG projects which provided me with some of the graphs shown is also acknowledged.

Selected Monographies, Books and Useful Internet Adresses

- **Solar Seismology from Space.** 1984. *R.K. Ulrich, J.W. Harvey, E.J. Rhodes Jr and J. Toomre eds.*, JPL Publ. 84-84, Pasadena. A good account on the state of the art of Helioseismology at the time when preparing the helioseismic experiments aboard SOHO.
- **Non-radial oscillations of stars.** 1989. *W. Unno, Y. Osaki, H. Ando, H. Saio and H. Shibahashi.* Univ. of Tokyo Press, 2nd edition. A theoretical approach to the problem of stellar oscillations.
- **The Interior of the Sun and Atmosphere.** 1991. *A.N.Cox, W.C. Livingstone and M. Matthews eds.* Spa. Sci. Ser., Univ. of Arizona Press. Tucson. A complete account of the solar structure and methods at the time.
- **Lecture Notes in Stellar Oscillations.** 1994. *J. Christensen-Dalsgaard.* D.f.I. print (3rd. edition). Aarhus. A comprehensive and extremely useful notes for students interested in the theory and interpretation of helioseismic data. Available from the author's web page: http://www.obs.aau.dk/ jcd.
- **The SOHO Mission.** 1995. *B. Fleck, V. Domingo and A. Poland eds..* Kluwer Ac. Pub. Dordrecht. A fully and complete scientific description of the SOHO mission.
- **The Structure of the Sun.** 1996. *T. Roca Cortés and F. Sánchez eds.* VIth Can. Islands Win. School. Camb. Univ. Press. Cambridge. Proceedings of a comprehensive course on Helioseismology (theory, instrumentation, data analysis techniques and interpretation), neutrinos and magnetic fields in the Sun.
- **GONG Results (SCIENCE special issue).** 1996. In Science (issue of 31 May 1996), vol. 272, pages (1233- 1388).
- http://sohowww.nascom.nasa.gov. ESA-NASA space SOHO mission.
- http://soi.stanford.edu. The SOI/MDI project.
- http://www.medoc-ias.u-psud.fr/golf/golf1.htm. The GOLF project.
- //ftp.estec.esa.nl/pub/loitenerife/VIRGO/virgo.html. The VIRGO project.
- http://helios.tuc.noao.edu. The GONG network.
- http://bison.ph.bham.ac.uk. The BiSON network project.
- http://boulega.unice.fr/iris/source/iris.html. The IRIS network project.
- http://www.obs.aau.dk. The TAC/Helioseismology branch.
- http://www.iac.es. The IAC's observatorio del Teide.

References

Anderson, E.R., Duvall Jr., T.R., Jefferies, S.M. (1990): ApJ **364**, 699

Bachmann, K.T. et al. (1995): *Helio- and Astero-seismology from the Earth and Space* , ASP Con. Ser. **76**, 156

Backus, G., Gilbert, F. (1970): Phil. Trans. Roy. Soc. London **A266**, 123

Bahcall, J.N. and Ulrich, R.K. (1988): Rev. Mod. Phys. **60**, 297

Basu, S. (1996): in *Sounding solar and stellar interiors.* (in press)

Basu, S., Christensen-Dalsgaard, J., Schou, J., Thompson, M.J., Tomczyk, S. (1996): ApJ **460**, 1064

Basu, S. et al. (1997): MNRAS, (in press)

Baudin, F. (1993): Ph.D. Thesis, University of Paris XI. Orsay. France

Baudin, F., Gabriel, A.H. and Gibert, D. (1993): A&A **276**, L1

Blamont, J.E. and Roddier, F. (1961) Phys. Rev. Lett. **7**, 437

Bogdan, T. et al. (1993): ApJ **406**, 723

Bogert, B.P., Healey, M.J. & Tuckey, J.W. (1963): in *Time series analysis*, 209. M. Rosenblat ed. Wiley.

Boumier, P. (1991): Ph. D. Thesis, University of Paris VII. Orsay. France

Boumier, P., van der Raay, H. B. and Roca Cortés, T. (1994): A&AS **107**, 177

Braun, D.C., Duvall, T.L. and LaBonte, B.J. (1988): ApJ **335**, 1015

Brookes, J.R., Isaak, G.R. and van der Raay, H.B., (1978): MNRAS **185**, 1

Brown, T. (1996): *The Structure of the Sun*, 1. (See above).

Cacciani, A., Fofi, M. (1978): Sol. Phys. **59**, 179

Cacciani, A. et al. (1988): in *Seismology of the Sun and Sun-like stars*, ESA-SP**286**, 181

Chaplin, W.J. et al. (1996): Sol. Phys. **168**, 1

Chou, D.Y. and the TON team (1995): Sol. Phys. **160**, 237.

Christensen-Dalsgaard, J. et al. (1986): Nature **315**, 378

Christensen-Dalsgaard, J., Gough D. O., Thompson, M. J. (1989): MNRAS **238**, 481

Christensen-Dalsgaard, J., Schou, J., Thompson, M.J. (1990): MNRAS **242**, 353

Christensen-Dalsgaard, J. (1996): in *The Structure of the Sun*, 47. (See above)

Christensen-Dalsgaard, J. et al. (1996): Science, 1286. (See above)

Craig, I.J.D., Brown, J.C. (1986): in *Inverse Problems in Astronomy: A Guide to Inversion Strategies for Remotely Sensed Data.* A. Hilger ed. Bristol

Domingo, V., Fleck, B. and Poland, A.I. (1995): *The SOHO mission*, 1. (See above)

Duvall, T.L. (1982): Nature **300**, 242

Duvall, T.L., Harvey, J.W., Pomerantz, M.A. (1986): Nature **321**, 500

Duvall, T.L. et al. (1993): Nature **362**, 430

Eff-Darwich, A. (1996): Ph.D. Thesis, Universidad de La Laguna. Tenerife. Spain

Fossat, E. and Ricort, G. (1971): Solar Phys. **28**, 311

Fossat, E. (1995): in *Helio- and Astero-seismology from the Earth and Space* , ASP Con. Ser. **76**, 387

Fröhlich, C. et al. (1988): in *Seismology of the Sun and sun-like stars* ESA SP-**286**, 359

Fröhlich, C. et al. (1995): *The SOHO Mission*, 101. (See above)

Gabriel, A.H. et al. (1995): *The SOHO Mission*, 61. (See above)

Goldreich, P. & Keeley, D.K., (1977): ApJ **212**, 243

Goldreich, P. & Kumar, P., (1988): ApJ **326**, 462

Gough, D.O. (1986): in *Hydrodynamic and magnetohydrodynamic problems in the Sun and stars*, 125. D.O. Gough ed.. Reidel.

Gough, D.O. (1990): in Lect. Notes in Phys. **367**, 283

Gough, D.O. (1996a): in *The Structure of the Sun*, 141. (See above)

Gough, D.O. et al. (1996b): Science **272**, 1296. (See above)

Grec, G, Fossat, E. and Pomerantz, M. (1980): Nature,**288**, 541

Grec, G, et al. (1991): Sol. Phys. **133**, 13

Grevesse, N. and Noels, A. (1993):in *Origin and Evolution of the Elements*,15. CUP. U.K.

Haber, D.A., et al. (1995): in *Helioseismology*. ESA SP-**376**, **2**, 141

Harvey, J. W. (1985): in *Probing the depths of a Star: the study of solar oscillations from space*, eds. R. W. Noyes y E. J. Rhodes Jr., JPL 400-327

Harvey, J. & the GONG Instrumental Development Team. (1988):in *Seismology of the Sun and Sun-like stars* ESA SP-**286**, 203

Hill, F. (1988): ApJ **333**, 996

Hill, H.A. (1985): ApJ, **290**, 765

Hill, H.A. and Gu, Y. (1988): Scientia Sinica A.

Hill, H.A. et al. (1991): in *Solar interior and Atmosphere*, 562. (See above)

Iglesias, C.A., Rogers, F.J., Wilson, B.G. (1992): ApJ **397**, 717

Isaak, G.R. (1961): Nature, **189**, 373

Jiménez, A. et al. (1987): A&A **172**, 323

Jiménez, A. (1988). Ph.D. Thesis. University of La Laguna. Tenerife. Spain

Jiménez, A. et al. (1988): A&A **192**, L-7

Jiménez, A. et al. (1990): Sol. Phys. **126**, 1

Kjeldsen, H. et al. (1995): AJ **109**, 1313

Korzennik, S. et al. (1988): in *Seismology of the Sun and sun-like stars* ESA SP-**286**, 117

Korzennik, S. (1990): Ph.D. Thesis. Univ. of California.

Kosovichev, A.G. et al. (1997): Sol. Phys. **170**, 43

LaBonte, B.J., Ronan, R., Kupke, R. (1995): Sol. Phys. **158**, 1

Libbrecht, K.G. (1988): in *Seismology of the Sun and sun-like stars* ESA SP-**286**, 3

Libbrecht, K.G. (1992): ApJ. **387**, 712

Mihalas, D., Däppen, W., Hummer, D.G. (1988): ApJ **331**, 815

Oppenheim, A.V., Schafer, R.W. & Stockham, T.G. (1968): IEEE **65**, 1264

Pallé, P.L. et al. (1986): A&A **169**, 313

Pallé, P.,L. (1991): Adv. Space Res. **11**, **4**, 29

Pallé, P.L. et al. (1995): ApJ **441**, 952

Pallé, P.,L. (1996): in *Sounding solar and stellar interiors*. (in press)

Patrón, J. (1994): Ph.D. Thesis. Universidad de La Laguna. Tenerife. Spain

Patrón, J. et al. (1995): ApJ **455**, 476

Pérez-Hernández, F. (1989): Ph.D. Thesis. Universidad de La Laguna. Tenerife. Spain

Pérez Hernández, F., Christensen-Dalsgaard, J., (1994): MNRAS **269**, 475

Rabello Soares et al. (1997): A&A **318**, 970

Régulo, C. et al. (1994): Ap.J., **434**, 384

Rhodes, Jr., E.J. et al. (1986): in *Seismology of the Sun and Distant Stars*, 309. D.O. Gough ed.. Reidel

Robillot, J.M., Bocchia, M. and Denis, N. (1993): Proceedings of the 5th IRIS Workshop and GOLF 93 Meeting. T. Roca Cortés ed., Tenerife. Spain

Rogers, F.J., Swenson, F.J. and Iglesias, C.A. (1996): ApJ **456**, 902

Rust, D.M., Burton, C.H., Leistner, A.J. (1986): in *Instrumentation in Astronomy* VI, SPIE-**627**, 39

Scherrer, P. et al. (1995): *The SOHO Mission*, 145. (See above)

Schou, J. (1992): Ph.D. Thesis. Aarhus University. Denmark

Takata, M. and Shibahashi, H. (1997): ApJ, (in press)

Thompson et al. (1996): Science, **272**. (See above)

Tomczyk, S. et al. (1995): Sol. Phys. **159**, 1

Toutain, T. (1996): in *Sounding solar and stellar interiors*. (in press)

Turck-Chièze, S., et al. (1993): Physics Reports **230**, 2

Turck-Chièze, S., et al. (1997): Sol. Phys., (in press)

Ulrych, T.J. (1971): Geophysics **36**, 650

Woodard, M., and Hudson, H. (1983): Nature **305**, 589

Woodard, M. (1984): Ph.D. Thesis. University of California. San Diego. USA

Solar Magnetic Fields: an Introduction

S.K. Solanki

Institute of Astronomy, ETH-Zentrum, CH-8092 Zürich, Switzerland

Abstract. The magnetic field of the Sun is thought to be produced by a dynamo in the solar interior and exhibits its greatest influence on the solar plasma in the tenuous outer layers of the solar atmosphere, where it lies at the heart of almost every major phenomenon. Most direct observations of the magnetic field are restricted to the solar surface, however. Both observational and theoretical methods have been employed heavily to obtain information on and an understanding of solar magnetism. It is the aim of these lecture notes to impart some of this knowledge to the reader: knowledge both of the magnetic field and of some of the methods used to investigate it.

First a short overview of the structure of magnetic fields in the observable layers of the Sun is given. This is followed by an introduction to the Zeeman effect and polarized light, which are fundamental to the measurement of the solar magnetic field, and to the magnetohydrostatic description of solar magnetic fields. Finally, current understanding of the small (magnetic elements) and large (sunspots) magnetic features observed in the solar photosphere is summarized.

1 Introduction: Overview of Solar Magnetic Fields

The magnetic field of the Sun has observable consequences ranging from changes in p-mode frequencies to the rapid acceleration of high speed streams in the solar wind. The vast variety of manifestations is due to the highly complex and dynamic nature of solar magnetism. These properties in turn result from the interaction of the magnetic field with solar differential rotation, convection, radiation, oscillations and waves.

On a large scale the complex structure of the solar magnetic field manifests itself by the presence of bipolar active regions and the magnetic network in the quiet Sun. This complexity continues down to the smallest observable scales, corresponding to a fraction of an arcsec, with the magnetic field being highly structured at all scales.

In the photosphere and interior of the Sun the magnetic field is concentrated into flux tubes or bundles of magnetic field lines. In the photosphere these flux tubes generally have field strengths of 1–3 kG and range in size from roughly 100 km (bright magnetic elements) to tens of thousands of km (dark sunspots). There is also some evidence for a weak, possibly turbulent magnetic field between the flux tubes. Although a considerable fraction of the magnetic flux may be in weak-field form, the flux tubes contain almost all the magnetic energy. Nevertheless, they cover only a very small fraction of the solar surface (on the order of 1%). This implies that in the solar photosphere the magnetic field is highly filamented.

The magnetic flux tubes expand with height. This expansion is particularly rapid in the mid chromosphere, where the field becomes almost horizontal, forming a magnetic canopy, i.e. a region of field overlying a field-free atmosphere. Above this layer basically the whole of the solar atmosphere is permeated by a magnetic field. Still higher up, in the corona, we need to distinguish between "open" field lines (i.e. field lines that reach out into interplanetary space) and "closed" (such as those forming loops or arcades).

The field becomes progressively more homogeneous in strength with height, although it is by no means completely homogeneous in the corona, in particular in active regions. Conversely, it becomes increasingly more inhomogeneous in direction. Whereas in the photosphere flux tubes are more or less vertical, almost any direction is possible in the corona and current sheets abound at tangential discontinuities of the field (at least according to theory).

In determining the magnitude and variety of ways in which the magnetic field manifests itself an important role is played by the approximate relative energy density of the field, $E_B = B^2/8\pi$, of the gas (thermal energy density), $E_t = \frac{3}{2}p$, and of the motions (kinetic energy density), $E_k = \rho v^2/2$. Here B is the magnetic field strength, p is the gas pressure, ρ is its density and v its velocity.

The relative values of these 3 quantitites are a strong function of height. In the solar convection zone $E_B \approx E_k$ and $E_B \ll E_t$, so that manifestations due to the field are expected to be small. As we go up in height E_B becomes increasingly important. Thus, in the photosphere $E_B \approx E_t \gg E_k$, in the chromosphere $E_B > E_t \gtrsim E_k$ and in the corona $E_B \gg E_t$, $E_B \gg E_k$. Consequently, the field plays an important role in the chromosphere and dominates the energetics and dynamics in the corona. Finally, in the solar wind outside the Alfvén radius $E_k > E_B$, so that the influence of the magnetic field is expected to diminish again in these layers.

This concludes my brief and incomplete introduction to the structure of the solar magnetic field. The field also evolves strongly as a function of time. The large scale part of this evolution is discussed by Proctor in his contribution to these proceedings.

2 Observation of Magnetic Fields

2.1 Zeeman Effect

In the absence of a magnetic field the energy E_J of the atomic levels of all except hydrogenic ions depends on the absolute value of the total angular momentum J [in the following, quantum numbers are often used instead of the operator eigenvalues for simplicity, e.g. J instead of $\hbar J(J+1)$], but not on any of its components, so that each level is $(2J+1)$-fold degenerate. This degeneracy disappears in the presence of a magnetic field and the energy of each level can then be written as

$$E_{J,M} = E_J + \mu_0 g M_J B. \tag{1}$$

The energy now also depends on M_J, the component of **J** parallel to the magnetic vector **B** $(-J \leq M_J \leq J)$.

In Eq. (1) $\mu_0 = e\hbar/(2mc)$ is the Bohr magneton (e is the electric charge, $\hbar = h/2\pi$ is Planck's constant, m the electron mass and c the speed of light) and g is the Landé factor given by

$$g = 1 + \frac{J(J+1) + S(S+1) - L(L+1)}{2J(J+1)} \tag{2}$$

in LS coupling (L and S are the total orbital and spin angular momentum quantum numbers, respectively). The main observational consequences of this splitting of the atomic levels is a splitting of the spectral lines formed between two such levels into three groups of lines (Zeeman, 1897) according to $\Delta M_J = 0, \pm 1$. The unshifted ($\Delta M_J = 0$) component is called the π-component, while the $\Delta M_J = \pm 1$ components are referred to as the $\sigma^\pm$-components.

For the special case of a transition between a $J = 1$ and a $J = 0$ level or between two levels with equal g-values, the line splits into exactly three components. In this case the line is called a *Zeeman-triplet* and the wavelength splitting of the line is given by

$$\Delta\lambda_H = (e/4\pi mc^2)gB\lambda^2 = 4.67\ 10^{-13}gB\lambda^2, \tag{3}$$

in cgs units, with $\Delta\lambda_H$ and λ in Angström and B given in G. The Zeeman splitting of an absorption line (Zeeman triplet) for a magnetic field parallel (upper frame) and perpendicular (lower frame) to the line of sight is illustrated in Fig. 1.

2.2   Polarized Light

In astronomy, polarized light is generally described by the four Stokes parameters I, Q, U and V, where I is the total intensity (i.e. the sum of the polarized and the unpolarized fractions of the light), and

$$Q = I_{\text{lin}}(\chi = 0) - I_{\text{lin}}(\chi = \pi/2), \tag{4}$$

$$U = I_{\text{lin}}(\chi = \pi/4) - I_{\text{lin}}(\chi = 3\pi/4), \tag{5}$$

$$V = I_{\text{circ}}(\text{right}) - I_{\text{circ}}(\text{left}). \tag{6}$$

Here $I_{\text{lin}}(\chi)$ refers to linearly polarized radiation whose electric vector makes an angle χ to some reference direction (defined, e.g., by the measuring apparatus). I_{circ} refers to circularly polarized light. The Stokes parameters are directly measurable quantities.

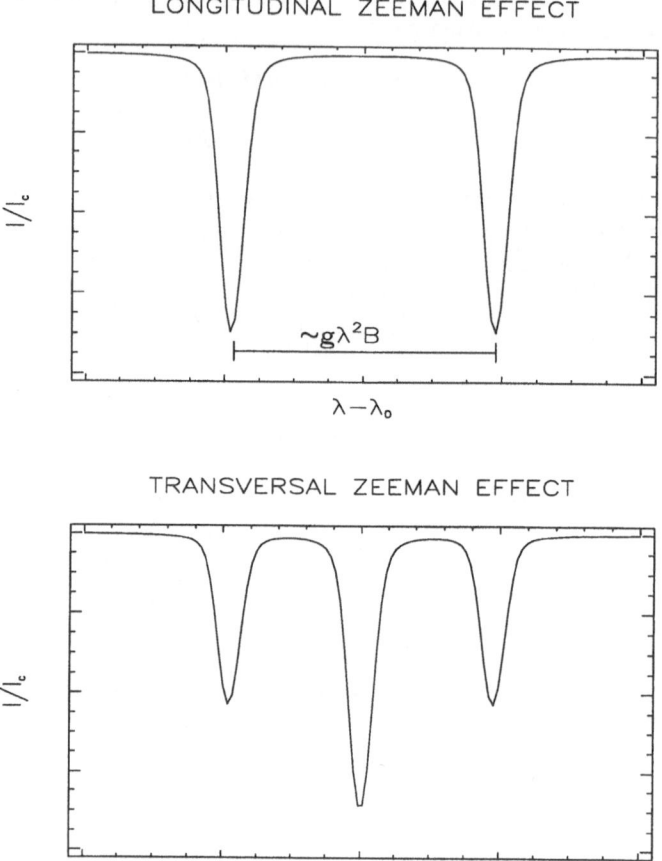

Fig. 1. Illustration of the longitudinal (upper frame) and transverse (lower frame) Zeeman effects for a Zeeman triplet. In the former case the magnetic field vector is directed along the line of sight, in the second case it is perpendicular to it. In the illustration the splitting (indicated by the horizontal line in the upper frame) is much larger than the line width in the absence of the field (courtesy of C. Frutiger).

In general $Q = U = V = 0$ if $B = 0$, since Q, U, V represent *net* linear and circular polarisation only and in the absence of a magnetic field $I_{\mathrm{circ}}(\text{right}) = I_{\mathrm{circ}}(\text{left})$, etc. This is an extremely useful property. By measuring Stokes Q, U, or V one obtains information exclusively on the magnetic features, even if these are spatially unresolved. Heuristically we can say that Stokes V is sensitive to the longitudinal or line-of-sight component of the magnetic field, while Stokes Q and U are sensitive to the transverse components.

2.3   LTE Radiative Transfer in a Magnetic Field

Here I exclusively discuss radiative transfer in a magnetic field. More on radiative transfer in general may be found in the lecture notes of Carlsson in these proceedings. A gentle introduction is given by Böhm-Vitense (1989).

In LTE (local thermodynamic equilibrium) the transfer equation for polarized radiation in the presence of a magnetic field can be written as

$$\frac{d\mathbf{I}_\nu}{d\tau} = (\mathbf{E} + \Omega_\nu)(\mathbf{I}_\nu - \mathbf{1}B_\nu), \tag{7}$$

where $\mathbf{I}_\nu$ is the Stokes vector at frequency ν, $\mathbf{E}$ is the unity matrix (not to be mistaken with the electric field vector introduced in Sect. 3), τ is the continuum optical depth along the line of sight ($d\tau = \kappa ds$, where κ is the absorption coefficient and ds is a geometrical path element along the line of sight), B_ν is the Planck function, $\mathbf{1} = (1,0,0,0)^{\mathrm{T}}$ (T signifies transposition) and Ω_ν is the absorption matrix

$$\Omega_\nu = \begin{pmatrix} \eta_I & \eta_Q & \eta_U & \eta_V \\ \eta_Q & \eta_I & \rho_V & -\rho_U \\ \eta_U & -\rho_V & \eta_I & \rho_Q \\ \eta_V & \rho_U & -\rho_Q & \eta_I \end{pmatrix}. \tag{8}$$

In Eq. (8)

$$\eta_I = \frac{\eta_0}{2}\sin^2\gamma + \frac{\eta_{+1} + \eta_{-1}}{4}(1 + \cos^2\gamma), \tag{9}$$

$$\eta_Q = \left(\frac{\eta_0}{2} - \frac{\eta_{+1} + \eta_{-1}}{4}\right)\sin^2\gamma \cos 2\chi, \tag{10}$$

$$\eta_U = \left(\frac{\eta_0}{2} - \frac{\eta_{+1} + \eta_{-1}}{4}\right)\sin^2\gamma \sin 2\chi, \tag{11}$$

$$\eta_V = \frac{\eta_{-1} - \eta_{+1}}{2}\cos\gamma, \tag{12}$$

$$\rho_Q = \left(\frac{\rho_0}{2} - \frac{\rho_{+1} + \rho_{-1}}{4}\right)\sin^2\gamma \cos 2\chi, \tag{13}$$

$$\rho_U = \left(\frac{\rho_0}{2} - \frac{\rho_{+1} + \rho_{-1}}{4}\right)\sin^2\gamma \sin 2\chi, \tag{14}$$

$$\rho_V = \frac{\rho_{-1} - \rho_{+1}}{2}\cos\gamma. \tag{15}$$

In Eqs. (9)–(15) γ is the angle between $\mathbf{B}$ and the line of sight and χ is the azimuthal angle of $\mathbf{B}$. The angles are illustrated in Fig. 2. The $\eta_{0,\pm1}$ are the ratios of the line to the continuum absorption coefficients of the π and $\sigma_{\pm}$-components, respectively, while $\rho_{0,\pm1}$ are the corresponding magneto-optical coefficients. The Zeeman splitting manifests itself in the strengths and shifts of the $\eta_{\pm 1,0}$ and $\rho_{\pm 1,0}$ values. For a Zeeman triplet η_0 is proportional to an unshifted Voigt function, ρ_0 to an unshifted Faraday-Voigt function, and

$$\eta_{\pm 1}(\lambda) = \eta_0(\lambda \pm \Delta\lambda_H), \tag{16}$$

$$\rho_{\pm 1}(\lambda) = \rho_0(\lambda \pm \Delta\lambda_H) \tag{17}$$

(cf. Landi Degl'Innocenti 1976, Kalkofen 1987). Eqs. (7)–(15) were first derived by Unno (1956) and Rachkovsky (1962).

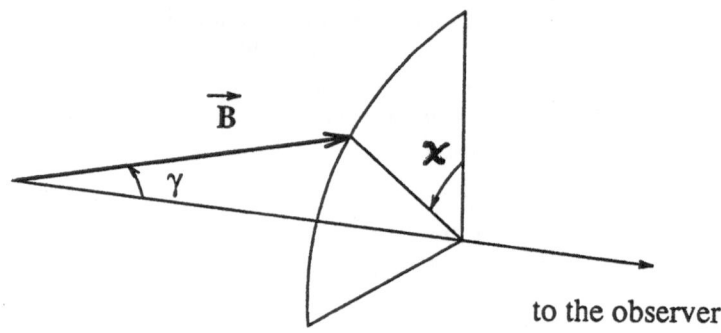

Fig. 2. Definition of the angles γ and χ used in the radiative transfer of Zeeman-split spectral lines.

For a plane-parallel atmosphere in which Ω_ν is height independent, i.e. **B** is also height independent, and in which the Planck function has the form

$$B_\nu = B_{\nu_0}(1 + \beta_0 \tau) \tag{18}$$

(Milne-Eddington model), an analytical solution of Eq. (7) may be obtained. In the absence of magneto-optical effects ($\rho_Q = \rho_U = \rho_V = 0$) the solution is particularly simple (Unno 1956):

$$I(\mu) = B_{\nu_0}\left(1 + \beta_0\mu\frac{1 + \eta_I}{(1 + \eta_I)^2 - \eta_Q^2 - \eta_U^2 - \eta_V^2}\right), \tag{19}$$

$$Q(\mu) = -B_{\nu_0}\beta_0\mu\frac{\eta_Q}{(1 + \eta_I)^2 - \eta_Q^2 - \eta_U^2 - \eta_V^2}, \tag{20}$$

$$U(\mu) = -B_{\nu_0}\beta_0\mu\frac{\eta_U}{(1 + \eta_I)^2 - \eta_Q^2 - \eta_U^2 - \eta_V^2}, \tag{21}$$

$$V(\mu) = -B_{\nu_0}\beta_0\mu\frac{\eta_V}{(1 + \eta_I)^2 - \eta_Q^2 - \eta_U^2 - \eta_V^2}, \tag{22}$$

where μ is the cosine of the angle between the LOS and the normal to the solar surface.

The denominator in Eqs. (19)–(22) is responsible for line saturation. Consider first the case $B = 0$. As the absorption in a line increases $(\eta_I \to \infty)$ $I(\mu)$ decreases as $1/(1 + \eta_I)$ and never drops below B_{ν_0}. A magnetic field causes the line to desaturate. For example, in a longitudinal field $\eta_Q = \eta_U = 0$ and η_V^2 varies between 0 for $\Delta\lambda_H = 0$ and $\eta_V^2 = \eta_I^2$ for complete splitting. Thus the denominator decreases from $(1 + \eta_I)^2$ to $(1 + 2\eta_I)$ as B increases. Consequently, spectral lines have a larger equivalent width when formed in the presence of a magnetic field.

Except for saturation effects described by the denominator the properties of Q, U and V are determined by η_Q, η_U and η_V. From Eqs. (10)–(12) it therefore follows that Q and U $(\sim \sin^2 \gamma)$ are only sensitive to the transverse component of the field, while V $(\sim \cos \gamma)$ reacts to the longitudinal field component. Equations (19)–(22) also show that at the limb ($\mu = 0$) Stokes $Q = U = V = 0$, while $I = B_{\nu_0}$.

2.4   Further Properties of Zeeman-split Stokes Parameters

In the following, some additional properties of the Stokes profiles are listed. Firstly, Stokes I and V do not depend on χ and Stokes Q and U are invariant to changes in χ that are multiples of 180°. Also, Stokes V changes sign if the longitudinal component of $\mathbf{B}$ changes its polarity, while Stokes I, Q and U do not. It is also worth noting that for lines formed in LTE, Stokes I, Q and U are symmetric around the line core wavelength in the absence of velocity gradients, while Stokes V is antisymmetric (Landi Degl'Innocenti and Landi Degl'Innocenti 1981). Not only is the symmetry of the profile shapes lost in the presence of longitudinal velocity gradients, but even the areas of the blue and red lobes of Stokes V may be unequal (Illing et al. 1975). The shape and field-strength dependence of Zeeman-split Stokes profiles is illustrated in Fig. 3 for an often-used spectral line.

For weak fields the Stokes I profile closely resembles the unsplit profile, while Q, U and V are very weak and increase in amplitude, but not in wavelength separation of the peaks, with B. At large B the line is completely split, i.e., the peak separation increases linearly with B and the amplitudes of Q, U and V become independent of B (often referred to as Zeeman saturation).

3   Description of Magnetic Fields

3.1   Multi-component Models Used to Interpret Observations

Magnetic features in the solar atmosphere generally possess a complex geometry. For the interpretation of observations, however, rather simple concepts are often used.

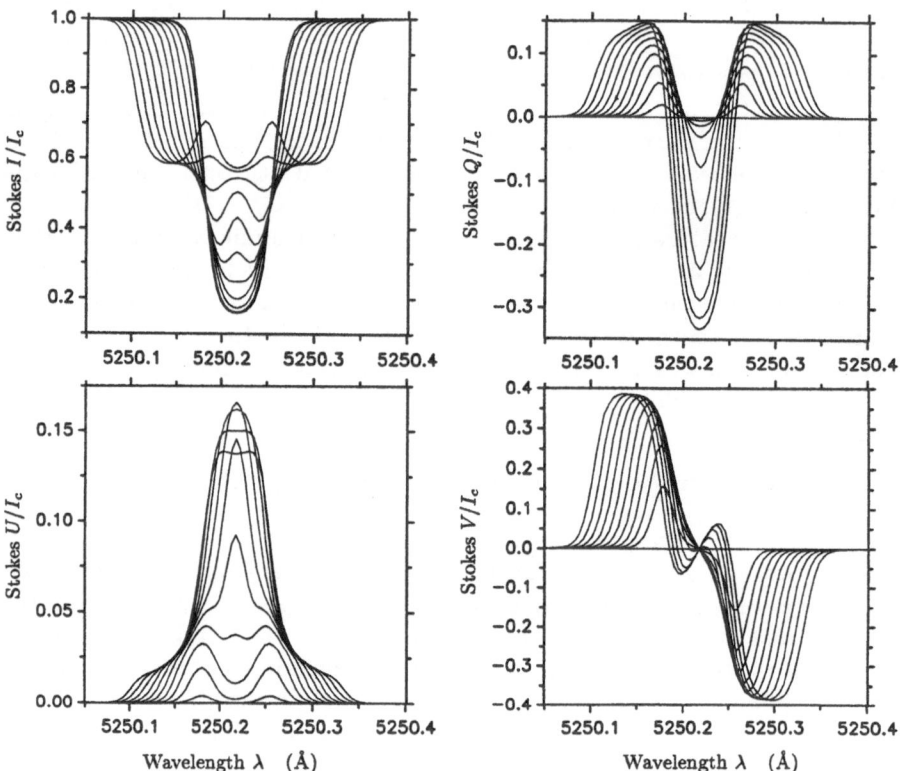

Fig. 3. Stokes I, Q, U and V profiles of Fe I 5250.2 Angström calculated for a quiet sun model atmosphere at solar disc centre ($\mu = 1$), with the magnetic parameters $\gamma = 45°$, $\chi = 0°$ and field strengths $B = 0$, 200, 400, 600, 800, 1000, 1200, 1400, 1600, 1800 and 2000 G. Note the different splitting regimes, e.g., in the Stokes V profile: For B between 0 and 400 G its amplitude increases almost linearly with field strength, while the peak separation remains almost unchanged (weak field regime with $\Delta\lambda_H/\Delta\lambda_D \ll 1$), for $B = 600$–1200 G $\Delta\lambda_H/\Delta\lambda_D \approx 1$, and for $B \gtrsim 1400$ G the Stokes V profile exhibits the behaviour of a completely split spectral line ($\Delta\lambda_H/\Delta\lambda_D \gg 1$). Here $\Delta\lambda_D$ is the Doppler width of the spectral line. The I and Q profiles begin to enter the fully split regime at 2000 G.

In recent years 2-component models have been commonly employed for the interpretation of observational data. They are based on the assumption that within a spatial resolution element two atmospheric components are present, one covering a fraction α of the surface, the other covering the remaining fraction $(1 - \alpha)$. In faculae (plages), or in the network, magnetic elements constitute the first component and the intervening field-free atmosphere the other. The magnetic component is generally assigned the surface fraction α, called the magnetic filling factor.

If the two components are spatially unresolved then the observed radiation is the weighted mean of the radiation coming from each, so that the observed Stokes parameters may be written as

$$\langle I \rangle = \alpha I_\mathrm{m} + (1 - \alpha) I_\mathrm{s}, \tag{23}$$

$$\langle Q \rangle = \alpha Q_\mathrm{m}, \tag{24}$$

$$\langle U \rangle = \alpha U_\mathrm{m}, \tag{25}$$

$$\langle V \rangle = \alpha V_\mathrm{m}. \tag{26}$$

The brackets $\langle \; \rangle$ denote averaging over the spatial resolution element, the subscript 'm' signifies light from the magnetic component, the subscript 's' light from the field-free surroundings. The great advantage of Stokes $\langle Q \rangle$, $\langle U \rangle$ and $\langle V \rangle$ over Stokes $\langle I \rangle$ is evident from Eqs. (23)–(26). In a 2-component model $\langle Q \rangle$, $\langle U \rangle$ and $\langle V \rangle$ only convey information on the magnetic component and are totally independent of the field-free component even if α is small, which in practice often is the case.

The multi-component model may also be applied to sunspot umbrae and penumbrae. For sunspot umbrae the dark umbral core and the brighter umbral dots constitute two components, for sunspot penumbrae the components may be the bright and dark filaments. In these cases both components have a magnetic field, so that all 4 Stokes parameters obtain contributions from both components. In addition, sunspots usually suffer from stray light, e.g., from the comparatively bright field-free surrounding atmosphere. The stray light can be treated as a third (let us assume for simplicity unpolarized) component. If we refer to the two sunspot components as bright (subscript 'b') and dark (subscript 'd'), assume that they cover fractions α_b and α_d of the resolution element, respectively, and denote the stray light component by I_s, then we must replace Eqs. (23)–(26) by the following expressions for the observed (spatially averaged) Stokes parameters

$$\langle I \rangle = \alpha_\mathrm{b} I_\mathrm{b} + \alpha_\mathrm{d} I_\mathrm{d} + (1 - \alpha_\mathrm{b} - \alpha_\mathrm{d}) I_\mathrm{s}, \tag{27}$$

$$\langle Q \rangle = \alpha_\mathrm{b} Q_\mathrm{b} + \alpha_\mathrm{d} Q_\mathrm{d}, \tag{28}$$

$$\langle U \rangle = \alpha_\mathrm{b} U_\mathrm{b} + \alpha_\mathrm{d} U_\mathrm{d}, \tag{29}$$

$$\langle V \rangle = \alpha_\mathrm{b} V_\mathrm{b} + \alpha_\mathrm{d} V_\mathrm{d}. \tag{30}$$

3.2 MHD Equations

For a realistic and physically consistent description of magnetic features, it is necessary to solve the magnetohydrodynamic (MHD) equations. For an inviscid and (on macroscopic scales) electrically neutral plasma, the MHD equations are composed of a mass, a momentum and an energy conservation equation (cf. Priest 1982),

$$\frac{\partial \rho}{\partial t} + \nabla(\rho \mathbf{v}) = 0 \, , \tag{31}$$

$$\rho \left(\frac{\partial \mathbf{v}}{\partial t} + (\mathbf{v} \cdot \nabla)\mathbf{v} \right) = -\nabla p + \rho \mathbf{g} + \frac{1}{c}\mathbf{j} \times \mathbf{B} \, , \tag{32}$$

$$\rho \left(\frac{\partial C_V T}{\partial t} + (\mathbf{v} \cdot \nabla)C_V T \right) = -p(\nabla \cdot \mathbf{v}) - \nabla \cdot (\mathbf{F}_R + \mathbf{F}_C + \mathbf{F}_O) + \eta \mathbf{j}^2 \tag{33}$$

Here ρ is the gas density, $\mathbf{v}$ is the gas velocity, p the gas pressure, $\mathbf{g}$ the grav-itational acceleration vector, $\mathbf{j}$ the electric current density, $\mathbf{B}$ the magnetic field (or, more precisely, the magnetic induction), C_V the specific heat of the gas at constant volume, T the temperature, $\mathbf{F}_R$ the radiative flux, $\mathbf{F}_O$ a term containing any other energy fluxes and η the electrical resistivity. Note that when deriving Eq. (33) the internal energy density was represented by $e = C_V T$ and substituted into the equation.

Equation (31) is in the well-known form of a continuity equation, which always signifies conservation of a quantity with time, in this case $\int \rho \, dV$, where dV is a volume element. It says that the density decreases (i.e., $\partial \rho / \partial t < 0$) when matter flows out of the considered volume element (i.e. divergence $\nabla(\rho \mathbf{v}) > 0$) and vice versa. The terms on the LHS of Eqs. (32) and (33) are material derivatives, i.e. they take into account time variations in momentum and energy density following the motion of the gas. The terms on the RHS of Eq. (32) describe forces, namely gas pressure gradients, gravity and magnetic forces. In Eq. (33) the RHS terms describe sources and sinks of energy density, such as gradients of energy fluxes or ohmic dissipation (except for the $p(\nabla \cdot \mathbf{v})$ term, which appears when transforming from the original equation for entropy to the current one for temperature).

Further equations are required to close the system (31)–(33). These in-clude the remaining Maxwell equations for quasistationary electromagnetic fields (i.e. neglecting $\frac{1}{c}\partial \mathbf{E}/\partial t$ terms and thus ruling out electro-magnetic radiation),

$$\nabla \times \mathbf{B} = \frac{4\pi}{c}\mathbf{j} \, , \tag{34}$$

$$\nabla \times \mathbf{E} = -\frac{1}{c}\frac{\partial \mathbf{B}}{\partial t} \, , \tag{35}$$

$$\nabla \cdot \mathbf{B} = 0 \, , \tag{36}$$

and Ohm's law

$$\mathbf{j} = \frac{1}{\eta}\left(\mathbf{E} + \frac{1}{c}\mathbf{v} \times \mathbf{B}\right).\tag{37}$$

Here $\mathbf{E}$ is the electric field vector. Note that in a time dependent problem, $\nabla \cdot \mathbf{B} = 0$ needs to be specified only as an initial condition. It can be shown by taking the divergence of Eq. (35) that Eq. (36) is then satisfied for all times.

Another equation missing so far is the equation of state. The ideal gas law is generally adequate in the solar atmosphere:

$$p = \frac{k}{m_p}\rho T,\tag{38}$$

where k is Boltzmann's constant and m_p is the mean particle mass.

In Eqs. (31)–(38) K, η, C_V and m_p are material properties, and in general depend on gas or electron density, temperature, etc. Their determination often requires the calculation of the complete ionization and excitation equilibria of all the atomic and molecular species in the solar atmosphere (in LTE this implies solving the Saha-Boltzmann equation). For example, when atoms, in particular hydrogen atoms, are ionized m_p is considerably lower due to the large number of free electrons (with almost negligible mass). The radiative energy flux $\mathbf{F}_R$ in Eq. (33) is related to the radiative intensity I_ν through

$$\mathbf{F}_R = \iint \mathbf{e}\, I_\nu \, d\Omega \, d\nu \ ,\tag{39}$$

where $\mathbf{e}$ is a unit vector describing the direction of the net radiative energy flow, $d\Omega$ is a solid-angle element and $d\nu$ is a frequency element. I_ν is in turn determined by solving the radiative transfer equation. Since the presence of magnetic fine structure often causes the geometry to depart from being plane-parallel, multi-dimensional radiative transfer must be carried out to determine $\mathbf{e}I_\nu$.

For the thermal conduction term we may write

$$\mathbf{F}_C = -K_C\,\nabla T,\tag{40}$$

where K_C is the thermal conductivity. Thermal conduction is mainly important in the solar corona. For more on the energy equation in the corona see the lecture notes by Gabriel.

The full MHD equations are rather complex and often simplified versions are used. In some situations the dynamics of the gas are relatively unimportant compared to other contributions. Such cases may be described by the magnetohydrostatic (MHS) approximation. The MHS equations can be obtained by setting $\mathbf{v} = 0$ and $\partial/\partial t = 0$ in Eqs. (31)–(38). Mass conservation is now trivially satisfied and momentum conservation reduces to force balance. We also introduce Eq. (34) into Eq. (32) and break up the magnetic force term, $(\nabla \times \mathbf{B}) \times \mathbf{B}$, into a magnetic pressure, $\nabla B^2/8\pi$ and a curvature $(\mathbf{B} \cdot \nabla)\mathbf{B}$ term.

$$-\nabla(p + \frac{\mathbf{B}^2}{8\pi}) + \rho\mathbf{g} + \frac{1}{4\pi}(\mathbf{B}\cdot\nabla)\mathbf{B} = 0, \tag{41}$$

$$-\nabla\cdot(\mathbf{F}_R + \mathbf{F}_C) + \eta\mathbf{j}^2 = 0, \tag{42}$$

$$\nabla\cdot\mathbf{B} = 0, \tag{43}$$

$$p = \frac{k}{m_p}\rho T. \tag{44}$$

In Eq. (41) the first term represents pressure balance, the second gravity and the last magnetic curvature forces. The $\mathbf{j}^2$ term in Eq. (42) may be expressed in terms of $\mathbf{B}$ using Eq. (34). In addition to these equations the condition of pressure equilibrium across any magnetic boundary with the coordinates $\mathbf{x}_b$, e.g. between a magnetic m and a non-magnetic n component of the atmosphere, must be fulfilled,

$$\frac{B_m^2}{8\pi}(\mathbf{x}_b) + p_m(\mathbf{x}_b) = p_n(\mathbf{x}_b). \tag{45}$$

3.3 Magnetohydrostatic Equilibrium of Flux Tubes: The Thin-Tube Approximation

The magnetic field in the solar interior and in the lower solar atmosphere is concentrated into discrete structures having a strong magnetic field, surrounded by a relatively field-free atmosphere. Although in reality the magnetic flux concentrations are probably devoid of symmetry (like most sunspots), for computational expedience they have generally been modelled as axially symmetric tubes.

Even for this idealized geometry the exact solution of the MHS equations can be a complex numerical undertaking and further simplifications have often been introduced. The subject has been reviewed with great insight by Schüssler (1986) for small flux tubes and by Pizzo (1987) mainly for sunspots. The equilibrium of flux tubes has also been discussed in the monographs by Parker (1979a) and Priest (1982).

For an axisymmetric flux tube in cylindrical coordinates (r, ϕ, z) with $\partial/\partial\phi = 0$, all the dependent variables can be expanded in a Taylor series according to the radial coordinate r if they are regular around $r = 0$ (Ferriz Mas and Schüssler 1989):

$$f(r, z) = \sum_{n=0}^{\infty} \frac{r^n}{n!}\left(\frac{\partial^n f}{\partial r^n}\right)\bigg|_{r=0}. \tag{46}$$

Here f represents each dependent variable in turn, i.e. B_z, B_r, p, T, ρ, etc. These expansions are then introduced into Eqs. (41)–(44).

If we now assume that the width of a flux tube is small compared to the scale of the vertical stratification, e.g. the pressure scale height

$$H_p = \frac{p}{\rho g} \, , \tag{47}$$

which is roughly proportional to temperature, then an approximate set of equations can be found by neglecting all terms higher than a certain order. The simplest set of equations is obtained if only zeroth order terms are included (Defouw 1976). This is referred to as the thin-tube or slender-tube approximation.

$$\frac{B_z^2(z)}{8\pi} + p_m(z) = p_s(z) \, , \tag{48}$$

$$\frac{dp_{m,s}}{dz} = \rho_{m,s} g \, , \tag{49}$$

$$\pi R^2(z) B_z(z) = \Phi = \text{const} \, , \tag{50}$$

$$p_{m,s} = \frac{k}{m_p} \rho_{m,s} T_{m,s} \, . \tag{51}$$

Index m describes quantities within the magnetic flux tube, index s describes quantities in the non-magnetic surroundings and Φ is the total magnetic flux. For Eqs. (48)–(51) to be valid R, the radius of the tube, has to be small compared to H_p. Note that in this approximation magnetic curvature terms are completely neglected (i.e. force balance reduces to pressure balance) and the vertical and horizontal components of the pressure balance decouple from each other (Eqs. 48 and 49). Magnetic flux conservation is now described (after integration of Eq. 43) by Eq. (50). Note that the energy equation cannot be as easily simplified as the other equations without additional assumptions, e.g., regarding $\mathbf{F}_R$. For simplicity it has not been included in the above set of equations. To first order B_r may be obtained as follows:

$$B_r = -\frac{r}{2} \frac{dB_z}{dz} \, . \tag{52}$$

3.4   Force-Free and Potential Fields

In many cases the magnetic field dominates over the gas pressure, gravity and dynamics. This is mainly the case in the upper solar atmosphere. Then the MHS equations can be further simplified. Let us start by rewriting Eq. (41).

$$-\nabla p + \frac{1}{4\pi}(\nabla \times \mathbf{B}) \times \mathbf{B} + \rho \mathbf{g} = 0 \, . \tag{53}$$

If the magnetic field is sufficiently strong to dominate over the other forces then the force-free approximation

$$(\nabla \times \mathbf{B}) \times \mathbf{B} = 0 \tag{54}$$

becomes valid. If, in addition, the current $\mathbf{j}$ vanishes then it follows from Eq. (34) that

$$\nabla \times \mathbf{B} = 0. \tag{55}$$

A curl-free field can always be written as

$$\mathbf{B} = -\nabla\varphi, \tag{56}$$

where φ is a scalar potential and the resulting $\mathbf{B}$ is called a potential field. Potential fields fill the whole atmosphere unless bounded by a current sheet. The field in a flux tube can be potential if it is axially symmetric and untwisted.

4   Magnetic Elements

Let me begin my description of the properties and physics of photospheric magnetic features at the small-scale end: the magnetic elements with diameters of less than a few hundred km. An introduction to their physics is given by Schüssler (1992) and detailed reviews by Schüssler (1990), Solanki (1993) and Stenflo (1994). Consider first their magnetic structure.

4.1   Magnetic Field of Magnetic Elements

Recent infrared observations and inversions have allowed accurate determinations of the field strength. Although the field strength averaged over the spatial resolution element $\langle B \rangle = \alpha B$ is often only a few 10s of G, the Zeeman splitting gives intrinsic B values of 1000–1500 G. A remarkable property of photospheric flux tubes is that the field strength averaged over the whole magnetic feature remains unchanged to within 50% (between 1000 and 1500 G in the mid photosphere) over 5–6 orders of magnitude in magnetic flux, i.e. from magnetic elements with an estimated diameter of 100 km to sunspots with diameters of many 10 000 km (Solanki & Schmidt 1993). For the smallest magnetic features (intranetwork elements) the field strength does decrease to a few 100 G, however (Keller et al. 1994, Lin 1995, Solanki et al. 1996).

The magnetic field of small-scale features is confined mainly by the internal gas-pressure deficit due to the partial evacuation of the flux tubes; curvature forces play only a small role in their photospheric layers. The field strength at a given height depends on the evacuation at that height. At a given geometrical height, z, the maximum field strength supported in this manner is obtained by completely evacuating the flux tube,

$$B_{\max}(z) = \sqrt{8\pi\, p_{\mathrm{s}}(z)}, \tag{57}$$

where p_{s} is the pressure in the field-free surroundings of the magnetic feature. However, an increased evacuation also implies a downwards shift of the continuum optical depth scale (there is less material to absorb light). This in turn signifies that the spectral lines used to measure the field are formed deeper

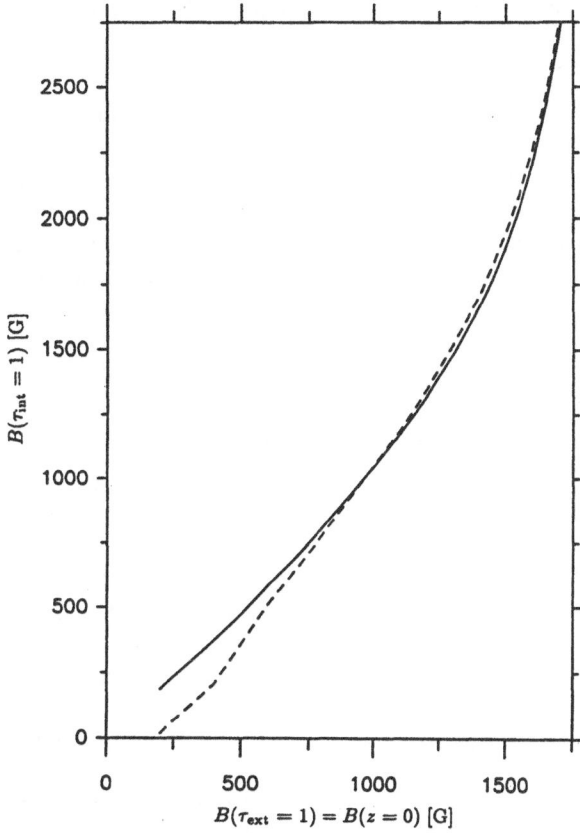

Fig. 4. Field strength at unit optical depth within the flux tube, $B(\tau_m = 1)$, vs. field strength at unit optical depth of the quiet sun, $B(\tau_s = 0) = B(z = 0)$. Due to the partial evacuation of the flux tubes the optical depth scales in the magnetic feature, τ_m, and in its surroundings, τ_s, are different, with $\tau_m = 1$ being reached at a deeper geometrical level than $\tau_s = 1$. The field strength is calculated in the thin-tube approximation for thermal structures corresponding to a plage flux-tube model (dashed curve) and for a quiet sun model (solid curve). See Solanki et al. (1992a) for a detailed discussion of this figure.

in the atmosphere, where p_s and consequently the field strength is higher. Figure 4 shows $B(\tau = 1)$ vs. $B(z = 0)$ for the simple case of a temperature structure that is independent of evacuation. Therefore, pressure balance by itself sets no useful limits on the observable field strength.

To understand the field strengths of magnetic elements we must consider the processes by which they are formed. After magnetic flux is swept into the downflow lanes of supergranules (on a larger scale) and granules (on a small scale) through the horizontal components of the convective motions, a process called flux expulsion, it gets further enhanced by the convective instability that leads to the convective collapse of the field.

The basis for the instability is the superadiabatic effect (Parker 1978). Let us start with a patch of relatively weak field that still allows some convective motions to occur within its confines. An adiabatic downflow within the magnetic patch makes it cooler than its superadiabatically stratified surroundings. A lower temperature implies a lower pressure scale height (Eq. 47), so that the gas pressure in the photosphere and the uppermost part of the convection zone is reduced. The field is then pressed together by the inward flowing gas, causing the magnetic pressure to increase until it cancels the gas pressure deficit. The whole process can become a runaway, driving a convective instability (Webb & Roberts 1978, Spruit & Zweibel 1979, Unno & Ando 1979), which can enhance the field strength even more.

How can the process be stopped? Spruit (1979) and Spruit & Zweibel (1979) suggest that a field with a strength above a critical value inhibits convection sufficiently to become stable. The observed dependence of magnetic field strength on flux per magnetic element is well reproduced by convective collapse models (Venkatakrishnan 1986).

Observations made in spectral lines formed at different heights show that B decreases with height. This is a natural consequence of the confinement of the field by excess gas pressure in the surroundings. Since the gas is gravitationally stratified its pressure decreases approximately exponentially with height, leading to a corresponding decrease of B. Since magnetic flux must be conserved the cross-sectional area of the flux tube increases exponentially with height, so that the upper atmosphere is more strongly filled with field. This expansion stops (or rather the field becomes purely radial) at the height at which the flux tubes merge together (above which the whole atmosphere is filled with field). Note that although above this height the field may be relatively homogeneous in strength, this does not necessarily imply that the heating is homogeneous. The observations suggest that in the chromosphere and the transition zone the heating (and hence the radiative losses) is (are) much more inhomogeneous than the field. See the lecture notes of Gabriel for the description of a model that reproduces this fact in the transition region (although an extension to the chromosphere is still problematic, but see Fiedler & Cally 1990).

On average the magnetic elements are nearly vertical. In active regions the peak in the distribution lies at inclinations of less than 10° to the vertical, although some elements with considerably larger inclinations are also observed (Martínez Pillet et al. 1997, Bernasconi 1997).

In the absence of other forces magnetic buoyancy works towards keeping magnetic flux tubes vertical. Recall that flux tubes are partially evacuated, i.e. have a lower gas density than their surroundings. Hence, for a sufficiently inclined flux tube denser gas overlies less dense material. In a gravitational field such a situation is of course unstable. Since one end of the flux tube is anchored at great depth a stable configuration is only reached when it is almost vertical (cf. Schüssler 1986, 1990 for more details).

4.2 Thermal Structure of Magnetic Elements

It is possible to determine the average temperature stratification within magnetic elements from observations of a group of spectral lines whose members have different temperature sensitivities. The main results may be summarized as follows.

1. Magnetic elements are cooler than the average quiet Sun at the same geometrical height in the continuum forming layers and below (i.e. in the lower photosphere), but are hotter in the higher layers.

2. Magnetic elements in the network (i.e. in regions in which the magnetic flux is relatively small) are hotter than in strong active region plages (where the magnetic filling factor is large).

3. The chromospheres of magnetic elements are particularly hot and bright: The chromospheric temperature rise starts 200–300 km below where it is thought to start in the average quiet sun.

Theory has to explain these observations, but most basically it has to explain why small flux tubes (magnetic elements) are bright while large flux tubes (sunspots) are dark.

There are basically 2 processes that determine the brightness of magnetic features in the visible continuum forming layers. Firstly, the magnetic field lines are dragged along by transverse flows (the field is said to be "frozen" into the plasma). Conversely, magnetic tension causes the field to brake localized transverse flows. A strong field, such as that in photospheric flux tubes, thus inhibits overturning convection, which is the main (and by far the most efficient) energy transport mechanism below the solar surface. Its inhibition therefore implies a blocking of the energy moving towards the solar surface, so that less energy reaches the surface in the magnetic features. The energy blocked in this manner is distributed throughout the solar convection zone due to its large thermal conductivity and thermal capacity (Spruit 1982). Hence, we don't expect prominent bright rings around flux tubes from the blocked energy flux.

The second process enhances the energy flux in magnetic features. Due to the lower gas density inside the flux tubes the absorption coefficient is also lower, so that we see deeper inside flux tubes (Wilson depression). This means that the surface through which radiation may escape is larger in flux tubes (in addition to the bottom we now also have the side walls). Hence flux tubes "channel" radiation to the surface. The efficiency of this process depends strongly on the ratio of the area of the walls to the bottom of the flux tube (i.e. on the Wilson depression and the size of the flux tube), as well as on the ratio of flux-tube size to photon horizontal mean free path (if the flux tube is horizontally optically thick, i.e. much wider than the photon mean free path then photons from the walls cannot contribute to the heating of the central part of the flux tube).

Whether a flux tube is bright or dark hence depends on the relative efficiency of these two processes. Since for sunspots the increase in surface area due to the Wilson depression (roughly 400–800 km) is small relative to the total surface and the spot diameter is much larger than the photon mean free path the inhibition of convection dominates and they are dark. In magnetic elements the *relative* increase in surface area is substantial and the diameter is on the order of the photon mean free path, so that the second process wins, making them bright.

In the lower photosphere simulations of magnetic features reproduce the observed temperature stratification, e.g. point 1 above, relatively well (Knölker et al. 1988, Knölker & Schüssler 1988). The models also reproduce point 2 quite well if the magnetic features in active region plage are on average somewhat larger than in the quiet Sun (Grossmann-Doerth et al. 1994). This assumption is supported by observations of the size of magnetic or brightness features (Spruit & Zwaan 1981, cf. Figures in Keller 1995). In the upper photosphere, however, theoretical models of magnetic elements are not hot enough. In particular, they do not produce a significant chromospheric temperature rise. Obviously another source of heating is missing. This source is probably related to waves, which transport energy from the solar interior along the flux tube and release it in the upper photosphere and chromosphere.

4.3 Velocities in Magnetic Elements

Although for many years magnetic elements were thought to harbour significant downflows (which posed serious problems to the requirement of mass conservation) they are currently thought to be free of steady flows greater than 200 m s^{-1} (Stenflo & Harvey 1985, Solanki 1986, Martínez Pillet et al. 1997). An exception are siphon flows. These are flows along (small) loops connecting opposite magnetic polarities on the solar surface, whose footpoints have different field strengths. Due to total pressure balance with the surroundings this implies that the gas pressure in the 2 footpoints is also different. If a flow starts along the loop from the footpoint with lower B (i.e. larger p) to the other, then the gas pressure difference will continue to drive it, and we obtain a siphon flow. Evidence has been found for siphon flows between magnetic elements and in sunspots (Rüedi et al. 1992, Martínez Pillet et al. 1994).

Of particular interest are waves travelling along flux tubes, since these are expected to transport the energy required to heat the upper photospheric and the chromospheric layers of flux tubes. Magnetic flux tubes can support a number of wave modes. In the thin-tube approximation these reduce to just 2, a longitudinal and a transverse mode (see e.g. Thomas 1985, Roberts 1986). If the flux tube can support twist, then an additional transverse mode is also allowed (often together with an additional longitudinal magneto-acoustic mode.

The latter will, however, not be discussed here any further). The geometry of these modes is illustrated in Fig. 5.

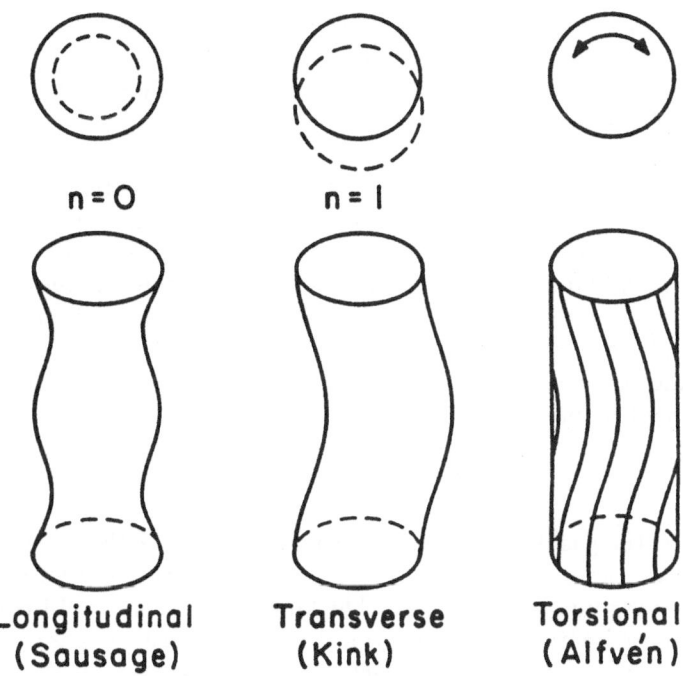

$n = 0$ $n = 1$

Longitudinal   **Transverse**   **Torsional**
(Sausage)   **(Kink)**   **(Alfvén)**

Fig. 5. Illustration of the three basic wave modes in a cylindrical and unstratified flux tube (adapted from Thomas 1985).

In the following I summarize some of the basic properties of these modes. More details on the propagation of these wave modes and their dissipation mechanisms are given by Ulmschneider in his lecture notes.

Longitudinal or sausage mode: This is a compressible magneto-acoustic wave, i.e. the gas is displaced mainly along the field lines. The dominating restoring forces are gas and magnetic pressure gradients. Above a critical frequency called the cutoff frequency this mode propagates with the tube speed,

$$c_T = \frac{c_s v_A}{\sqrt{c_s^2 + v_A^2}}, \tag{58}$$

where c_s is the sound speed (proportional to temperature) and v_A is the Alfvén speed,

$$v_A = \frac{B}{\sqrt{4\pi\rho_m}}. \tag{59}$$

In Eq. (59) ρ_m is the gas density within the flux tube. Note that $c_T < c_s$ and $c_T < v_A$. Below the cutoff the wave is evanescent and can transport no energy. For typical parameters of observed magnetic elements the cutoff frequency of this mode $\omega_T \approx 1.2\omega_s$, where ω_s is the cutoff frequency of acoustic waves. This is equivalent to saying that tube modes propagate only when their period is smaller than approximately 160 s. Thus only relatively short period waves propagate (their periods are far shorter than the typical lifetimes of granules).

The second wave mode, the torsional mode is incompressible and is in fact a pure Alfvén wave, whose restoring force is the magnetic curvature force. It propagates with the Alfvén speed v_A. Note that this mode has no cutoff frequency. Even very slowly excited torsional waves propagate.

Finally, the kink wave is incompressible and is thus Alfvén-like in character. It is not a pure Alfvén wave, however, since in addition to magnetic curvature, buoyancy also acts as a restoring force (in the presence of this wave parts of the evacuated flux tube are bent over, which thus become susceptible to buoyancy). It propagates at the kink speed

$$c_k = v_A \sqrt{\frac{\rho_m}{(\rho_m + \rho_s)}}, \tag{60}$$

where ρ_s is the gas density in the surroundings. Due to the enhanced inertia of this wave, produced by the need to displace external (field free) material by the swaying flux tube, $c_k < v_A$. The kink wave has a cutoff frequency, ω_k, that is approximately $0.4\omega_s$ for typical magnetic element parameters. This corresponds to a cutoff period of 480 s or 8 min, which is of the same order as the typical lifetime of granules.

Relatively little direct observational evidence exists for waves in magnetic elements. Most observations exhibit significant oscillations only at periods close to 5 minutes (Giovanelli et al. 1978, Fleck 1991). Volkmer et al. (1995), however, find 100 s periods in Stokes V, with a 280 m s^{-1} amplitude. According to linear theory this period corresponds to propagating longitudinal tube waves. Volkmer et al. (1995) estimate an energy flux of 1.6–2.3×10^7 erg cm^{-2} s^{-1}, sufficient to heat the associated chromospheric network if the wave is not significantly radiatively or acoustically damped on the way. Unfortunately, only a single magnetic feature was found to exhibit such oscillations and confirmation is important.

Observations of a more indirect nature, such as those of line broadening or of the spatial rms of Stokes V wavelength shifts indicate larger amplitudes. The rms shifts have values of 300–400 m s^{-1} (Martínez Pillet et al. 1997). The association of such seemingly random shifts with waves is unclear.

Observations of Stokes V asymmetry and width show values that suggest the presence of even stronger (2 km s^{-1}) non-stationary motions within magnetic elements. But again, which fraction of these is due to propagating waves (i.e. those that can carry energy into the upper atmosphere) is at present completely open.

5   Sunspots

Sunspots are magnetic structures that appear dark on the solar surface. Each sunspot is characterized by a dark core, the umbra, and a less dark halo, the penumbra. The presence of a penumbra distinguishes sunspots from the usually smaller pores, which are dark magnetic features intermediate in size between magnetic elements and sunspots.

The magnetic field, reaching up to 3500 G in the umbra, is the root cause of the sunspot phenomenon. Overviews of the structure and physics of sunspots are given in the proceedings edited by Thomas and Weiss (1992) and Schmieder et al. (1997).

5.1   General Properties of Sunspots

Sunspots exhibit a considerable size distribution, with the smallest being the commonest and the largest the rarest. Bogdan et al. (1988) conclude from Mt. Wilson white-light images that the umbral area distribution is well described by a log normal function; i.e. the logarithm of the umbral area is normally distributed. Very large sunspots can occasionally reach diameters of 60 000 km.

Sunspots live for days to months. The lifetime increases with size. Unfortunately lifetimes are often not very certain, since the decay of sunspots is very gradual and sunspots are usually born and die overnight or behind the solar limb (Murphy's law applies to sunspots as well).

The brightness and thus the temperature of a sunspot is a function of spatial position within the spot. It changes on large scales (spots are composed of an umbra or multiple umbrae and a penumbra) and small (bright umbral dots, bright and dark penumbral filaments). On a large scale the temperature is related to the field strength (Kopp & Rabin 1992, Martínez Pillet & Vázquez 1993, Solanki et al. 1993) in the sense that it increases along with B, but in a non-linear manner.

The most obvious fine-scale structures are the radial dark and bright filaments forming the penumbra. In the umbra bright, dot-like structures with a horizontal extent well below $1''$ are often seen and are referred to as umbral dots.

In the photosphere and immediate subsurface layers the temperature structure of umbrae is close to radiative equilibrium, suggesting that most of the energy is transported by radiation. The effective temperature of umbrae lies between 4200 and 4800 K, that of penumbrae roughly between 5300 and 5600 K (as compared to approximately 5800 K for the quiet Sun). The transition from umbrae to penumbrae is relatively sharp in the thermal structure (but is practically invisible in the magnetic field). The penumbra covers 70–80% of the surface area of a large, mature sunspot.

The heating of the outer atmosphere appears to be significantly reduced above sunspot umbrae relative to the strongest plage, in the sense that chromospheric and transition region emission (Gurman 1993), as well as the X-ray flux (Sams et al. 1992) is reduced above the umbra. Also, the chromospheric temperature rise starts at a greater height above sunspots than in the quiet sun (e.g. Avrett 1981, Maltby et al. 1986, Lites et al. 1987).

5.2   Magnetic Structure

Sunspots have a field strength of $B = 2500$–3500 G in their darkest portions (umbral cores) and $B = 700$–1000 G at their outer edges. At the centre of a spot the field is vertical ($\gamma' = 0$), while it is inclined by $\gamma' = 70$–$80°$ to the vertical at the visible sunspot boundary. Some recent measurements of $B(r/r_p)$ and $\gamma'(r/r_p)$, where r_p is the outer penumbral boundary in regular, i.e. almost circular sunspots are shown in Figs. 6 and 7 (cf. Keppens & Martínez Pillet 1996). For comparison, the edge of the umbra, r_u/r_p lies at 0.4–0.5. Such regular, isolated sunspots do not appear to show significant global azimuthal twist of the field (e.g. Landolfi et al. 1984, Lites & Skumanich 1990). Dipole fits to such data (e.g., Lites & Skumanich 1990) are reasonable, but not perfect (see Fig. 8).

The observations also suggest that sunspots are bounded by current sheets, i.e. at the sunspot boundary B falls off rapidly within a radial distance that is small compared to the size of the sunspot (Solanki & Schmidt 1993).

From the known $B(r)$ and $\gamma'(r)$ it is possible to determine the relative amounts of magnetic flux emerging through the umbra and the penumbra (Solanki & Schmidt 1993). The observations imply that over half the flux emerges in the penumbra. Consequently, sunspot penumbrae are deep, i.e. in the penumbra the magnetic field doesn't simply lie flat on the solar surface (as in the case of a shallow penumbra), but extends considerably below it. This is important, since it has consequences for the energy transport mechanism in the penumbra (a thin or shallow penumbra may be heated by radiation, but an additional mechanism must be acting in a deep penumbra).

The magnetic field of a sunspot continues well beyond its white-light boundary as an almost horizontal canopy with a base in the mid photosphere (Giovanelli & Jones 1982, Solanki et al. 1992b, 1994, Bruls et al. 1995). The field strength above the height of the canopy base continues to decrease steadily for increasing $r/r_p > 1$, i.e. for increasing distance from the white-light sunspot.

Within the visible outline of the sunspot the field strength decreases with height. At photospheric levels in the penumbra $dB/dz \approx 0.5$–3 G km^{-1} (Balthasar & Schmidt 1993, Bruls et al. 1995). When averaged over a height range of 2000 km or more $dB/dz \approx 0.3$–0.6 G km^{-1} in the umbra (Henze et al. 1982, Lee et al. 1993, Rüedi et al. 1995). The measured umbral values are in good agreement with simple theoretical predictions of 0.5–1 G km^{-1} (Yun 1972).

At small scales the penumbral magnetic field appears to be filamented into two components, an inclined component and a horizontal component.

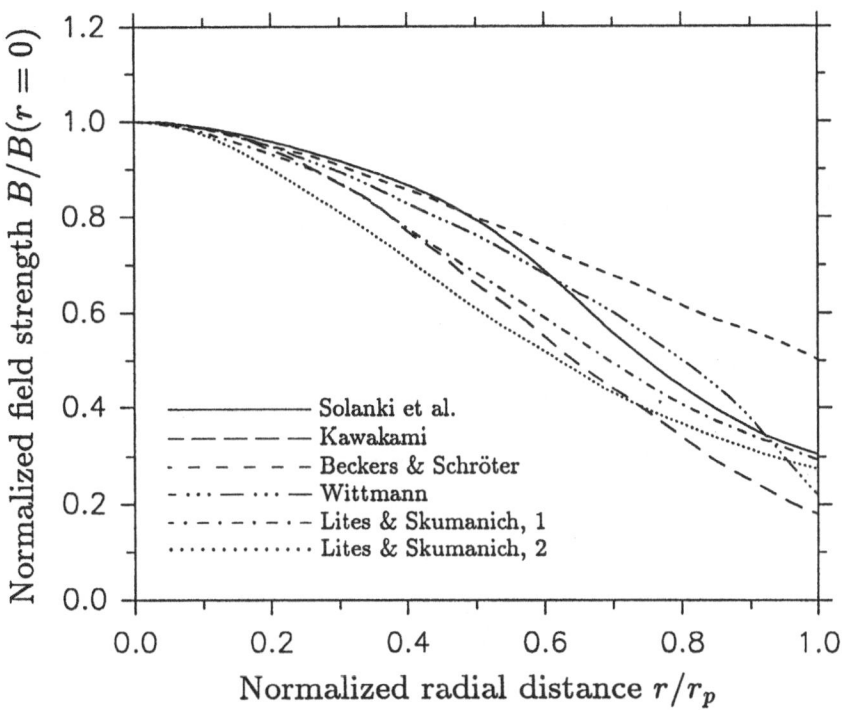

Fig. 6. Field strength normalized to its central value $B/B(r = 0)$ vs. r/r_p, the radial distance from the geometric centre of the sunspot normalized to the outer penumbral radius. The curves represent the smoothed means of the data of Beckers & Schröter (1969), Wittmann (1974), Kawakami (1983), Lites & Skumanich (1990) and Solanki et al. (1992b).

First hints of this fluted or uncombed magnetic structure have been reported by Degenhardt & Wiehr (1991), Schmidt et al. (1992) and Title et al. (1993). The horizontal magnetic component is restricted in height and is probably best described by horizontal flux tubes (Solanki & Montavon 1993, see Fig. 9).

5.3   Dynamic Structure

The dominant signature of photospheric dynamics in sunspots is the Evershed effect, named after its discoverer J. Evershed. It is composed of a shift and asymmetry of spectral lines, with opposite signs in the limbward and discward sides of the penumbra (Evershed 1909, Maltby 1964, Schröter 1965, Wiehr 1995, etc.). The discovery of the "uncombed fields" in the penumbra (see Sect. 5.2) has solved a long-standing problem, namely that the Evershed flow is very nearly horizontal (e.g. Maltby 1964, Shine et al. 1994), whereas the

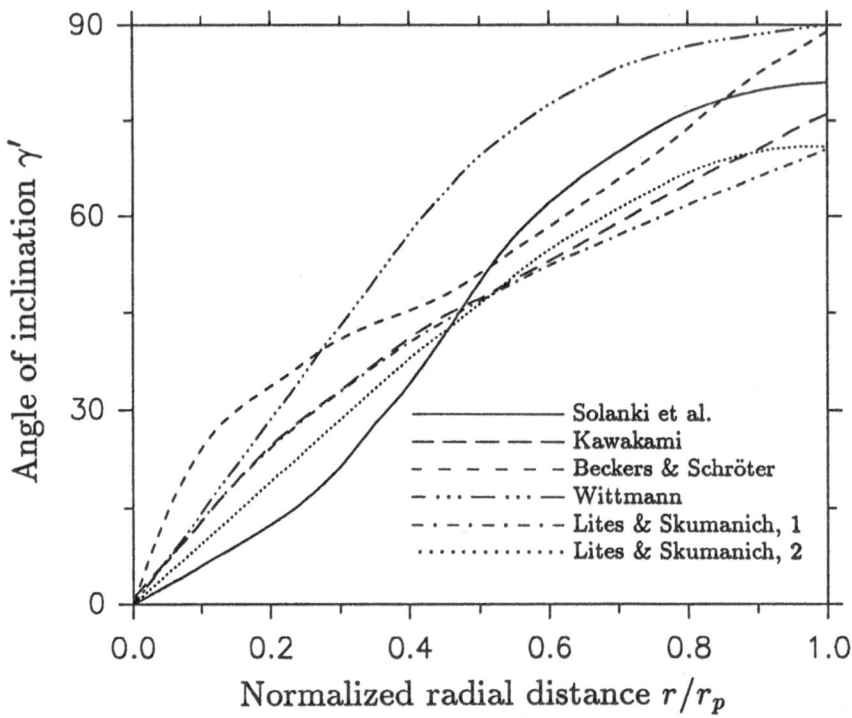

Fig. 7. Inclination γ' to the vertical vs. r/r_p for the same observations as in Fig. 6.

penumbral field is on average inclined. If the Evershed flow is restricted to the horizontal component of the field then matter can flow and field lines can still be frozen into the plasma. The Evershed effect does continue outside the sunspot boundary, but it does so only above the base of the magnetic canopy. In addition, the mass flux in the penumbra is found to be much larger than in the superpenumbral canopy (Solanki et al. 1994). Recently it has been suggested that much of the outflowing material flows down again at the outer penumbral boundary (Rimmele 1995, Westendorp Plaza et al. 1997).

The Evershed effect is restricted in height. The line shifts decrease rapidly with height of line formation (e.g. St. John 1913, Börner & Kneer 1992) and at sufficiently large heights (above the temperature minimum) even change sign, so that in the chromosphere the flow is directed inwards (so-called inverse Evershed effect, St. John 1913, Haugen 1967, Alissandrakis et al. 1988).

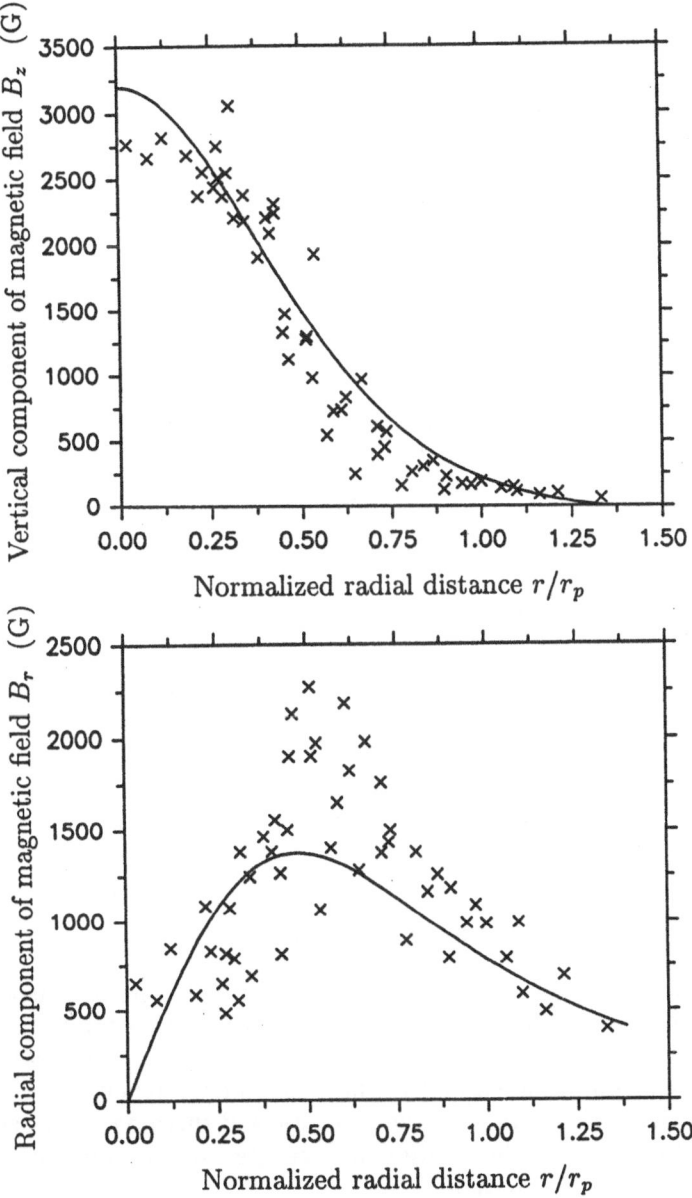

Fig. 8. Fit of a buried dipole (solid curves) to the measured $B_z(r/r_p)$ in Fig. 8a and to $B_r(r/r_p)$ in Fig. 8b. The observations are represented by crosses. Here B_z is the vertical and B_r is the radial component of the magnetic vector (from Walther 1992).

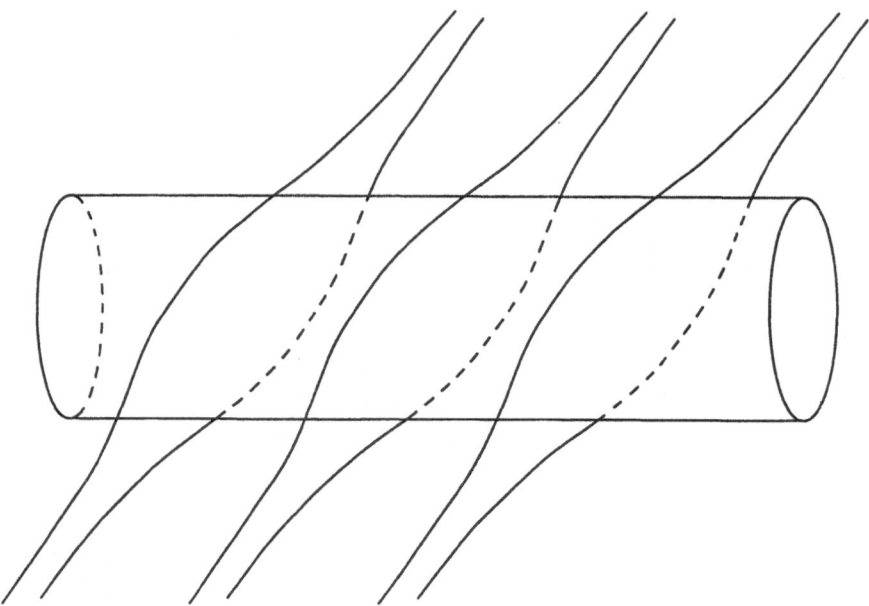

Fig. 9. Detail of the fine-scale magnetic structure of the penumbra, as derived from observations. Small-scale horizontal flux tubes are surrounded by inclined field lines.

5.4   Theory of the Magnetic Structure of Sunspots

Monolith vs. Cluster Models: The magnetic stucture of sunspots is generally modelled assuming axial symmetry and a static equilibrium. Usually significant additional assumptions are also made, since the computation of the magnetic configuration without further assumptions requires the simultaneous and consistent solution for the magnetic and thermodynamic structures, which in turn requires the solution of a realistic energy equation in addition to the force balance equation.

Basically four types of magnetic solutions have been investigated, similarity solutions, return-flux models, current-sheet models and models with arbitrary, but prescribed horizontal pressure distributions.

One basic assumption underlying all attempts to quantitatively model the global magnetic structure of sunspots is the assumption that the sunspot is monolithic below its surface. Since these layers are not directly accessible to observations (although helioseismology is providing some insight), this assumption cannot be directly tested.

Parker (1979a, b, c) proposed that just below the surface the magnetic field of a sunspot breaks up into many small flux tubes due to the fluting or interchange instability (Parker 1975, see Fig. 10). Basically, since magnetic curvature forces are smaller in small flux tubes (even non-existant in the thin-tube approximation) the total energy in the magnetic field is lowered if a large flux tube (monolithic spot) breaks up into many small flux tubes (spaghetti). Later it was found, however, that magnetic buoyancy can save sunspots ($\phi > 10^{20}$ Mx) from going unstable to fluting (Meyer et al. 1977), so that this argument does not favour the production of subsurface 'spaghetti'. Due to its buoyancy any small flux tube that breaks away from the spot is pushed upward and since the sunspot expands with height, it hits the spot again and is reabsorbed. Nevertheless, a cluster or spaghetti or jellyfish model of sunspots (see Fig. 10) has the advantage that it can readily explain the high thermal flux seen in the umbra, as well as umbral dots (as field-free intrusions into the sunspot, Parker 1979c, Choudhuri 1986). One argument against the cluster model comes from Spruit (1981a), who pointed out that the potential field in the observable layers produced by a collection of buried magnetic monopoles held together at great depth has a maximum strength that increases too strongly with the number of monopoles. It increases much more rapidly with the size of the spot than shown by the observations (e.g. Brants & Zwaan 1982, Kopp & Rabin 1992). Another reason why the cluster cannot be too loosely constructed is that the observed radial dependence of the field is not too different from a buried dipole (Fig. 8), which is prototypical of a monolithic tube.

Self-similar Models: In such models the radial dependence of **B** is prescribed. It is the same at all heights, hence the name similarity or self-similar models. Self-similar models, although straightforward to construct, are too restrictive to describe sunspots accurately. For example, self-similar models have great difficulty reproducing the observed continuum structure of sunspots composed of a central umbra surrounded by a penumbra (Landman & Finn 1979, Murphy 1990).

Return-flux Models: Osherovich (1982) extended self-similar models to include field lines emerging in the outer part of the sunspot which return to the solar surface just outside the visible sunspot (return flux). In this model the emerging return flux is said to constitute the penumbra. The model, however, finds little support from observations, since no net opposite polarity flux is seen in the immediate surroundings of sunspots (e.g. Solanki & Schmidt 1993, Lites et al. 1993). The presence of some return flux inside the penumbra been deduced by Westendorp Plaza et al. (1997), but it probably corresponds only to a small fraction of the flux emerging in the penumbra.

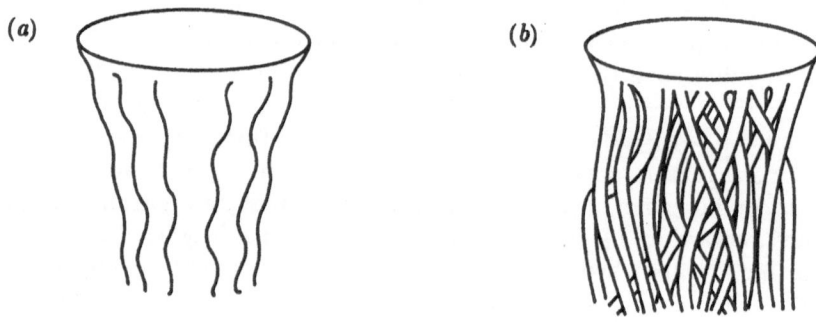

Fig. 10. Sketch of the monolith (left) and cluster (right) models of sunspots (from Thomas & Weiss 1992 by permission).

Models Without Current Sheets: A set of such models was calculated by Pizzo (1986). The pressure is specified throughout the numerical box, but not simply as a function of spatial position. Rather, it partly depends on the distribution of the field lines themselves (hydrostatic equilibrium acts along each field line). Such models are considerably more sophisticated than the self-similar or return-flux models described above and consequently also come closer to describing the observations adequately. The main problem that they face is the lack of a clear sunspot boundary, whose presence is suggested by various observations (Beckers & Schröter 1969, Solanki & Schmidt 1993).

Models With Current Sheet: Wegmann (1981) proposed a solution to the free boundary problem, i.e. the determination of the position of the boundary of, e.g., a flux tube whose interior is potential, i.e. current free, so that all the current is concentrated in a boundary current sheet. Schmidt & Wegmann (1983) first applied this technique to sunspots. Jahn (1989) extended it to include body currents in his model, i.e. the field in the sunspot is not purely potential. The body currents were restricted to the outer part of the sunspot and chosen such that the surface field matches the observations of Beckers & Schröter (1969). Jahn (1992) extended the treatment by introducing an internal magnetic boundary corresponding roughly to the boundary between the umbra and the penumbra.

Pure current sheet models fail to reproduce the observations (Jahn 1989), but combined sheet and body current models (Jahn 1989, 1992) provide relatively good fits to the observations of the global magnetic structure of sunspots and are the most promising for future study.

Uncombed Fields: Basically, the ideas regarding the origin of the un-combed fields in sunspot penumbrae consider the small-scale magnetic structure to be dynamic and its complexity to result from an instability.

Spruit (1981b) and Jahn (1992) propose that the complex structure is due to the convective exchange of flux tubes. So far, however, only a highly qualitative description of this mechanism has been provided. One possible such scenario is: A tube near the bottom of the (deep) penumbra is heated by the non-magnetic convective gas with which it comes into contact. The heated tube is buoyant and rises, coming to the surface as an inclined tube. At the surface it radiates away its excess energy, loses its buoyancy, becomes more horizontal and sinks again. The rise of a buoyant penumbral flux tube from the outer (subsurface) boundary of the sunspot to the solar surface has been modelled by Jahn & Schmidt (1994) and Schlichenmaier et al. (1997). According to these simulations the tube is horizontal and bright (at least in its inner part) and supports an outward flow of material (Evershed effect?) when it reaches the solar surface.

The other proposal comes from Wentzel (1992). He starts with an inclined field and produces a horizontal field component by invoking a density inversion with height. To produce the density inversion he postulates a spatially localized impulsive upflow in the inclined magnetic component. In the almost vertical field of the umbra he expects such an upflow to be visible as an umbral dot. In the penumbra, at a larger inclination of the field to the vertical, the flow, which fills a small flux tube up to a certain height, now causes dense material to overlie less dense material. An instability develops which, according to Wentzel (1992), causes the filled flux tube to fall and become horizontal. Furthermore, he expects the surplus material, which has flown up into the inclined flux tube, to now flow outward along the horizontal tube. A detailed numerical treatment of the collapse of a flux tube would be of great value. Also needed are observations, in particular ones that follow the evolution and dynamics of the uncombed field.

5.5   Brightness and Thermal Structure

Biermann (1941) first proposed that the reduced brightness of sunspots is due to the inhibition of convective motions by the magnetic field. Since convection dominates energy transport below the observable layers, a quenching of convection makes a sunspot look like an obstacle to the outward heat flux through the convection zone. This leads to a diversion of energy away from the sunpot, which reduces the energy flux through the spot and produces a darkening (Sect. 4.2). The diverted energy flux is mainly stored in the convection zone and released only over a time scale of 10^5 years, much longer than the lifetime of a spot (Spruit 1977, 1982). A complete quenching of convection reduces the heat flux so efficiently that the question to ask is not why umbrae and penumbrae are so dark, but rather why they are so bright, in particular the penumbrae. As discussed in Sect. 4.2 sunspots are too large and their Wilson depression too small for radiation from the walls to be important.

Basically two approaches have been taken to solve this problem. The first has been to form the sunspot out of a cluster of small flux tubes (Spaghetti model, see Sect. 5.4). Due to the tapered shape of each small flux tube, increasingly more non-magnetic gas is present between them at increasing depth. Consequently, below the sunspot convection can penetrate relatively unhindered until close to the surface. In addition, for a sunspot composed of n small flux tubes the surface area of the side walls of the flux tubes, over which the convective gas can radiate into the magnetized gas, is $\sqrt{n}$-times larger than for a simple monolithic sunspot. A larger side-wall surface compared to the horizontal cross-sectional area leads to a much more efficient heating of the tubes. This theory, like many other aspects of the cluster model, has not been quantitatively worked out in detail.

The second proposal incorporates convective transport within a simple model of a monolithic sunspot. Magnetoconvection is a vast subject in itself and I refer the reader to reviews devoted specifically to this topic (e.g. Hughes & Proctor 1988, Weiss 1991, Proctor 1992). Here I only mention two results. The current picture is that oscillatory convection dominates in the 2000 km immediately below the surface, while overturning convection (such as seen in granules) dominates in the deeper layers.

Nordlund & Stein (1990) carried out a 3-D, fully compressible calculation of a box including the solar surface near its top. They treated radiative transfer consistently. First, granular convection is allowed to develop. Then a strong, homogeneous and vertical field is added. The field immediately begins to inhibit the convection and the surface intensity falls rapidly with time, until it becomes as low as 10% of the original granular (i.e. quiet sun) intensity. At this point the temperature gradient of the atmosphere becomes so large that a short episode of convection is started, which raises the emergent intensity for a brief period of time. After this the intensity again gradually drops until another convective episode is triggered. The convection in these models is overturning. It appears that episodic convection can produce and maintain the correct umbral temperature in a monolithic model of the umbra.

5.6   Theory of the Evershed Effect

The simplest explanation of the Evershed effect assumes that the line shifts and asymmetries are produced by a steady, almost radial outflow. This scenario was put on a solid physical footing by Meyer & Schmidt (1968), who presented a siphon flow model of the Evershed effect (cf. Thomas & Montesinos 1993). The field strength in the outer penumbra (700–900 G) is smaller than in typical small-scale magnetic elements (1000–1500 G), so that the gas pressure is expected to be larger in the penumbra. If the field lines from the outer penumbra are connected to magnetic elements, then the pressure difference at equal gravitational potential drives a flow from the penumbra to the magnetic elements. One nice feature of this model is that it naturally explains the inverse Evershed effect. The field strength in the umbra (2000–3000 G) is larger than in small magnetic elements, so that a field line connecting the umbra with a magnetic element should support an inflow into the umbra. If, however, the outflowing matter returns to the solar interior within the sunspot the siphon flow model requires the field strength in the outer penumbra to be larger than in the inner penumbra, which is not what is generally observed.

Recent high resolution observations further suggest that the Evershed effect is not steady but episodic (Rimmele 1994, Shine et al. 1994), suggestive of a time-dependent phenomenon. Models such as those of Schlichenmaier et al. (1997) and of Wentzel (1992) would then be closer to the observations.

6   Conclusions

This concludes the current brief introduction to solar magnetism. Much has been left unsaid, not just about the magnetic field in the solar interior (e.g., dynamos, see Proctor, this volume) and in the chromosphere and corona (see the contributions of Ulmschneider and of Walsh), which have practically not been touched upon here. The surface has only been scratched even as far as the photospheric layers of flux tubes are concerned.

Inspite of the large amount of knowledge of solar magnetism the number of open questions is immense. I expect SOHO to provide considerable new insight into solar magnetism, in particular its connection with solar heating, but also its large-scale structure and evolution and its influence on p-modes and solar irradiance variations. Nevertheless, the resolution of many open questions, in particular those connected with the physics of photospheric magnetic features, will probably have to wait for higher spatial resolution spectropolarimetric observations, either from the ground or from space (e.g. as planned by the japanese Solar-B mission).

References

Alissandrakis C.E., Dialetis D., Mein P., Schmieder B., Simon G., 1988, *Astron. Astrophys.* **201**, 339

Avrett E.H., 1981, in *The Physics of Sunspots*, L.E. Cram, J.H. Thomas (Eds.), National Solar Obs., Sunspot, NM, p. 235

Balthasar H., Schmidt W., 1993, *Astron. Astrophys.* **279**, 243

Beckers J.M., Schröter E.H., 1969, *Solar Phys.* **10**, 384

Bernasconi P.N., 1997, *Stokes vector polarimetry: Observation and analysis of solar magnetic fields*, Ph.D. Thesis, ETH, Zürich

Biermann L., 1941, *Vierteljahrsschrift Astron. Ges.* **76**, 194

Bogdan T.J., Gilman P.A., Lerche I., Howard R., 1988, *Astrophys. J.* **327**, 451

Böhm-Vitense E., 1989, *Introduction to Stellar Astrophysics. Vol. 2: Stellar Atmospheres*, , Cambridge University Press, Cambridge

Börner P., Kneer F., 1992, *Astron. Astrophys.* **259**, 307

Brants J.J., Zwaan C., 1982, *Solar Phys.* **80**, 251

Bruls J.H.M.J., Solanki S.K., Carlsson M., Rutten R.J., 1995, *Astron. Astrophys.* **293**, 225

Choudhuri A.R., 1986, *Astrophys. J.* **302**, 809

Defouw R.J., 1976, *Astrophys. J.* **209**, 266

Degenhardt D., Wiehr E., 1991, *Astron. Astrophys.* **252**, 821

Evershed J., 1909, *Monthly Notices Royal Astron. Soc.* **69**, 454

Ferriz Mas A., Schüssler M., 1989, *Geophys. Astrophys. Fluid Dyn.* **48**, 217

Fiedler R.A.S., Cally P.S., 1990, *Solar Phys.* **126**, 69

Fleck B., 1991, *Rev. Mod. Astron.* **4**, 90

Giovanelli R.G., Jones H.P., 1982, *Solar Phys.* **79**, 267

Giovanelli R.G., Livingston W.C., Harvey J.W., 1978, *Solar Phys.* **59**, 49

Grossmann-Doerth U., Knölker, M., Schüssler M., Solanki S.K., 1994, *Astron. Astrophys.* **285** 648

Gurman J.B., 1993, *Astrophys. J.* **412**, 865

Haugen E., 1967, *Solar Phys.* **2**, 227

Henze W., Jr., Tandberg-Hanssen E., Hagyard M.J., Woodgate B.E., Shine R.A., Beckers J.M., Bruner M., Gurman J.B., Hyder C.L., West E.A., 1982, *Solar Phys.* **81**, 231

Hughes D.W., Proctor M.R.E., 1988, *Ann. Rev. Fluid Mech.* **20**, 187

Illing R.M.E., Landman D.A., Mickey D.L., 1975, *Astron. Astrophys.* **41**, 183

Jahn K., 1989, *Astron. Astrophys.* **222**, 264

Jahn K., 1992, in *Theory of Sunspots*, J.H. Thomas, N. Weiss (Eds.), Cambridge University Press, p. 139

Jahn K., Schmidt H.U., 1994, *Astron. Astrophys.* **290**, 295

Kalkofen W., 1987, (Ed.), *Numerical Radiative Transfer*, Cambridge University Press, Cambridge

Kawakami H., 1983, *Publ. Astron. Soc. Japan* **35**, 459

Keller C.U., 1995, *Reviews in Modern Astronomy* **8**, 27

Keller C.U., Deubner F.-L., Egger U., Fleck B., Povel P., 1994, *Astron. Astrophys.* **286**, 626

Keppens R., Martínez Pillet V., 1996, *Astron. Astrophys.* **316**, 229

Knölker M., Schüssler M., 1988, *Astron. Astrophys.* **202**, 275

Knölker M., Schüssler M., Weisshaar E., 1988, *Astron. Astrophys.* **194**, 257

Kopp G., Rabin D., 1992, *Solar Phys.* **141**, 253

Landi Degl'Innocenti E., 1976, *Astron. Astrophys. Suppl Series* **25**, 379

Landi Degl'Innocenti E., Landi Degl'Innocenti M., 1981, *Il Nuovo Cimento B* **62**, 1

Landman D.A., Finn G.D., 1979, *Solar Phys.* **63**, 221

Landolfi M., Landi Degl'Innocenti E., Arena P., 1984, *Solar Phys.* **93**, 269

Lee J.W., Hurford G.J., Gary D.E., 1993, *Solar Phys.* **144**, 45

Lin H., 1995, *Astrophys. J.* **446**, 421

Lites B.W., Skumanich A., 1990, *Astrophys. J.* **348**, 747

Lites B.W., Skumanich A., Rees D.A., Murphy G.A., Carlsson M., 1987, *Astrophys. J.* **318**, 930

Lites B.W., Elmore D.F., Seagraves P., Skumanich A., 1993, *Astrophys. J.* **418**, 928

Maltby P., 1964, *Astrophys. Norvegica* **8**, 205

Maltby P., Avrett E.H., Carlsson M., Kjeldseth-Moe O., Kurucz R.L., Loeser R., 1986, *Astrophys. J.* **306**, 284

Martínez Pillet V., Vázquez M., 1990, *Astrophys. Space Sci.* **170**, 75

Martínez Pillet V., Lites B.W., Skumanich A., Degenhardt D., 1993, *Astron. Astrophys.* **270**, 494

Martínez Pillet V., Lites B.W., Skumanich A., 1997, *Astrophys. J.* **474**, 810

Meyer F., Schmidt H.U., 1968, *Z. Angew. Math. Mech.* **48**, 218

Meyer F., Schmidt H.U., Weiss N.O., 1977, *Monthly Notices Royal Astron. Soc.* **179**, 741

Murphy G.A., 1990, NCAR Cooperative Thesis No. 124

Nordlund A., Stein R.F., 1990, in *Solar Photosphere: Structure, Convection and Magnetic Fields*, J.O. Stenflo (Ed.), Kluwer, Dordrecht, *IAU Symp.* **138**, 191

Osherovich V.A., 1982, *Solar Phys.* **77**, 63

Parker E.N., 1975, *Solar Phys.* **40**, 291

Parker E.N., 1978, *Astrophys. J.* **221**, 368

Parker E.N., 1979a, *Cosmical Magnetic Fields*, Clarendon Press, Oxford

Parker E.N., 1979b, *Astrophys. J.* **230**, 905 –913. Sunspots and the Physics of Magnetic Flux Tubes. I. The General Nature of Sunspots.

Parker E.N., 1979c, *Astrophys. J.* **234**, 333 –347. Sunspots and the Physics of Magnetic Flux Tubes. IX. Umbral Dots and Longitudinal Overstability.

Pizzo V.J., 1986, *Astrophys. J.* **302**, 785

Pizzo V.J., 1987, in *Theoretical Problems in High Resolution Solar Physics II*, G. Athay, D.S. Spicer (Eds.), NASA Conf. Publ. 2483, p. 1

Priest E.R., 1982, *Solar Magnetohydrodynamics*, Reidel, Dordrecht

Proctor M.R.E., 1992, in *Sunspots: Theory and Observations*, J.H. Thomas, N.O. Weiss (Eds.), Kluwer, Dordrecht, pp. 221

Rachkovsky D.N., 1962, *Izv. Krymsk. Astrofiz. Obs.* **28**, 259

Rimmele T.R., 1994, *Astron. Astrophys.* **290**, 972

Rimmele T.R., 1995, *Astrophys. J.* **445**, 511

Roberts B., 1986, in *Small Scale Magnetic Flux Concentrations in the Solar Photosphere*, W. Deinzer, M. Knölker, H.H. Voigt (Eds.), Vandenhoeck & Ruprecht, Göttingen, p. 169

Rüedi I., Solanki S.K., Livingston W., Stenflo, J.O., 1992, *Astron. Astrophys.* **263**, 323

Rüedi I., Solanki S.K., Livingston W., 1995, *Astron. Astrophys.* **293**, 252

Sams B.J. III, Golub L., Weiss N.O., 1992, *Astrophys. J.* **399**, 313

Schlichenmaier R., Jahn K., Schmidt H.U., 1997, in *Advances in the Physics of Sunspots*, B. Schmieder, J.C. del Toro Iniesta, M. Vázquez (Eds.), Astron. Soc. Pacific Conf. Ser. Vol. 118, p. 140

Schmidt H.U., Wegmann R., 1983, in *Dynamical Problems in Mathematical Physics*, B. Brosowski, E. Martensen (Eds.), Verlag P. Lang, Frankfurt, p. 137

Schmidt W., Hofmann A., Balthasar H., Tarbell T.D., Frank Z.A., 1992, *Astron. Astrophys.* **264**, L27

Schmieder B., del Toro Iniesta J.C., Vázquez M., (Eds.)1997, *Advances in the Physics of Sunspots*, Astron. Soc. Pacific Conf. Ser. Vol. 118

Schröter E.H., 1965, *Z. Astrophys.* **62**, 228

Schüssler M., 1986, in *Small Scale Magnetic Flux Concentrations in the Solar Photosphere*, W. Deinzer, M. Knölker, H.H. Voigt (Eds.), Vandenhoeck & Ruprecht, Göttingen, p. 103

Schüssler M., 1990, in *Solar Photosphere: Structure, Convection and Magnetic Fields*, J.O. Stenflo (Ed.), Kluwer, Dordrecht, *IAU Symp.* **138**, 161

Schüssler M., 1992, in *The Sun — a Laboratory for Astrophysics*, J.T. Schmelz, J.C. Brown (Eds.), Kluwer, Dordrecht, p. 191

Shine R.A., Title A.M., Tarbell T.D., Smith K., Frank Z.A., 1994, *Astrophys. J.* **430**, 413

Solanki S.K., 1986, *Astron. Astrophys.* **168**, 311

Solanki S.K., 1993, *Space Sci. Rev.* **61**, 1

Solanki S.K., Montavon C.A.P., 1993, *Astron. Astrophys.* **275**, 283

Solanki S.K., Schmidt H.U., 1993, *Astron. Astrophys.* **267**, 287

Solanki S.K., Rüedi I., Livingston W., 1992a, *Astron. Astrophys.* **263**, 312

Solanki S.K., Rüedi I., Livingston W., 1992b, *Astron. Astrophys.* **263**, 339

Solanki S.K., Walther U., Livingston W., 1993, *Astron. Astrophys.* **277**, 639

Solanki S.K., Montavon C.A.P., Livingston W., 1994, *Astron. Astrophys.* **283**, 221

Solanki S.K., Zuffrey D., Lin H., Rüedi I., Kuhn J., 1996, *Astron. Astrophys.* **310**, L33

Spruit H.C., 1977, *Solar Phys.* **55**, 3

Spruit H.C., 1979, *Solar Phys.* **61**, 363

Spruit H.C., 1981a, in *Physics of Sunspots*, L.E. Cram, J.H. Thomas (Eds.), National Solar Obs., Sunspot, NM, p. 98

Spruit H.C., 1981b, in *Physics of Sunspots*, L.E. Cram, J.H. Thomas (Eds.), National Solar Obs., Sunspot, NM, p. 359

Spruit H.C., 1982, *Astron. Astrophys.* **108**, 348

Spruit H.C., Zwaan C., 1981, *Solar Phys.* **70**, 207

Spruit H.C., Zweibel E.G., 1979, *Solar Phys.* **62**, 15

St. John C.E., 1913, *Astrophys. J.* **37**, 322

Stenflo J.O., 1994, *Solar Magnetic Fields: Polarized Radiation Diagnostics*, Kluwer, Dordrecht

Stenflo J.O., Harvey J.W., 1985, *Solar Phys.* **95**, 99

Thomas J.H., 1985, in *Theoretical Problems in High Resolution Solar Physics*, H.U. Schmidt (Ed.), Max Planck Inst. f. Astrophys., Munich, p. 126

Thomas J.H., Montesinos B., 1993, *Astrophys. J.* **407**, 398

Thomas J.H., Weiss N., 1992, in *Sunspots: Theory and Observations*, J.H. Thomas, N. Weiss (Eds.), Kluwer, Dordrecht, p. 3

Title A.M., Frank Z.A., Shine R.A., Tarbell T.D., Topka K.P., Scharmer G., Schmidt W., 1993, *Astrophys. J.* **403**, 780

Unno W., 1956, *Publ. Astron. Soc. Japan* **8**, 108

Unno W., Ando H., 1979, *Geophys. Astrophys. Fluid Dyn.* **12**, 107

Venkatakrishnan P., 1986, *Nature* **322**, 156

Volkmer R., Kneer F., Bendlin C., 1995, *Astron. Astrophys.* **304**, L1

Walther U., 1992, *Diplomarbeit*, Institute of Astronomy, ETH, Zürich

Webb A.R., Roberts B., 1978, *Solar Phys.* **59**, 249

Wegmann R., 1981, in *Numerical Treatment of Free Boundary Value Problems*, J. Albrecht (Ed.), Birkhäuser, Stuttgart, p. 335
Weiss N.O., 1991, *Geophys. Astrophys. Fluid Dyn.* **69**, 229
Wentzel D.G., 1992, *Astrophys. J.* **388**, 211
Westendorp Plaza C., del Toro Iniesta J.C., Ruiz Cobo B., Martínez Pillet V., Lites B.W., Skumanich A., 1997, *Nature* **389**, 47
Wiehr E., 1995, *Astron. Astrophys.* **298**, L17
Wittmann A.D., 1974, *Solar Phys.* **36**, 29
Yun H.S., 1972, *Solar Phys.* **22**, 137
Zeeman P., 1897, *Phil. Mag.* **43**, 226

Heating of Chromospheres and Coronae

Peter Ulmschneider

Institut für Theoretische Astrophysik der Universität Heidelberg, Tiergartenstr. 15, D–69121 Heidelberg, Germany
ulmschneider@ita.uni-heidelberg.de

Abstract. Chromospheres and coronae owe their existence to mechanical heating. In the present work the mechanisms which are thought to provide steady mechanical heating are reviewed. These mechanisms can be classified as *hydrodynamical–* and *magnetic heating mechanisms* and each of these can be subdivided further on basis of the fluctuation frequency. Rapid fluctuations generated by the turbulence in the convection zones lead to acoustic waves and to mhd waves (AC–mechanisms), slow fluctuations to pulsational waves and to stressed fields with current sheets (DC–mechanisms).

Solar and stellar observations, as well as acoustic and mhd wave generation rate computations on basis of convection zone models and a Kolmogorov-type energy spectrum representation for the turbulence, provide great progress towards the understanding of the complete dependence of the chromospheres and coronae on the properties of the underlying stars.

1   Chromospheres and coronae

Chromospheres and coronae are hot outer layers of stars where the temperatures are much higher than the stellar effective temperature T_{eff}. In the chromospheres the temperatures rise in outward direction to values of around 10000 K, while in the coronae the temperatures are in the million degree range. These layers are observed in the Radio, UV and X-ray spectral ranges. That the Sun has a million degree corona was first discovered by Grotrian in 1939 and Edlén in 1941 by identifying the coronal lines (observed since 1869) as transitions from low lying metastable levels of the ground configuration of highly ionized metals. The green coronal line at 5303 Å for instance emanates from the ion Fe XIV and the red line at 6374 Å from Fe X. Observing the Sun in the UV from 300 to 1400 Å the OSO satellites in the 1960's then showed the full sequence of ions e.g. from O II to O VI, the latter emitted at a temperature of about 10^5 K, as well as MgX which indicates 10^6 K. The Lyman, Carbon and Silicon continua from these observations were later used to construct models (like by Vernazza, Avrett & Loeser 1981) of the outer solar atmosphere. The IUE and Einstein satellites in the 1970's and 1980's, measuring UV and X-ray emissions from stars, subsequently showed that the Sun is not exceptional, but that probably all nondegenerate stars have such hot outer shells.

Einstein observations (Vaiana et al. 1981) showed that *O- and B-stars*, which do not have surface convection zones, have strong X-ray emission which today is attributed not to contiguous layers around those stars, but to shocks generated by rapidly moving individual gas blobs driven by radiative instability (see Sec. 5 below) in the outer stellar envelope.

F-, G-, K- and M-stars have chromospheres and often coronae much like our Sun. These are attributed to the wave- and magnetic energy generation in the surface convection zones of these stars. Late giants and supergiants do not have coronae (Linsky 1985). Finally, *A-stars* appear to have neither chromospheres nor coronae. There is a conspicuous X-ray gap for A-stars (Vaiana et al. 1981) and no UV-emission indicating an outwardly rising temperature. However, theoretically it can be argued that these stars still have a strong radiation field where photospheric motions can be initiated and thus acoustic disturbances produced. It is well known that by wave energy conservation even a very small disturbance, propagating over many density scale heights, will generate shock waves which lead to dissipation rates proportional to the pressure (or density) (Ulmschneider 1991). As the radiative cooling goes with the square of the density it is seen that eventually heating will overwhelm cooling, leading to hot outer layers. However, this might be so far away from the star that no appreciable emission is generated and thus observed. Thus, theoretically at least, A-stars should also have hot outer envelopes. This supports our above stated conclusion that very likely all nondegenerate stars have hot outer layers.

Chromospheres and coronae have another very important property which should never be overlooked. Except for T-Tau type stars, where the chromospheric emission results from mass-infall from accretion disks, the typical star with a chromosphere and corona does not receive energy from infinity. Thus the detailed physics of chromospheres and coronae is completely dependent on the underlying star. We are faced with the great challenge to understand the dependence of chromospheres and coronae on the stellar interior. For this reason, in the present review, not only the individual heating mechanisms are discussed but an attempt is made to relate heating and mechanical energy fluxes to the structure of the convection zone. One day we expect that a careful simulation of the surface convection zone of a given star will result in an accurate prediction of its chromosphere and corona.

2   Necessity of mechanical heating

Consider a gas element in the chromosphere. An amount of heat dQ (erg/cm^3) flowing into the element across its boundaries raises the entropy in the element by

$$dS = \frac{dQ}{\rho T} \quad , \tag{1}$$

where T is the temperature and ρ the density. As the entropy in the gas element (moving in a wind with velocity v) is conserved, one has in the laboratory (Euler) frame

$$\frac{\partial S}{\partial t} + v\frac{\partial S}{\partial z} = \frac{dS}{dt}\Big|_R + \frac{dS}{dt}\Big|_J + \frac{dS}{dt}\Big|_C + \frac{dS}{dt}\Big|_V + \frac{dS}{dt}\Big|_M \quad . \tag{2}$$

Here t is time and z is the height in a plane-parallel atmosphere. The first four terms on the right hand side stand for radiative, Joule, thermal conductive and viscous heating. We have added a term $dS/dt|_M$ which increases the entropy due to *mechanical heating*. Mechanical heating comprises all processes which convert nonradiative, nonconductive hydrodynamic or magnetic energy (henceforth called mechanical energy) flowing through the element into microscopic random thermal motion.

As the chromosphere exists on the Sun for billions of years, one usually (for an average model) can neglect the term $\frac{\partial S}{\partial t}$ in Eq. (2). With a solar wind mass loss rate of $\dot{M} = 10^{-14} M_\odot/y$ one can compute a wind flow speed

$$v = \frac{\dot{M}}{4\pi\rho R_\odot^2} = \frac{10^{-14} 2 \cdot 10^{33}}{3 \cdot 10^7 4\pi \left(7 \cdot 10^{10}\right)^2 \rho} \approx \frac{1.1 \cdot 10^{-11}}{\rho} \quad . \tag{3}$$

Typical flow velocities in the chromosphere from this formula on basis of the Vernazza et al. (1981) model C, henceforth called VAL81, are shown in Tab. 1. Because the wind velocity in the chromosphere and transition layer is small

z (km)	0	500	1000	1500	2100	2543
T (K)	6400	4200	5900	6400	10000	447000
ρ (g/cm^3)	$2.7 \cdot 10^{-7}$	$6.0 \cdot 10^{-9}$	$4 \cdot 10^{-11}$	$2.5 \cdot 10^{-12}$	$1.2 \cdot 10^{-13}$	$2.3 \cdot 10^{-15}$
c_S (cm/s)	$8.3 \cdot 10^5$	$6.7 \cdot 10^5$	$9 \cdot 10^5$	$8.3 \cdot 10^5$	$1.5 \cdot 10^6$	$1.0 \cdot 10^7$
v (cm/s)	$4.0 \cdot 10^{-5}$	$1.8 \cdot 10^{-3}$	0.27	4.3	90	$4.7 \cdot 10^3$

Table 1. Temperature T, density ρ, sound speed c_S and wind velocity v in the VAL81 model as a function of height z.

compared to the other characteristic velocity, the sound speed c_S (see Tab. 1), the entire left hand side of Eq. (2) can be neglected. Below we will show that $dS/dt|_J$, $dS/dt|_C$, $dS/dt|_V$ can also be neglected in the chromosphere (while $dS/dt|_C$ is important only in the transition layer and corona). In the chromosphere we thus find

$$\frac{dS}{dt}\Big|_R + \frac{dS}{dt}\Big|_M = 0 \quad . \tag{4}$$

Conclusion: *In stellar chromospheres the main energy balance is between radiation and mechanical heating.* For simplicity of the argument we now assume gray radiation. With the mean intensity J (erg/(cm^2 s sterad Hz)), the frequency-integrated Planck function B (erg/(cm^2 s sterad Hz)) and the gray absorption coefficient $\bar{\kappa}$ (1/cm) one finds

$$\frac{4\pi\bar{\kappa}}{\rho T}\left(J - B\right) + \left.\frac{dS}{dt}\right|_M = 0 \quad . \tag{5}$$

In the special case of *radiative equilibrium* one has $dS/dt|_M = 0$ and obtains $J = B$. With $J = \sigma T_{eff}^4/2\pi$, valid for the surface of a star, roughly represented by a black body of effective temperature T_{eff} (the factor 1/2 comes from the fact that there is only radiation going away from the star), and $B = \sigma T^4/\pi$, one gets for the outer stellar layers a boundary temperature

$$T = \left(\frac{1}{2}\right)^{1/4} T_{eff} \approx 0.8\, T_{eff} \quad . \tag{6}$$

That is, in absence of mechanical heating we would expect that the regions above the stellar surface have temperatures of the order of the boundary temperature. However, as a chromosphere is a layer where the temperature rises in outward direction to values $T \gg T_{eff}$, it is clear that one must have $B \gg J$ and therefore $dS/dt|_M \gg 0$. This shows that for chromospheres mechanical heating is essential. In addition, as the energy loss of the transition layer and corona cannot be balanced by thermal conduction from a reservoir at infinity, but must ultimately be supplied from the stellar interior, we conclude that *for the existence of chromospheres and coronae mechanical heating is essential.*

Moreover, chromospheres and coronae can only be maintained if mechanical heating is supplied *constantly*. The time scale, in which an excess temperature will cool down to the boundary temperature if the mechanical heating were suddenly disrupted, is given by the the *radiative relaxation time* for which we have

$$t_{Rad} = \frac{\Delta E}{-\Phi_R} = \frac{\rho c_v \Delta T}{16\bar{\kappa}\sigma T^3 \Delta T} = \frac{\rho c_v}{16\bar{\kappa}\sigma T^3} \approx 1.1 \cdot 10^3 \; s \quad . \tag{7}$$

Here we have expanded the radiative cooling rate $-\Phi_R = 4\bar{\kappa}\sigma(T^4 - T_{eff}^4/2) \approx 16\bar{\kappa}\sigma T^3 \Delta T$ for $T_{eff}^4/2 \approx T^4$ and at $z = 1280$ km in the VAL81 model, used temperature $T = 6200$ K, pressure $p = 4.4$ dyn/cm^2, opacity $\bar{\kappa}/\rho = 4.1 \cdot 10^{-4}$ cm^2/g from Eq. (20), specific heat $c_v = 9.6 \cdot 10^7$ erg/gK, Stefan-Boltzmann constant $\sigma = 5.6 \cdot 10^{-5}$ $erg/cm^2 s$ K^4. It is seen that in timescales of a fraction of an hour the chromosphere would cool down to the boundary temperature if mechanical heating would suddenly stop.

3   Overview of the heating mechanisms

Tab. 2 summarizes the mechanisms which are thought to provide a steady supply of mechanical energy to balance the chromospheric and coronal losses

(Kuperus, Ionson & Spicer 1981, Wentzel 1981, Ulmschneider, Priest & Rosner 1991, Narain & Ulmschneider 1990, 1996). Here occasional transient and localized heating events like large flares are not considered, because they do not contribute appreciably to the persistent chromospheric and coronal heating. The term heating mechanism comprises three physical aspects, the *generation* of a carrier of mechanical energy, the *transport* of mechanical energy into the chromosphere and corona and the *dissipation* of this energy in these layers.

energy carrier	*dissipation mechanism*
hydrodynamic heating mechanisms	
acoustic waves, P<P_A pulsational waves, P>P_A	shock dissipation shock dissipation
magnetic heating mechanisms	
1. alternating current (AC) or wave mechanisms	
slow mode mhd waves, longitudinal mhd tube waves	shock dissipation
fast mode mhd waves	Landau damping
Alfvén waves (transverse, torsional)	mode-coupling resonance heating compressional viscous heating turbulent heating Landau damping
magnetoacoustic surface waves	mode-coupling phase-mixing resonant absorption
2. direct current (DC) mechanisms	
current sheets	reconnection (turbulent heating, wave heating)

Table 2. Heating mechanisms

Tab. 2 shows the various proposed energy carriers which can be classified into two main categories as *hydrodynamic* and *magnetic* heating mechanisms. Both the hydrodynamic and the magnetic mechanisms can be subdivided further by frequency. Acoustic waves are high frequency hydrodynamic fluctuations with periods $P < P_A$ (here P_A is the acoustic cut-off period, c.f.

Eq. (11)) and pulsational waves have periods $P > P_A$. The magnetic mechanisms are subdivided into high frequency wave- or AC(alternating current)-mechanisms and current sheet- or DC(direct current)-mechanisms where one has time variations of low frequency. Also in Tab. 2 the mode of dissipation of these mechanical energy carriers is indicated.

Ultimately the mechanical energy carriers derive their energy from the nuclear processes in the stellar core from where the energy is transported by radiation and convection to the stellar surface. In late-type stars the mechanical energy generation arises from the gas motions of the surface convection zones. These gas motions are largest in the regions of smallest density near the top boundary of the convection zone. Due to this, the mechanical energy carriers, particularly the waves, are generated in a narrow surface layer.

As the gas motions in the convection zone can be described by a common temporal and spatial turbulence spectrum consisting of a characteristic distribution from large to small cell size and from long to short time scales of the gas elements, it is clear that different parts of that spectrum are correlated with each other. We thus expect to find correlations between the various energy carriers owing to the common mode and region of generation.

4   Elementary heating processes

In the dissipation process, mechanical energy is converted into heat. That is, organized motion or potential energy is converted into random thermal motion. As will be shown below, an efficient conversion process is almost always associated with the generation of large variations of the physical variables over very small scales. For instance, it has been known for a long time that an efficient way to dissipate acoustic waves is the formation of shocks, where the physical variables abruptly vary over distances of a molecular mean free path.

Consider a typical acoustic or magnetohydrodynamic disturbance in the solar chromosphere with characteristic values, size $L = 200\ km$, temperature $\Delta T = 1000\ K$, velocity $\Delta v = 3\ km/s$ and magnetic field perturbation $\Delta B = 10\ G$. Using appropriate values for the thermal conductivity $\kappa_{th} = 10^5\ erg/cm\ s\ K$, viscosity $\eta_{vis} = 5 \cdot 10^{-4}\ dyn\ s/cm^2$ and electrical conductivity $\lambda_{el} = 2 \cdot 10^{10}\ s^{-1}$ we find for the *thermal conductive heating rate* $(erg\ cm^{-3}\ s^{-1})$

$$\Phi_C = \frac{d}{dz}\kappa_{th}\frac{dT}{dz} \approx \frac{\kappa_{th}\Delta T}{L^2} \approx 3 \cdot 10^{-7} \quad , \tag{8}$$

the *viscous heating rate* $(erg\ cm^{-3}\ s^{-1})$

$$\Phi_V = \eta_{vis}\left(\frac{dv}{dz}\right)^2 \approx \frac{\eta_{vis}\Delta v^2}{L^2} \approx 1 \cdot 10^{-7} \quad , \tag{9}$$

and the *Joule heating rate* $(erg\ cm^{-3}\ s^{-1})$

$$\Phi_J = \frac{j^2}{\lambda_{el}} = \frac{c_L^2}{16\pi^2 \lambda_{el}} (\nabla \times B)^2 \approx \frac{c_L^2 \Delta B^2}{16\pi^2 \lambda_{el} L^2} \approx 7 \cdot 10^{-5} \quad . \qquad (10)$$

Here j is the current density and c_L the light velocity. The three heating rates show that normally these processes are inadequate to balance the empirical chromospheric cooling rate of $10^{-1} erg\ cm^{-3}\ s^{-1}$ from the VAL81 model. Only when the length scale L is considerably decreased, can the heating rates be raised to acceptable levels. For acoustic waves as well as slow mode mhd- and longitudinal mhd tube waves, this is accomplished by shock formation. For magnetic cases, by the formation of current sheets.

5   Hydrodynamic heating mechanisms

There are two types of hydrodynamic mechanisms, *acoustic waves* and *pulsational waves*. Acoustic waves have periods smaller than the acoustic cut-off period

$$P_A = \frac{4\pi c_S}{\gamma g} \quad , \qquad (11)$$

where c_S is the sound speed, g the gravity and $\gamma = 5/3$ the ratio of specific heats. Pulsational waves have periods $P > P_A$. Typical values for the acoustic cut-off period for the Sun ($g = 2.74 \cdot 10^4\ cm/s^2$, $c_S = 7\ km/s$) are $P_A \approx 190\ s$ and for Arcturus ($g = 50\ cm/s^2$, $c_S = 6\ km/s$) $P_A \approx 1.5 \cdot 10^5\ s$.

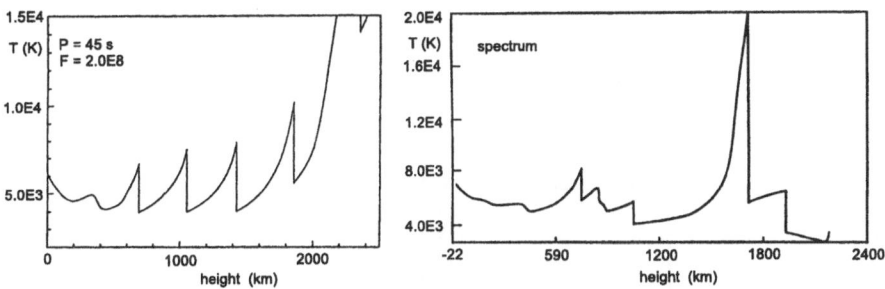

Fig. 1. Left panel: monochromatic, radiatively damped acoustic wave with period $P = 45\ s$ and initial energy flux $F_M = 2 \cdot 10^8\ erg\ cm^{-2}\ s^{-1}$. Right panel: propagating acoustic wave spectrum of the same initial energy.

For *late-type stars* of spectral type F to M, acoustic waves are generated by turbulent velocity fluctuations near the top boundary of the stellar convection zones. This process is discussed in detail in Sec. 8. Propagating down the steep density gradients of the outer stellar layers the acoustic waves suffer amplitude growth due to wave energy conservation. This amplitude growth occurs despite of strong radiation damping. As the wave crests of large amplitude waves move faster than the wave troughs, shocks form by which the waves dissipate their energy.

Fig. 1 (left panel) shows a calculation of a monochromatic, radiatively damped acoustic wave of period $P = 45\ s$ and initial energy flux $F_M = 2 \cdot 10^8\ erg\ cm^{-2}\ s^{-1}$. It is seen that the wave grows from very small amplitude up to a point where sawtooth shocks form which attain a limiting strength. This limiting strength for a given star depends only on the wave period (larger periods give stronger shocks). Also shown in Fig. 1 (right panel) is a calculation with an acoustic spectrum. Here the wave energy likewise is dissipated by shocks.

In *early-type stars* of spectral type O to A, where the surface convection zones are absent, it is the intense radiation field of these stars which generates acoustic disturbances and amplifies them to strong shocks. This mechanism works as follows (c.f. Fig. 2). Consider a gas blob in the outer atmosphere of an early-type star. The line-opacity κ_ν as function of frequency ν of the gas in the blob at rest is shown dashed in Fig. 2. Also shown is the radiation intensity I_ν of the photospheric stellar radiation field which has an absorption line where κ_ν is large. If by chance the gas blob acquires a slight outward velocity, then its line-opacity κ_ν (drawn) becomes Doppler shifted towards the violet and photons from the violet side of the stellar absorption line will be absorbed in the blob, imparting their momentum and accelerate the blob even more (see Fig. 2). This results in a line-opacity κ_ν shifted even more out of the region of the photospheric absorption line etc. This process is called *radiative instability* and results in a powerful acceleration of the gas blobs which leads to the formation of strong shocks with x-ray emitting post-shock regions and intense local heating.

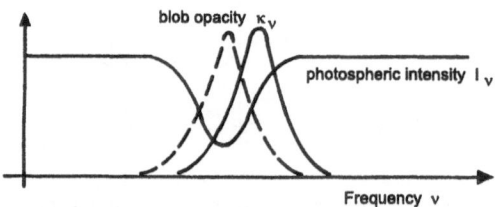

Fig. 2. Radiative instability for an accelerating gas blob in an intense radiation field.

The other hydrodynamic heating mechanism is by *pulsational waves*. Pulsational waves are prominent in Mira-star pulsations, but also in other late-type giants. These pulsations are generated by the κ-*mechanism*. The κ-mechanism (here κ refers to the opacity) functions similarly as the internal combustion engine in motorcars (see Fig. 3). In the internal combustion engine a reactive gas mixture is compressed in a pulsational motion and is ignited at the moment of strongest compression, resulting in a violent decompression. The timing of the ignition ensures that the pulsational motion is amplified. In the κ-mechanism the opacity of stellar envelope material increases (due to the adiabatic temperature and pressure increase) when the star contracts in a pulsational motion. The opacity increase leads to an increased absorption of radiation energy and thus to a large heat input into the contracted envelope layers. The overheated envelope layers subsequently react by rapid expansion, thus driving the pulsation. These pulsational waves propagate to the outer stellar atmosphere where they form shocks.

This and related processes, in principle and possibly with different drivers (e.g. the $\epsilon - mechanism$ where the nuclear energy generation is increased, see e.g. Kippenhahn & Weigert 1990), works also for nonradial oscillations. Any process which kicks on the basic pulsational and vibrational modes of the outer stellar envelope belongs to the category pulsational wave mechanisms. For the Sun the 3 min oscillation is such an example of a basic resonance which is generated by transient events produced in the convection zone. A systematic study of this heating mechanism for late-type stars is missing at the present time.

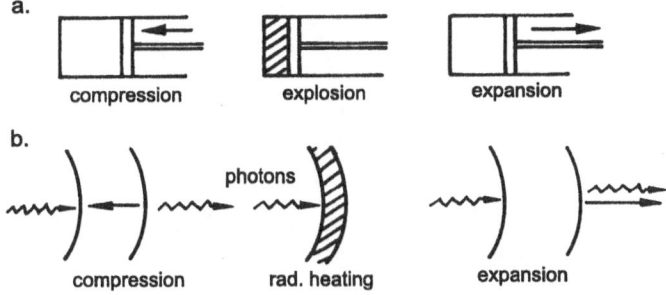

Fig. 3. a: Generation of pulsation by a gasoline engine, b: by the κ-mechanism.

6   Magnetic field structure

The magnetic fields of the Sun and their relation with the granulation and supergranulation cells, as well as the chromospheric network are discussed in other sections of this book. Reviews of the solar magnetic field structure have also been given by Mehltretter (1974), Zwaan (1978) and Stenflo (1978). It is found that the magnetic fields often appear in the shape of flux tubes both in the chromosphere and in the corona. At about 1500 km height the

individual flux tubes fill out the entire space and form the *magnetic canopy*.
In the corona the field strength is $B \approx 10 - 100 \ G$. At the surface of the Sun
the field strength in an isolated flux tube of the chromospheric network is
roughly $B \approx 1500 \ G$. One has horizontal pressure balance

$$p_i + \frac{B^2}{8\pi} = p_e \quad , \tag{12}$$

where p_i is the gas pressure inside the flux tube and p_e the gas pressure
in the non-magnetic region outside. At the solar surface at $z = 0$ one finds
$p_e = 1.2 \cdot 10^5 dyn/cm^2$ in the VAL81 model. If the tube were empty, that is,
$p_i = 0$, one would have $B = \sqrt{8\pi p_e} = 1740 \ G$. This is called *equipartition
field strength*. Actually the tube is not empty but has a gas pressure of about
1/4 to 1/5 of the outside pressure. Fig. 4 shows the three wave modes (Spruit
1981) which are thought to propagate along magnetic flux tubes. Longitudinal
tube waves produce cross-sectional variations of the tube, they are essentially
acoustic tube waves. The transverse and torsional Alfvén waves do not show
a cross-sectional variation of the tube.

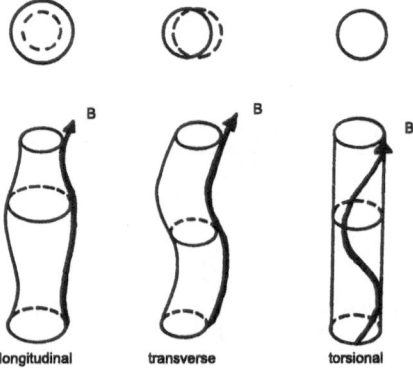

Fig. 4. The three possible wave modes in magnetic flux tubes.

7 Magnetic heating mechanisms

Consider a cylindrical magnetic flux tube of length $l_\parallel$ and diameter $l_\perp$, where
the magnetic field B is along the axis of the tube. The convective gas motions
outside the tube lead to magnetic field perturbations δB either in tangential
or in radial direction. One then has $\delta B \approx B l_\perp / l_\parallel \approx B u \tau / l_\parallel$, where u is the
velocity and τ the characteristic time of the convective flow. From this the
energy density of the perturbation is (where $c_A = B/\sqrt{4\pi\rho}$ is the Alfvén
speed)

$$E = \frac{\delta B^2}{4\pi} \approx \frac{B^2}{4\pi} \left(\frac{u}{l_\parallel}\right)^2 \tau^2 = \rho_o c_A^2 \left(\frac{u}{l_\parallel}\right)^2 \tau^2 \quad . \tag{13}$$

1. For slow tube motion ($\tau > l_\parallel/c_A$) one has a generated energy flux for the DC-mechanism

$$F_{DC} = E \frac{l_\parallel}{\tau} = \rho_o c_A^2 \frac{u^2}{l_\parallel} \tau \quad , \tag{14}$$

2. for fast tube motion ($\tau < l_\parallel/c_A$), with an effective length $l_\parallel = c_A \tau$, one has a generated energy flux for the AC-mechanism

$$F_{AC} = \rho_o u^2 c_A \quad . \tag{15}$$

A typical time scale is the Alfvén transit time t_A:

$$t_A = l_\parallel/c_A = l_\parallel \frac{\sqrt{4\pi\rho}}{B} \quad . \tag{16}$$

The magnetic waves are generated by *fast* ($t << t_A$) *turbulent velocity fluctuations* outside the tube. There velocity fluctuations are produced in the convection zone and affect the tube by compressional, transverse and torsional perturbations, but also by sudden events (see Sec. 7.8). Above the canopy the waves encounter a more or less *homogeneous medium* and other wave modes are possible. *Fast* and *slow mode* magnetohydrodynamic (MHD) waves propagate there.

If the motions of the convection zone are *slow* ($t >> t_A$), then instead of waves, stressed magnetic structures are built up which contain a large amount of energy. Here often magnetic fields of opposite polarity are brought together and form current sheets. The energy of the stressed field is then released by *reconnection*, where the field lines break open and reconnect in such a way that the field geometry afterwards is simpler. These reconnection processes usually occur suddenly like in a *flare* where the magnetic field energy of a large spatial region is released in seconds. Smaller reconnection events are called *microflares*. The local release of energy generates waves in turn. However, same as for waves the ultimate *source* of the DC-heating mechanism is the *convection zone*.

Let us now list the main magnetic heating processes.

7.1   Mode-coupling

This mechanism is not a heating process by itself, but converts wave modes, which are difficult to dissipate by non-linear coupling into other modes, where the dissipation is more readily achieved. Typical cases are the conversion of transverse or torsional Alfvén waves into acoustic-like longitudinal tube waves which dissipate their energy by shock heating. Fig. 5 shows an example of such a process. It is seen that the magnetic tension force which is directed towards the center of curvature can be split into longitudinal and transverse

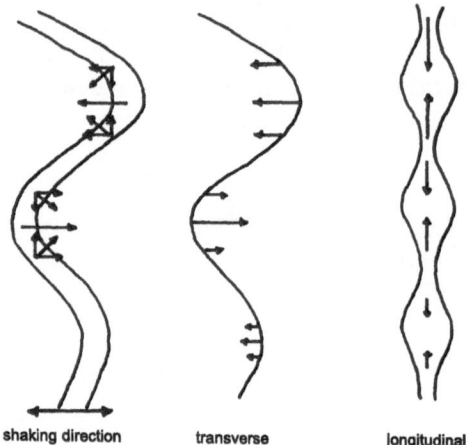

Fig. 5. Mode-coupling between transverse and longitudinal waves.

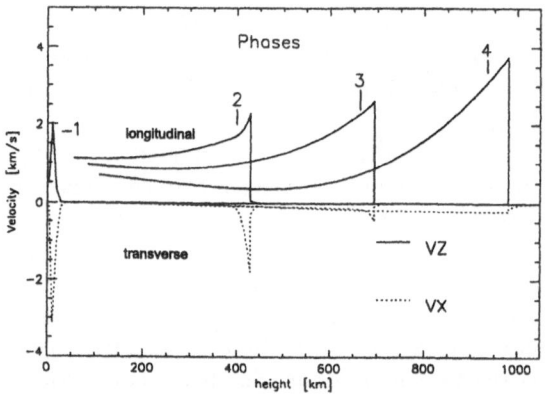

Fig. 6. Longitudinal wave pulse and shock generated by mode-coupling from an applied transverse wave pulse, after Zhugzhda et al. (1995).

components. The longitudinal force components act to compress the gas in the tube such that a longitudinal wave of twice the frequency is generated.

Mode-coupling is particularly efficient when the transverse waves are very stochastic in nature as is expected from observation (Muller et al. 1994) and from wave generation calculations (see Figs. 21 and 22 below). Fig. 6 (after Zhugzhda et al. 1995) shows the generation and development of a longitudinal shock wave pulse produced by mode-coupling from an applied transverse wave pulse. These calculations have to be taken with some caution because of two reasons. First, the Zhugzhda et al. (1995) computations are performed with a one-dimensional time-dependent code using the thin tube approximation. There is presently an extensive not yet fully conclusive discussion in the literature about how to take into account for the swaying tube the back-reaction of the external fluid. Second, the recent three-dimensional time-dependent work by Ziegler & Ulmschneider (1997) on swaying magnetic flux tubes in the solar atmosphere shows that there is extensive leakage of the transverse wave energy into the outside medium. Thus the true value of the longitudinal wave energy generation by mode-coupling is presently not well determined.

7.2   Resonance heating

Resonance heating occurs, when upon reflection of Alfvén waves at the two foot points of the coronal loops, one has constructive interference (see Fig. 7). For a given loop length $l_\parallel$ and Alfvén speed c_A, resonance occurs, when the wave period is $mP = 2l_\parallel/c_A$, m being a positive integer. Waves which fulfill the resonance condition are trapped and after many reflections are dissipated by Joule-, thermal conductive or viscous heating.

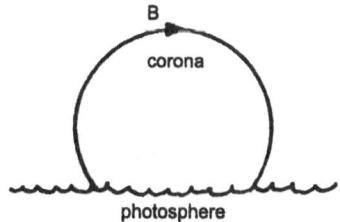

Fig. 7. Resonance heating in coronal loops.

7.3   Compressional viscous heating

Compressional viscous heating, recently proposed by Strauss (1991), is a very promising mechanism for coronal regions, where the gyro-frequency is much larger than the collision-frequency. Swaying an axial magnetic flux tube sideways with velocity $\mathbf{v}_\perp$ results in a transverse Alfvén wave which is incompressible ($\nabla \cdot \mathbf{v}_\perp = 0$) to first order. This is different for tubes with helicity, where one has $\nabla \cdot \mathbf{v}_\perp \approx \dot{\rho}/\rho$ (see Fig. 8). With an increase of the density, the magnetic field is compressed and the gyro-frequency increased. Gyrating around the field lines more quickly, the ions after colliding with each other, generate larger velocities in non-perpendicular directions as well, which constitutes the heating process.

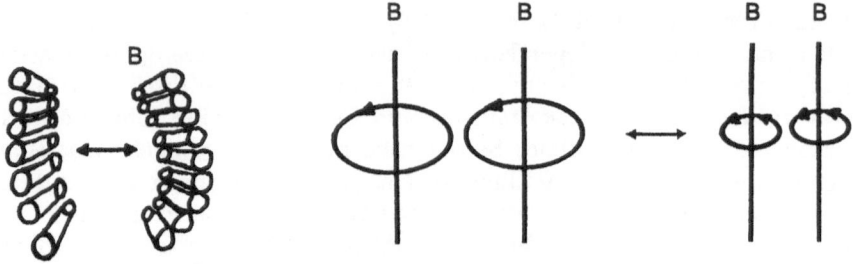

Fig. 8. Compressional viscous heating in helical fields.

7.4 Turbulent heating

In a turbulent flow field with high Reynolds number there are bubbles of all sizes. The energy usually is put into the largest bubbles. Because of the large inertial forces, the big bubbles are ripped apart into smaller bubbles, and these in turn into still smaller ones etc. This process is called turbulent cascade. A turbulent flow field can be described by three characteristic quantities, density ρ, bubble scale $l_k = 2\pi/k$, and the mean velocity u_k of such bubbles. k is the wavenumber. It is easily seen, that from these three quantities only one combination for a heating rate can be formed

$$\Phi_k = \rho \frac{u_k^3}{l_k} \quad \left[\frac{erg}{cm^3 s}\right] \quad . \tag{17}$$

If there are no other losses, like by radiation, all the energy which is put in at the largest bubbles must reappear in the smaller bubbles etc. Thus if $k1$, $k2$, $\cdots$ represents a series of smaller and smaller bubbles one must have $\Phi_{k1} = \Phi_{k2} = \cdots = \text{const}$. This implies

$$u_k \sim l_k^{1/3} \quad , \tag{18}$$

which is the *Kolmogorov law*. The range $l_{k1} \cdots l_{kn}$ of validity of this law is called the *inertial range*. Consider what happens if l_k becomes very small. From Eq. (9) one finds for the viscous heating rate, $\Phi_V = \eta_{vis}(du/dl)^2 \approx \eta_{vis} u_k^2/l_k^2 \approx \eta l_k^{-4/3}$, which goes to infinity for $l_k \to 0$. Thus at some small enough scale, viscous heating sets in and the inertial range ends. It is seen that turbulent heating lives from the formation of small scales. One can visualize the process as follows. Because of the continuous splitting of bubbles into smaller sizes, with the velocities decreasing much less rapidly, one eventually has close encounters of very small bubbles with large velocity differences where viscous heating dominates.

As the fluctuations generated in the turbulent convection zone produce acoustic and mhd waves it is of interest to deduce from the inertial range an estimate of the frequency range of the generated waves. If $k_0 \approx 2\pi/H$ is the wavenumber of the scale where the energy is put into the turbulence, with H being the scale height, we have for the size l_k of the smallest bubble (where viscosity ends the cascade) $\eta_{vis} u_k^2 / l_k^2 = \rho u_{k_0}^3 / l_{k_0}$ and $u_k^3 / l_k = u_{k_0}^3 / l_{k_0}$. From this we obtain

$$l_k = \left(\frac{\eta_{vis} l_{k_0}^{1/3}}{\rho u_{k_0}} \right)^{3/4} . \tag{19}$$

With $l_{k_0} = H = 150 \ km$, $u_{k_0} = 1 \ km/s$, $\rho = 3 \cdot 10^{-7} \ g/cm^3$, $\eta_{vis} = 5 \cdot 10^{-4} \ dyn \ s/cm^2$, one finds $l_k = 2.9 \ cm$ as well as $u_k = 290 \ cm/s$ and derives a maximum frequency of $\nu_k = u_k/l_k = 100 \ Hz$ or a period of $P = 1/100 \ s$. This estimate is somewhat idealized as small bubbles become transparent to radiation. In this case the temperature excess, which drives the convection, is exchanged directly via radiation. Thus it is expected that the optical depth limits the bubble size. Assuming that the smallest bubble has an optical depth of $\tau = l_k \overline{\kappa} = 0.1$, where

$$\frac{\overline{\kappa}}{\rho} = 1.38 \cdot 10^{-23} p^{0.738} T^5 \quad cm^2/g \quad , \tag{20}$$

is the gray H^- opacity, $T = 8320 \ K$ the temperature and $p = 1.8 \cdot 10^5 \ dyn/cm^2$ the gas pressure, one finds $l_k = 6.3 \cdot 10^4 \ cm$, $u_k = 1.6 \cdot 10^4 \ cm/s$, $\nu_k = 0.25 \ Hz$ and $P = 3.9 \ s$ for the smallest bubble.

So far we have discussed turbulent heating in a non-magnetic environment. When there is a magnetic flux tube one has to take into account the tube geometry, the frozen-in condition and the wave modes. Fig. 9 shows how the turbulent dissipation of a torsional Alfvén wave is pictured (after Heyvaerts & Priest 1983, Hollweg 1983). Shearing motions in azimuthal direction generate closed magnetic loops (similarly to the growth of Helmholtz-Kelvin-type instabilities) over the tube-cross-section, which decay into smaller tubes etc. and are ultimately dissipated by reconnection.

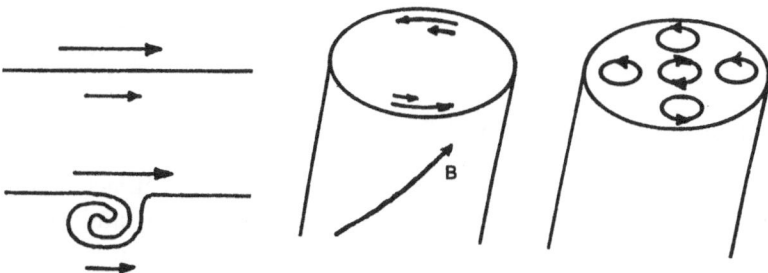

Fig. 9. Turbulent heating in magnetic flux tubes.

7.5   Landau damping

Landau damping occurs at coronal heights, where the collision rate becomes small. As Chen (1974) has explained, this process is analogous to surfing on ocean waves (see Fig. 10). When surfing, a surfboard rider launches himself in propagation direction into the steepening part of an incoming wave and gets further accelerated by this wave. In Landau damping, the propagating wave accelerates gas particles which, due to their particle distribution function, happen to have similar direction and speed as the wave. Because a distribution function normally has many more slower particles than faster ones, the wave looses energy to accelerate the slower particles. This gained energy is eventually shared with other particles in the process to reestablish the distribution function, which constitutes the heating mechanism.

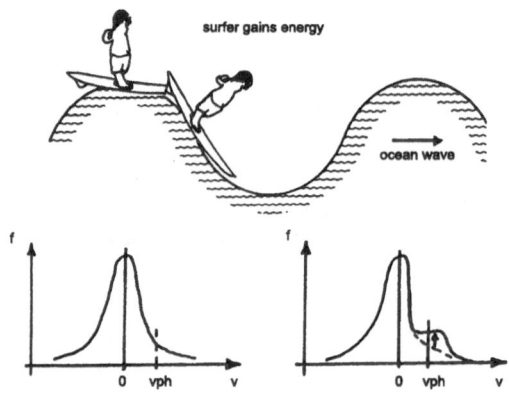

Fig. 10. Landau damping and the analogy to surfing.

7.6   Resonant absorption

In the process of resonant absorption one considers magnetoacoustic surface waves in a magnetic field $\mathbf{B}$ which points in z-direction, and varies from $\mathbf{B}_1$ to $\mathbf{B}_2$ in x-direction (see Fig. 11, panel a). The surface wave, with its field perturbation $\delta B = B_x'$ in x-direction, has a phase speed $v_{ph} = ((B_1^2 + B_2^2)/(4\pi(\rho_1 + \rho_2)))^{1/2}$, such that at an intermediate position x_o, the phase speed becomes equal to the local Alfvén speed $c_{Ao} = B(x_o)/\sqrt{4\pi\rho(x_o)}$. In panel b of Fig. 11 consider the wave fronts of the peak (drawn) and trough (dotted) of a surface wave. Because to the right of x_o, the Alfvén speed is

larger and to the left smaller, the wave fronts at a later time get tilted, relative
to the phase, propagating with speed c_{Ao} (see panel c). At a still later time
(panel d) the wave fronts get tilted even further and approach each other
closely at the position x_o. This leads to small scales and intense heating by
reconnection at that field line.

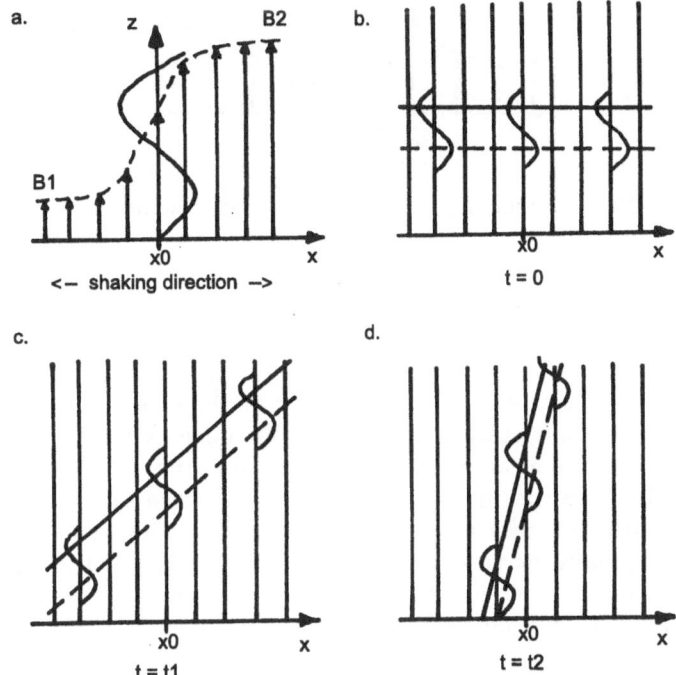

Fig. 11. Resonant absorption. In a field pointing in z-direction, where the field
strength varies in x-direction. a: resonant absorption of a surface wave (shaking in
x-direction), b: wave fronts at time $t = 0$, c and d: at later times t_1 and t_2.

7.7 Phase-mixing

For phase-mixing (c.f. Fig. 12) one considers a magnetic field geometry similar
to that in Fig. 11, however, the field perturbation $\delta B = B'_y$ of the wave is
now in y-direction, perpendicular to the x- and z-directions. As the Alfvén
speeds of two closely adjacent regions x_1 and x_2 are different, it is seen that
after propagating some distance Δz, the fields $B'_y(x_1)$ and $B'_y(x_2)$ will be
very different, leading to a current sheet and strong dissipation. Here again
it is the appearance of small scale structures which lead to dissipation.

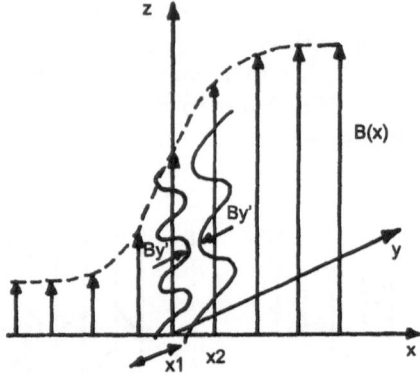

Fig. 12. Phase-mixing of a surface wave (shaking in y-direction).

7.8   Reconnection in current sheets

As example of the DC heating mechanisms, Fig. 13, after Parker (1991), shows how slow foot point motions caused by the convection zone at both legs of a coronal loop, starting from a minimum energy axial field state (left panel), will build up a tangled and braided field configuration of high energy (right panel). Note that in Fig. 13 the two legs of the tube ($z = 0$, $z = L$) are drawn such that they appear at the bottom and top. As the foot point motions perpetually increase the energy buildup, the system tries increasingly hard to return to a lower energy state. Because of the hopelessly complicated tangling this can only be achieved via reconnection. At many locations in the web of field lines, oppositely directed fields occur, giving rise to local current sheets, which by reconnection (in the form of microflares) release the magnetic field energy. The energy is dissipated both directly and via the generation of waves and turbulence. Note that similarly to the wave mechanisms, reconnection happens in small scale regions. These reconnective events of different magnitude have been observed on the Sun by Brueckner & Bartoe (1983) as sudden velocity shifts in the C IV ($T \approx 10^5$ K) transition layer line with velocities of 250 km/s and even 400 km/s. The authors have termed these phenomena turbulent events and high velocity jets, respectively.

The question, whether microflares is a significant coronal heating mechanism and what its importance is as compared to wave heating (DC– versus AC–heating), has recently been studied by Wood, Linsky & Ayres (1996) by observing C IV and Si IV transition layer lines (see Fig. 14). They found that the total line profile can be explained as a combination of a very broad profile, attributed to microflares, and a narrow profile, attributed to wave heating. A careful study of observations of this type, particularly with respect to the three stellar parameters T_{eff}, gravity and rotation, will allow to

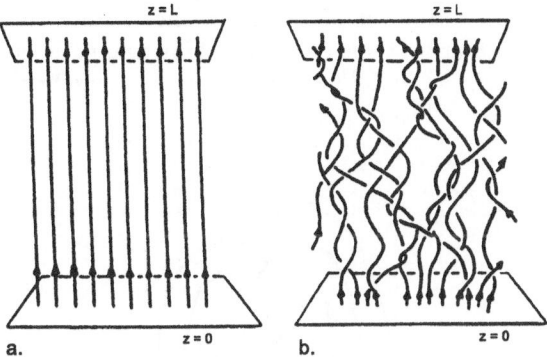

Fig. 13. a: Magnetic fields in coronal loops, initial axial field (left), b: tangled fields after considerable foot point motions (right), after Parker (1991).

make great progress in the identification of the individual heating processes and generally in the understanding of coronal heating as a function of the basic parameters of the underlying stars.

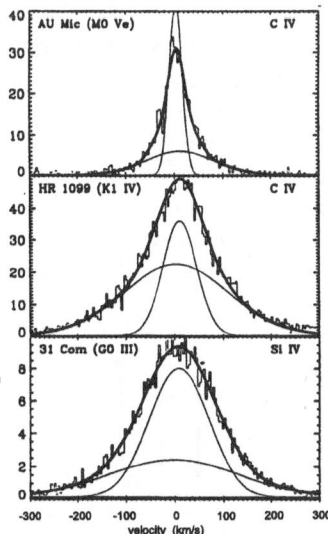

Fig. 14. Surface fluxes in transition layer lines (in 10^4 erg cm^{-2} s^{-1} A^{-1}) of giants and main sequence stars, after Wood et al. (1996).

Another example of current sheet formation and heating is seen in Fig. 15, after Priest (1991). It shows an arcade system, which by slow motion is laterally compressed and develops a current sheet. Here oppositely directed fields reconnect. Similar systems of approaching magnetic elements of opposite polarity and large scale field annihilation are thought to be responsible for the heating of X-ray bright points.

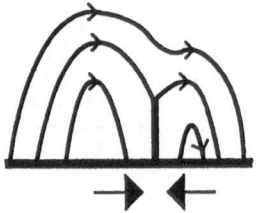

Fig. 15. Current sheets in arcade systems, after Priest (1991).

8   Acoustic energy generation

8.1   The process of quadrupole sound generation

Typically in stellar convection zones, sound is generated by pressure and density fluctuations produced by Reynolds stresses (Fig. 16). Such sound generation by free turbulence, called *quadrupole sound generation*, is the most efficient way to generate sound even in terrestrial situations, when mass injections and rigid vibrating surfaces are absent. Chopped mass injection like in a rotating perforated disk siren is the most efficient way to produce sound and is termed *monopole sound generation*, while vibrating surfaces like in a tuning fork leads to the less efficient *dipole sound generation*.

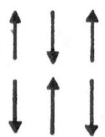

Fig. 16. Reynolds stresses.

Using the linearized continuity equation and the equation of motion one has

$$\frac{\partial \rho'}{\partial t} + \nabla \cdot \rho_0 \mathbf{v} = 0 \quad , \tag{21}$$

$$\frac{\partial \rho_0 \mathbf{v}}{\partial t} + c_S^2 \nabla \rho' = -\nabla \cdot \rho_0 \mathbf{v}\mathbf{v} \quad . \tag{22}$$

Here we retain the Reynolds stress term $\rho_0 \mathbf{v}\mathbf{v}$ which usually in the linearized equations is *neglected* as normally (21), (22) contain only first order terms. We expect therefore that this source term is very small. Operating with $\partial/\partial t$ on (21) and with $-\nabla\cdot$ on (22) and adding, a wave equation with a source term is obtained

$$\left(\frac{\partial^2}{\partial t^2} - c_S^2 \nabla^2 \right) \rho' = \nabla\nabla : \rho_0 \mathbf{v}\mathbf{v} \quad . \tag{23}$$

From the LHS of this equation we see that $\nabla \sim \frac{1}{c_S} \frac{\partial}{\partial t}$. Applying this to the RHS is incorrect because in a stellar convection zone the velocity $\mathbf{v}$ in the source term is given by the convective velocity u and the flow time scales are much slower $\partial/\partial t \to u/c_S \partial/\partial t$. On the RHS one thus has

$$\nabla \to \frac{u}{c_S{}^2} \frac{\partial}{\partial t} \quad . \tag{24}$$

With this we find from Eq. (23)

$$\rho' \approx \rho_0 \frac{u^4}{c_S^4} \quad . \tag{25}$$

The acoustic energy flux F_M is twice the kinetic energy density (because of equipartition of kinetic and potential energy) times the sound speed:

$$F_M = \rho_0 v^2 c_S \quad . \tag{26}$$

Using the amplitude relations for sound waves

$$\frac{\rho'}{\rho_o} = \frac{v}{c_S} = \frac{p'}{\gamma p_o} \quad , \tag{27}$$

one finds for the acoustic flux by *quadrupole sound generation:*

$$F_M^{quad} = \frac{1}{\rho_0} c_S^3 \rho'^2 = \rho_0 \frac{u^8}{c_S^5} \quad . \tag{28}$$

Considering a Kolmogorov turbulence spectrum (like our Eq. (31), below), Lighthill has derived from (28) the expression

$$F_M = \int 38 \frac{\rho_0 u^8}{c_S^5 H} dz \quad , \tag{29}$$

called *Lighthill- or Lighthill-Proudmann formula*, where H is the scale height. Fig. 17 taken from Goldstein (1976) shows how well Lighthill's u^8-formula is observed to work in terrestrial applications. Here gas from jet engines is directed against a wire mesh, behind which turbulent flows develop. As there is neither mass injection nor a vibrating surface, one has quadrupole sound generation.

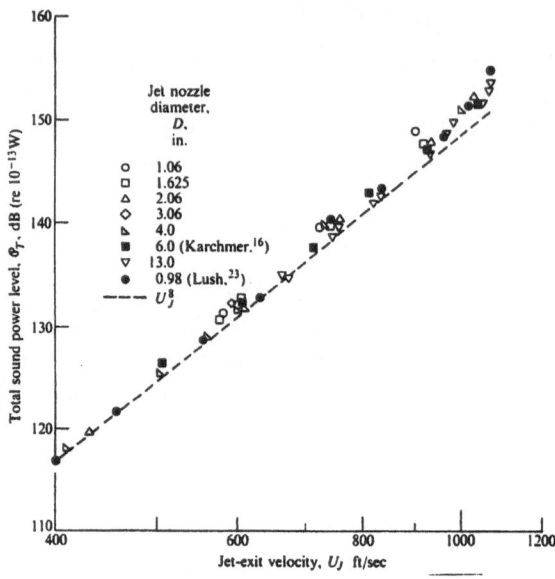

Fig. 17. Noise from jet engines, after Goldstein (1976).

8.2   Acoustic flux computations for stars

Quadrupole sound generation from turbulence was originally developed by
Lighthill (1952, 1954) as well as Proudman (1952) and was further extended
by Stein (1967, 1968) to stellar convection zones. The Lighthill-Stein theory
has been rediscussed in detail by Musielak et al. (1994). These authors found
that the stellar turbulence is best represented by a Kolmogorov-type energy
spectrum (see Eqs. (31) to (34), below).

To calculate acoustic fluxes of the Sun and of stars *three parameters have
to be specified*: T_{eff}, g and the mixing-length parameter α, where α is the
ratio of the mixing-length to the scale height. Typically one has $\alpha \approx 1.5$ to
2.0. These three parameters uniquely specify a *convection zone model*. This
model provides ρ_0, u, c_S and H versus z and allows to calculate the acoustic
fluxes F_M using the described methods. Acoustic fluxes as well as acoustic
energy spectra computed this way for a large number of late-type stars in
the HR-diagram (after Ulmschneider, Theurer & Musielak 1996) are shown
in Figs. 18 and 19. These authors find that the total acoustic fluxes using the
very elaborate Lighthill-Stein theory are fairly close to the values given by
the simple Lighthill formula (see dotted values in Fig. 18). This fact is due
to the use of the Kolmogorov spectrum in both methods.

While in terrestrial situations, monopole, dipole and quadrupole sound
generation is a sequence of progressively less important ways to produce

acoustic waves, this is different in stellar situations, where external mass injections and rigid surfaces are absent. Here quadrupole sound generation is largest and, except for late-type dwarf stars, even dominant. The acoustic

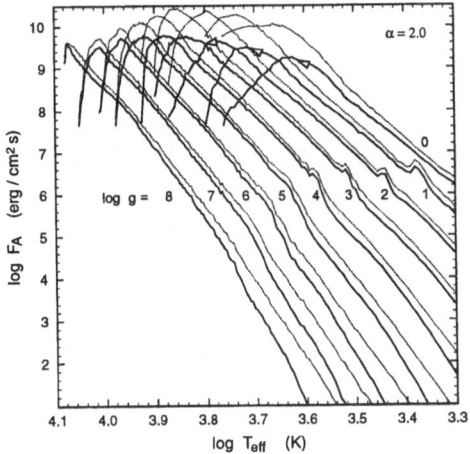

Fig. 18. Acoustic fluxes F_M for stars versus T_{eff} for given $\log g$ and $\alpha = 2.0$.

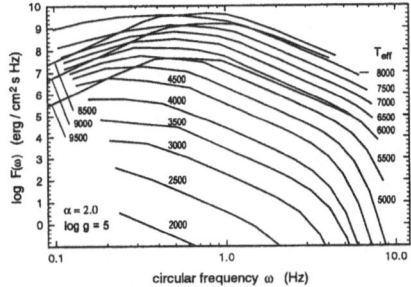

Fig. 19. Acoustic spectra $dF_M(\omega)/d\omega$ for stars versus circular frequency ω for given T_{eff} and $\alpha = 2.0$, for $\log g = 5$.

spectra extend roughly over the range $\omega_A < \omega < 20\omega_A$ (see Fig. 19), where $\omega_A = 2\pi/P_A = \gamma g/2c_S$ is the acoustic cut-off frequency and P_A the acoustic cut-off period. $\omega = 2\pi\nu = 2\pi/P$ is the circular frequency and P the wave period. For the Sun the acoustic spectrum has a maximum at a period

$$P_{max} \approx \frac{P_A}{4} \approx 60 \ s \quad .$$ (30)

At this point it is interesting to show an application of the acoustic wave

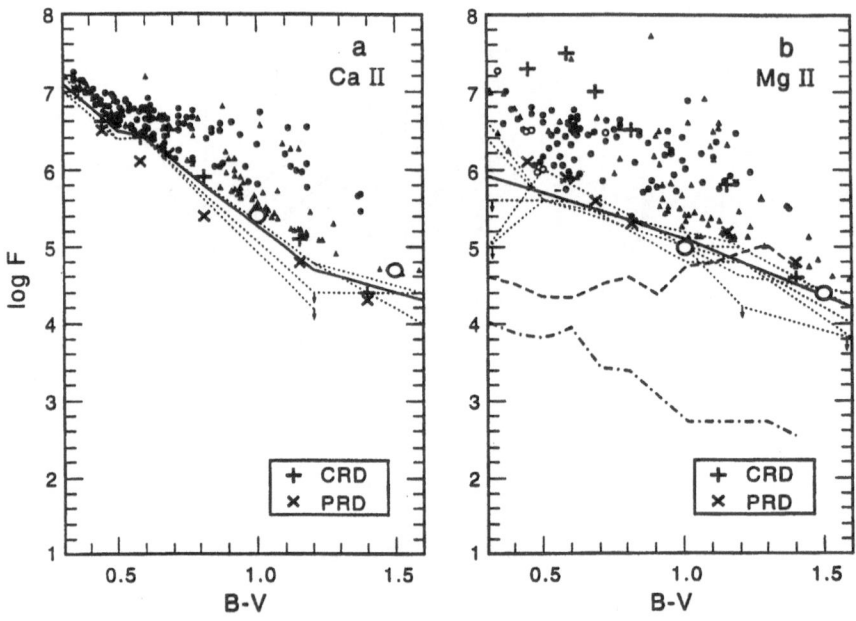

Fig. 20. Theoretical and empirical chromospheric emission fluxes in the Ca II and Mg II lines versus colour for late-type stars, after Buchholz et al. (1998).

generation computations which allowed to identify acoustic waves as the main heating mechanism for magnetic field free regions in late-type stars. Fig. 20 shows the chromospheric emission flux in the Ca II H+K and the Mg II h+k lines of late-type stars versus the color index B-V after Buchholz, Ulmschneider & Cuntz (1998). Here the dots represent main-sequence and the triangles giant stars. It is seen that the stars have a minimum emission (given by the drawn empirical line, called *basal flux line*) which is interpreted as due to pure acoustically heated chromospheres. Dots and triangles above this line represent stars which have additional magnetic heating, correlated with rotation. Also shown are simulations (x's for main-sequence and circles for giant stars) obtained from theoretical chromosphere model calculations based on the generated acoustic wave fluxes. The nice agreement of the theoretical and empirical (basal) emission fluxes for a wide range of stars with different T_{eff} (equivalent to B-V) and gravity shows the validity of acoustic heating. The conclusion, that acoustic heating is the main mechanism for magnetic field free regions of stars, could not have been drawn from solar observations alone, as here the magnetic and hydrodynamic heating mechanisms are very difficult to disentangle. Varying the T_{eff} and gravity dependences of the different heating mechanisms allows to identify the important mechanisms using stellar observations.

9 Magnetic energy generation

The radiative emission of the solar chromosphere is particularly concentrated in the network region surrounding supergranulation cells. Here strong vertically directed magnetic flux tubes have been collected by the convective granular and supergranular flows and were strengthened by the convective collapse. The turbulent non-magnetic environment outside these flux tubes strangles and displaces the magnetic tubes which leads to the generation of longitudinal and transverse mhd waves. These tubes, moreover, are thought to occupy the centers of cyclonic downflow channels where turbulent motions give rise to torsional perturbations of the tubes which lead to the generation of torsional waves. Thus all three wave types, longitudinal, transverse and torsional modes are generated by the turbulent convection zone.

Both solar observations and theoretical convection zone simulations show that the turbulent energy spectrum is reasonably well described by a Kolmogorov type law. As shown by Musielak et al. (1994) the turbulence in the solar convection zone can very likely be represented by an extended Kolmogorov spectrum $E(k)$ with a modified Gaussian frequency factor $\Delta(\frac{\omega}{ku_k})$. The extended Kolmogorov spectrum (which describes the energy distribution among the different spatial scales k) is given by

$$
E(k) = \begin{cases} 0 & : \quad 0 < k < 0.2k_t \\ a\frac{u_t^2}{k_t}\left(\frac{k}{k_t}\right) & : \quad 0.2k_t \leq k < k_t \\ a\frac{u_t^2}{k_t}\left(\frac{k}{k_t}\right)^{-5/3} & : \quad k_t \leq k \leq k_d \end{cases} \quad , \tag{31}
$$

where the factor $a = 0.758$ is determined from the normalization condition

$$
\int_0^\infty E(k)dk = \frac{3}{2}u_t^2 \quad . \tag{32}
$$

The modified Gaussian frequency factor (describing the temporal behaviour of the turbulence) is given by

$$
\Delta\left(\frac{\omega}{ku_k}\right) = \frac{4}{\sqrt{\pi}}\frac{\omega^2}{|ku_k|^3}e^{-\left(\frac{\omega}{ku_k}\right)^2} \quad , \tag{33}
$$

where ω is the circular frequency and u_k the velocity of bubbles of scale k. u_k is computed from

$$
u_k = \left[\int_k^{2k} E(k')dk'\right]^{1/2} \quad . \tag{34}
$$

Here $k_t = 2\pi/H$ (H being the scale height) is the characteristic scale where the main energy input occurs. u_t is the root mean square velocity in one spatial direction (see Eq. (36) below).

Assume that at a given height z, the horizontal turbulent velocity fluctuation in one spatial direction can be described by a large number of ($N \approx 100$) superposed partial waves of given frequencies ω_n

$$v_x(z,t) = \sum_{n=1}^{N} u_n \sin(\omega_n t + \varphi_n) \,, \tag{35}$$

as function of time t. Here $\varphi_n = 2\pi r_n$ is an arbitrary but constant phase angle and r_n a random number in the interval $[0, 1]$. The amplitude of the partial waves u_n is then determined from the normalization condition of the assumed turbulent energy spectrum:

$$\frac{3}{2}u_t^2 = \frac{3}{2}\overline{v_x^2} = \frac{3}{4}\sum_{n=1}^{N} u_n^2 = \int_0^\infty d\omega \int_0^\infty dk \; E(k)\Delta\left(\frac{\omega}{ku_k}\right) = \sum_{n=1}^{N} E'(\omega_n)\Delta\omega \,. \tag{36}$$

To compute transverse tube wave fluxes we let the external velocity fluctuations $v_x(z,t)$ shake the tube, while for longitudinal tube waves we let the external turbulent pressure fluctuations p_{turb} work on the tube. For a thin magnetic flux tube sticking in a convection zone, the pressure balance of Eq. (12) is modified and now given by

$$p + \frac{B^2}{8\pi} = p_e + p_{turb} \,, \tag{37}$$

where

$$p_{turb} = \rho_e \left(v_x^2(\mathbf{r},t) + v_y^2(\mathbf{r},t) + v_z^2(\mathbf{r},t)\right) \,. \tag{38}$$

Upon time averaging one finds for homogeneous isotropic turbulence

$$p_0 + \frac{B_0^2}{8\pi} = p_e + 3\rho_e u_t^2 \,, \tag{39}$$

where p_0 is the average internal gas pressure and B_0 the average magnetic field strength. This equation shows that when one adds a turbulent flow outside the magnetic flux tube, then the mean external pressure is increased. This is due to the fact that in Eqs. (37), (38) the external turbulent pressure is a quantity which cannot be negative and thus must fluctuate around an average positive value. This additional turbulent pressure is small, typically one finds $\overline{p_{turb}} = 6 \cdot 10^{-2} p_e$.

Subtracting the averaged equation from Eq. (37) one obtains a perturbation equation

$$p' + \frac{B_0 B'}{4\pi} = p_{turb} - 3\rho_e u_t^2 = p'_{turb} \,, \tag{40}$$

where $p' = p - p_0$ is the internal gas pressure fluctuation and $B' = B - B_0$ the magnetic field strength perturbation. In a thin tube p' and B' are related and one finds

$$p' = \frac{\beta}{2/\gamma + \beta} \, p'_{turb} \quad , \tag{41}$$

where $\beta = 8\pi p_0/B_0^2$ is the plasma β. Making use of the amplitude relations for longitudinal tube waves, and the tube speed $c_T = c_S c_A / \sqrt{c_S^2 + c_A^2}$ one finds for the internal longitudinal velocity fluctuations

$$v_{\parallel} = \frac{c_S^2}{c_T} \frac{p'}{\gamma p_0} = \frac{\beta}{2/\gamma + \beta} \frac{p'_{turb}}{\rho_0 c_T} = \frac{\beta}{2/\gamma + \beta} \frac{3\rho_e(v_x^2 - u_t^2)}{\rho_0 c_T} \quad . \tag{42}$$

The velocity fluctuations v_x and $v_{\parallel}$ constitute the boundary conditions for a time-dependent one-dimensional magnetohydrodynamic wave code with which the transverse and longitudinal tube wave fluxes (Huang, Musielak & Ulmschneider 1995, Ulmschneider & Musielak 1998) can be computed employing the expressions

$$F_{trans} = -\frac{B}{4\pi} B_x v_x \quad , \tag{43}$$

and

$$F_{long} = v_{\parallel} p' \quad . \tag{44}$$

Note that the magnetic field perturbation B_x and the pressure perturba-

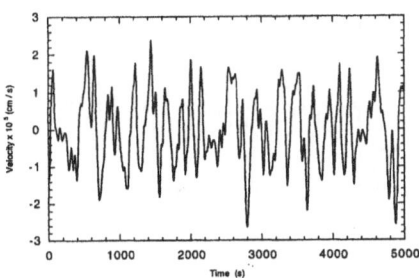

Fig. 21. Fluctuating turbulent velocity v_x as function of time for an extended Kolmogorov energy spectrum with a modified Gaussian frequency factor for an assumed rms convective velocity $u_t = 1 km/s$.

tion p' are automatically evaluated (including their phase-shifts against the velocity fluctuations) in the wave code by the mhd equations, once the time-dependent velocity boundary conditions are specified.

Fig. 21 shows the fluctuating horizontal velocity v_x computed using Eq. (35) for a prescribed rms turbulent velocity of $u_t = 1 \, km/s$. It is seen that due to the many partial waves, v_x is very stochastic in nature. u_t is a typical mean excitation velocity which is about one third of the maximum convective velocity obtained from a convection zone code for the Sun using a mixing-length parameter $\alpha = 2$. Fig. 21 shows that occasionally large velocity spikes occur which can approach 3 km/s, which is in good agreement with observations. The instantaneous and time-averaged transverse and longitudinal

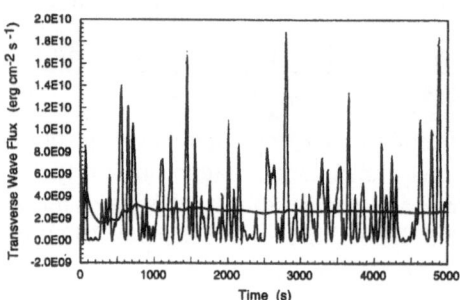

Fig. 22. Instant and time-averaged transverse tube wave energy fluxes versus time.

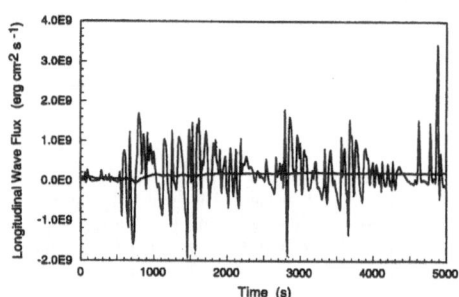

Fig. 23. Instant and time-averaged longitudinal tube wave energy fluxes versus time.

wave energy fluxes resulting from these excitations are shown in Figs. 22 and 23. From several computations using mean excitation velocities $u_t = 1.0$ to 1.5 km/s and varying the shaking height from -150 to +150 km height in the solar atmosphere one finds that the longitudinal tube wave fluxes F_{long} are of the order of $3 \cdot 10^8 erg \ cm^{-2} \ s^{-1}$ and the transverse tube wave fluxes are about a factor of 20 larger. Figs. 24 show the computed mhd tube wave energy spectra. It is seen that these spectra are dominated in their frequency behaviour by the Kolmogorov input spectrum.

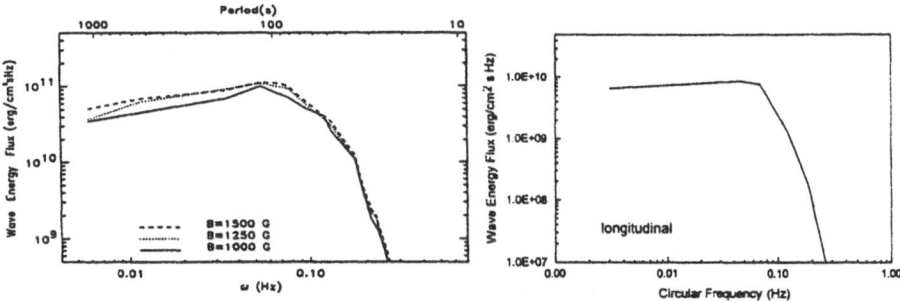

Fig. 24. Left: Fourier spectrum of the transverse, right: of the longitudinal energy flux.

10   Conclusions

Probably all nondegenerate stars have chromospheres and coronae. These hot outer layers are physically completely dependent on the properties of the underlying star and it is our ultimate aim to elucidate this physical dependence. Chromospheres and coronae owe their existence to mechanical heating. There are *hydrodynamical–* and *magnetic heating mechanisms*, each of which, on basis of the frequency of fluctuations, can by further classified into subcategories (acoustic– and pulsational waves as well as AC– and DC–mechanisms).

Rapid fluctuations generated by the turbulence in the convection zones lead to acoustic waves and to mhd waves (AC–mechanisms), slow fluctuations to pulsational waves and to stressed fields with current sheets (DC–mechanisms). Acoustic and mhd wave generation rates can presently be computed on basis of convection zone models and the (Kolmogorov–type) energy spectrum of the turbulence for a large number of stars.

Nonmagnetic chromospheric regions of late-type stars appear to be heated by shock dissipation of *acoustic waves*. For slowly rotating stars, which have weak or no magnetic fields, acoustic waves are the dominant chromospheric heating mechanism.

Except for F-stars and giants, the chromospheric heating of rapidly rotating late-type stars is dominated by *magnetic heating*. This heating very likely is by acoustic-like longitudinal mhd tube waves in the lower chromosphere up to the transition layer, and by Alfvén waves in the transition layer and corona. But these higher layers and the corona are also efficiently heated by DC–mechanisms (microflares). Note that microflares lead to both direct dissipation and wave generation.

Which of the various proposed mechanisms is the prime candidate for the dissipation of Alfvén waves is presently not well understood. This is also true for the question in what situation DC– or AC–heating is more important.

Here mhd modeling using the complete fluctuation spectrum of the convection zone and more solar and stellar observations will bring the answer.

References

Brueckner G.E., Bartoe J.D.F.: 1983, ApJ 272, 329
Buchholz B., Ulmschneider P., Cuntz, M.: 1998, ApJ, in press
Chen F.F., 1974, *Introduction to Plasma Physics*, Plenum Press, New York
Goldstein M.E.: 1976, *Aeroacoustics*, Mc Graw Hill, New York
Heyvaerts J., Priest E.R.: 1983, A&A 117, 220
Hollweg J.V.: 1983, in *Solar Wind V*, M. Neugebauer Ed., NASA CP-2280, p. 5
Huang P., Musielak, Z.E., Ulmschneider P., 1995, A&A 279, 579
Kippenhahn R., Weigert A.: 1990, *Stellar Structure and Evolution*, Springer, Berlin
Kuperus M., Ionson J.A., Spicer D.S.: 1981, Ann. Rev. Astr. Ap. 19, 7
Lighthill M.J.: 1952, Proc. Roy. Soc. London A211, 564
Lighthill M.J.: 1954, Proc. Roy. Soc. London A222, 1
Linsky J.L., 1985, Solar Phys. 100, 333
Mehltretter P.: 1974, Solar Phys. 38, 43
Muller R., Roudier Th., Vigneau J., Auffret H.: 1994, A&A 283, 232
Musielak Z.E., Rosner R., Stein, R.F., Ulmschneider P., 1994, ApJ 423, 474
Narain U. & Ulmschneider P.: 1990, Space Sci. Rev. 54, 377
Narain U. & Ulmschneider P.: 1996, Space Sci. Rev. 75, 453
Parker E.N.: 1991, in *Mechanisms of Chromospheric and Coronal Heating*, Springer, Berlin, p. 615
Priest E.R.: 1991, in *Mechanisms of Chromospheric and Coronal Heating*, Springer, Berlin, p. 520
Proudmann I.: 1952, Proc. Roy. Soc. London A214, 119
Spruit H.: 1981, A&A 98, 155; 102, 129
Stein R.F.: 1967, Solar Phys. 2, 385
Stein R.F.: 1968, Ap.J. 154, 297
Stenflo J.O.: 1978, Rep. Progr. Phys. 41, 865
Strauss H.R.: 1991, Geophys. Res. Let. 18, 77
Ulmschneider P., Musielak Z.E. 1998, A&A, submitted
Ulmschneider P., Theurer J., Musielak Z.E.: 1996, A&A 315, 212
Ulmschneider P., Priest E.R., Rosner R.: 1991, *Mechanisms of Chromospheric and Coronal Heating*, Springer, Berlin
Ulmschneider P., 1991, in *Mechanisms of Chromospheric and Coronal Heating*, Springer, Berlin, p. 328
Vaiana G.S. et al.: 1981, ApJ 245, 163
Vernazza J.E., Avrett E.H., Loeser R., 1981, ApJ Suppl 45, 635
Wentzel D.G.: 1981, *The Sun as a Star*, NASA-CNRS Series
Wood B.E., Linsky J.L., Ayres T.R.: 1996, ApJ 478, 745
Zhugzhda Y., Bromm V., Ulmschneider P.: 1995, A&A 300, 302
Ziegler U., Ulmschneider P.: 1997, A&A 327, 854
Zwaan C.: 1978, Solar Phys. 60, 213

Solar Wind

Eckart Marsch

Max-Planck-Institut für Aeronomie, Max-Planck-Straße 2, D-37191 Katlenburg-Lindau, Germany

Abstract. A concise tutorial review is given of solar wind observations and theory with emphasis on the more recent findings from the Ulysses and SOHO missions, in particular about the plasma state and structure of the polar coronal holes, and on theoretical efforts to model the coronal hole flows and the fast solar wind emanating therefrom. The structure of the large-scale corona is discussed and the microstate of the solar wind in terms of particle and wave observations is analysed. Observational evidence is mounting that the magnetically open coronal holes are far away from a state of local thermodynamic equilibrium, as is the associated interplanetary solar wind, and that they differ substantially in their plasma parameters from the low-latitude streamers with closed magnetic fields, which only open intermittently to release the slow solar wind. The coronal sources and their plasma boundary conditions as well as the interplanetary constraints on the wind models are presented and discussed. Modern theories and models of the solar wind are reviewed. First the basic concept of the single-fluid Parker-type model is outlined, and then two-fluid models are described, in particular those incorporating strong heating of protons close to the Sun, which yields fast acceleration with the terminal wind speed being attained within 10 $R_\odot$. Finally, the most recent modelling efforts to generate the wind plasma through ionization in the chromospheric network are presented.

1 Introduction

The solar wind is the supersonic outflow of completely-ionized gas from the solar corona. It consists of protons and electrons with an admixture of a few percent of alpha particles and heavy, much-less-abundant, ions in different ionization stages. The hot corona at temperatures of typically 1-2 MK expands radially into interplanetary space. The flow becomes supersonic at a few solar radii (Parker 1958). The solar magnetic field lines are dragged away by the wind, because of its high conductivity, and transform into heliospheric field lines, which attain the form of spirals due to solar rotation (at a siderial rate of 25 days, or 27 days seen from Earth). Mass conservation and spherical geometry imply that for constant wind speed the density must decrease in proportion to r^{-2}. This yields a radially declining ram pressure of the wind which is brought on collision with the interstellar gas, through a shock transition to a full stop at the heliopause. Here the weak pressure of the local instellar medium is capable of breaking the wind flow and the heliosphere ends. The heating of the solar corona and the acceleration of the solar wind are among the important unsolved problems of space plasma and solar physics.

At the outset, I would like to point out the limited scope of this paper, which is the writeup of key notes and compilation of some figures from two lectures given at this summer school. It is entirely impossible to cope here with the vast material accumulated over many decades of research on the solar corona and solar wind. For the reader who wants to study the literature in detail some books and reviews are listed below, which are of interest from the historical point of view and contain in-depth studies of the topics only touched upon in this paper. The books of Parker (1965) and Hundhausen (1972) cover the early years of solar wind research. The Helios epoch is reviewed in Schwenn and Marsch (1990,1991). The results anticipated and obtained from Ulysses are composed in the proceedings edited by Marsden (1985, 1995). MHD turbulence is reviewed by Tu and Marsch (1995a), and the ongoing SOHO mission is described in the book edited by Fleck et al. (1995). Concerning the developments in the recent past years, several reviews are available: the reports of Isenberg (1991), McKenzie (1991), Barnes (1992), and Marsch (1994). Many aspects of solar wind research are dealt with in the solar wind conference proceedings, most recently the ones edited by Winterhalter et al. (1996).

2   Large-scale structure of corona and wind

2.1   The structure of the minimum-activity corona

The large-scale structure of the corona changes over the 11-year solar cycle and becomes particularly simple near the minimum period, where it can be described essentially by a dipolar and quadrupolar component supplemented by contributions from the near-Sun current sheet, features which can clearly be inferred from Fig. 1 showing coronal images obtained in the green Fe XIV 5303 Å emission line and in white light with the LASCO coronagraph on SOHO (e.g., Schwenn 1997). Banaszkiewicz et al. (1997) provided a simple algebraic model for the field near minimum, which describes the global field properties and also allows one to derive the expansion factor of magnetic flux tubes required in models of the fast wind originating in the polar coronal holes. Of course, the low-latitude coronal holes, as typically seen during the Helios primary mission (Schwenn 1990), also gave rise to recurrent fast streams, a fact already known for a long time from the famous Italy-boot-shaped coronal hole seen on Skylab in the mid-seventies.

Whereas it is generally agreed upon that the polar coronal holes are the main source regions of the fast solar wind, and that the equatorial streamer belt is the birth place of the slow and transient wind (Sheeley et al. 1997), it was suggested by Habbal at al. (1997) that fast streams might also originate at low latitudes, where they flow along ray-like structures which pervade the corona everywhere (Woo and Habbal 1997). Yet, is the equatorial corona fully closed with the streamer belt inhibiting fast outflows, or are the magnetic field "surfaces" patchy and punctured with isolated flux tubes leaking out

Fig. 1. Composite image of the solar corona taken by the C1 (in the green iron emission line) and C2 (in white light) LASCO coronagraphs on February 1st, 1996, with a spatial resolution of 11.4 arcsec/pixel (Schwenn et al. 1997). Note the high-latitude streamers merging beyond 2 $R_\odot$ into the heliospheric current sheet.

into the heliosphere even at low latitudes? Some UVCS coronal images seem to indicate flows along open rays even near the equator, and the transitions from fast to slow flows seem rather sharp, with the slow flows being restricted to within only $\pm(10 - 20)°$. The general flow pattern is clearly visible in Fig. 2 (Woch et al. 1997), which presents in a polar diagram the wind speed versus latitude and the polarity of the magnetic field measured on the Ulysses spacecraft. Note the overall steadiness of the fast wind and variability and intermittent nature of the slow wind (intermingled with fast recurrent streams from low-latitude coronal holes).

2.2   Fine-scale structures in coronal holes and fast wind

We now turn our attention to the solar wind meso-scale variations. Although the wind speed, as shown in Fig. 2, appears at high latitudes to be quite constant on average, there remain clear meso-scale variations in velocity and other parameters, which indicate the existence of fine structures of coronal origin that survive the transit to the in-situ point of observation. Some years ago Thieme et al. (1990) provided evidence from Helios in-ecliptic observations for solar wind structures of a few degrees in extent, which seem to be related with or to originate from the chromospheric network in CHs. Koutchmy et al. (1991) have reviewed what is known from eclipse observations about fine-scale structures, discontinuities and filaments in the corona. Many of these structures have their remnants seen in the near-Sun solar wind. There seems to exist a whole hierarchy of structures ranging from large CHs (fast wind) and streamers (slow wind) to degree-sized coronal features, which leave imprints on the wind as variations in speed, density and temperature, or e.g. as pressure-balanced structures (Tu and Marsch 1995a).

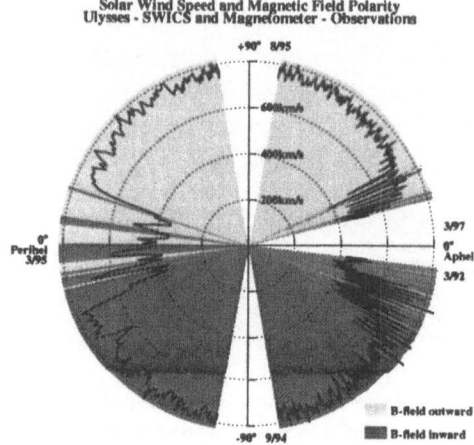

Fig. 2. A polar diagram of the solar wind speed variation with heliospheric latitude as obtained from the SWICS instrument on Ulysses (Woch et al. 1997). Speed corresponds to the radius and latitude to the polar angle, with 0° aphelion (perihelion) to the right (left). Shading corresponds to the polarity of the magnetic field. The the sign of the one-day averages of the radial component is shown.

The most conspicuous bright features in the otherwise dark polar coronal holes are the plumes, visible in the EIT, LASCO, UVCS and SUMER images obtained from SOHO. There are many questions as to the nature of these bright ray-like structures, being of the angular size of the supergranules, i.e. 2-3 degrees as seen from Sun center. Where are plumes rooted, are they dynamic or static, are there plasma flows within them? What is their magnetic structure? Apparently, plumes are not static but may support compressive oscillations, as reported by DeForest and Gurman (1997), who showed evidence for magnetoacoustic waves propagating through the plumes with wave periods of 10-15 minutes. De Forest et al. (1997) showed EIT images combined with MDI magnetograms, which illustrate that plumes arise from morphologically-unipolar magnetic footpoints. Interestingly enough, Lamy et al. (1997) provided from LASCO images evidence that plumes come and go, yet tend to reappear above the same foot points, which last much longer and clearly corrotate with the Sun. Obviously, much work is still required to unravel the mysteries of plumes.

2.3   The three-dimensional heliosphere

The large-scale structure of the heliosphere in three dimensions has been described in several articles in the book edited by Marsden (1995). The Ulysses observations analysed by Woch et al. (1997) resulted in a polar plot of the wind speed and magnetic field magnitude, which was already shown in Fig. 2, providing a concise overview of the three-dimensional heliosphere. A study of the magnetic field as a function of distance, longitude and latitude in the inner heliosphere was made by Balogh (1995), showing that the distant radial field was rather monopolar, with large compressive deviations occurring near the heliospheric current sheet. The combined image from LASCO in Fig. 1 (Schwenn et al. 1997) nicely illustrates the corona near solar activity minimum. The current sheet becomes more complicated and highly structured and warped during increased activity of the Sun, being associated with a complex multipole field and many small activity regions. Hoeksema (1995) has monitored the evolution of the current sheet by means of plots of the source field polarity at 2.5 $R_\odot$, as reconstructed through potential field extrapolations from the surface fields of the Sun.

Barnes et al. (1995) clearly showed that the mass flux and Helium abundance of the solar wind are remarkably stable and constant all over the solar surface with the exception of the equatorial streamer belt, where transients and coronal mass ejections (CMEs) render the mass flux more varibale. Yet, its daily average seems to be even invariant in time over a full magnetic solar cycle. Helios (Schwenn 1990) found 22 years earlier nearly the same mass flux than Ulysses! The SWAN experiment on SOHO allowed us recently to derive global averages of the mass flux from sky maps in $Ly \alpha$. Bertaux et al. (1997) quoted a mean value of 2.2×10^8 cm^{-2}s^{-1}, consistent with in-situ values, assuming that 1/3 of space is filled by slow wind (within a latitudinal band of $\pm 20°$) and 2/3 by fast wind. In the past, there has been a controversal debate as to what determines the mass flux and keeps it so constant, given the consideration that it should depend sensitively on the coronal temperature (Leer and Holzer 1980). Peter and Marsch (1997) have convincingly argued that the proton flux is intrinsically coupled to the photon flux of the Lyman recombination continuum below the wavelength of 911 Å, and that in fact these fluxes are equal. The generation of the coronal and solar wind plasma takes place in a thin ionization layer of Hydrogen in the chromosphere, where the mass flux is ultimately determined, and thus no mass flux problem may arise in the corona.

2.4   Transients and coronal mass ejections

Besides the steady fast and unsteady slow wind there is a third class of wind related to coronal magnetic activity, for instant in prominence eruptions, causing interplanetary disturbances (e.g. Schwenn 1983), such as magnetic clouds (reviewed by Burlaga 1990). Spectacular events of this kind were frequently registered and visualized by the LASCO coronagraph. Coronal mass ejections drive usually shock waves in the wind, which when hitting the Earth's magnetosphere can thus cause magnetic storms and other perturbations. Here we will concentrate in the remainder of the paper on the fast wind, which is the basic mode of solar wind. It is perhaps most easily modelled, since its explanation will not involve large-scale reconnections on the Sun, which are needed to explain the origin of the slow and transient wind.

The initial velocity profiles of these types of flow have been inferred from white-light coronagraph difference images and are shon in Fig. 3 after Sheeley et al. (1997). The moving coronal features, marking the slow wind flow, cluster in speed around a parabolic radial path characterized by a constant acceleration of about $4 \, \mathrm{ms}^{-2}$. This profile is consistent with an isothermal expansion at a temperature of about 1.1 MK and a sonic point near 5 $R_\odot$, in good agreement with the simple Parker model to be discussed below in

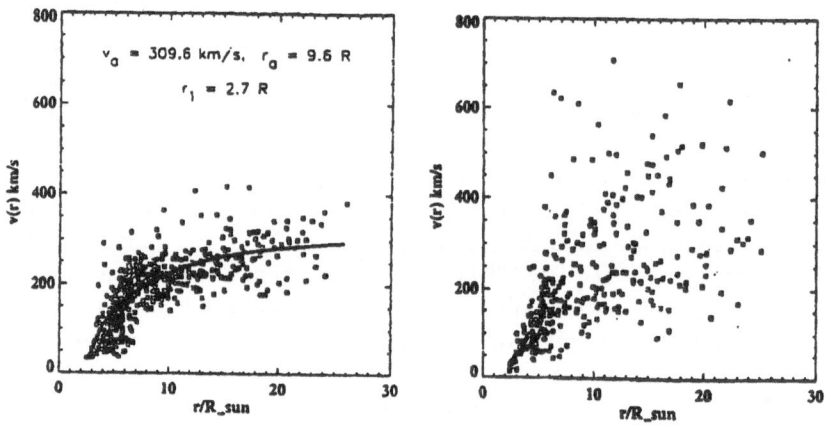

Fig. 3. Speed versus height profiles obtained by tracking measurements of coronal density enhancements observed by the LASCO coronagraph (Sheeley et al. 1997). Left: slow wind, with points clustering around a smooth curve with a parabolic shape. This solid curve represents the best fit to the unweighted data points; right: coronal mass ejections, with points showing a wider distribution.

3   Microstate of the solar wind: Particles and waves

3.1   Introduction

An overview is given of the microscopic state of the solar wind with emphasis on recent Ulysses high-latitude observations and previous Helios in-ecliptic observations. Emphasis is placed on the connection of interplanetary kinetic scale phenomena with their generating microscopic processes in the corona.

Some nonthermal particle features, such as proton-ion differential streaming, ion beams, temperature anisotropies, and skewed electron distributions
associated with collisionless heat conduction, and the related wave-particle
interactions are discussed. At all heliospheric latitudes the radial evolution of
the internal state of the wind resembles a complicated relaxation process, in
the course of which free (as compared with LTE and MHD equilibrium conditions) particle kinetic energy is converted into wave and turbulence energy
on a wide range of scales. This leads to intermittent wave-particle scattering
and unsteady anomalous transport, mixed with the weak effects of the rare
Coulomb collisions. Spherical expansion and large-scale inhomogeneity force
the wind to attain a complex state of dynamic statistical equilibrium between
the particle species and field fluctuations.

3.2   Velocity distributions and kinetic properties

Besides Coulomb collisions, kinetic wave-particle interactions induced by
small-scale waves and fluctuations play a key role in particle scattering and
related transport and largely shape velocity distributions together with the
mean large-scale interplanetary forces. As a consequence, the observed distributions of ion and electrons come in various shapes and vary widely with the
solar wind speed, heliographic coordinates and the phase of the solar cycle.
Typical proton distributions, ranging from Maxwellian in the current sheet
to highly nonthermal in fast streams, which have temperature anisotropies
with $T_\perp > T_\parallel$ and field-aligned beams with a drift speed larger than the local
Alfvén speed. Ion differential streaming also occurs, which is known to fade
away with heliocentric distance (Marsch 1991a), yet sometimes is still seen
on Ulysses beyond 1 AU at shocks and stream interaction regions.

The shape variations of the ion distribution functions in fast wind have
been investigated systematically. Whereas high-energy extensions are a uni-
versal property of the protons they are not seen in the alpha particles which
tend to move faster than the protons and surf on the ubiquitous Alfvén waves
(for a review of this phenomenon see Marsch 1991a). Some typical proton
distributions are presented in Fig. 4, together with the radial profile of the
average magnetic moment $\mu_p = T_p/B$ of the particles as a function of radial
distance from the Sun. The radially increasing μ_p indicates continuous ion
heating perpendicular to the field in the interplanetary medium.

Heavy ions have often been used to trace wave effects in the wind and
as test particles to probe the plasma. One of the most prominent features of
minor ion is that they all travel at the same speed (about by V_A faster than
the protons) and show temperature scaling in proportion to their masses, with
the exception of Helium (Neugebauer 1981). Ulysses observations showed,
now with very good statistical significance, that all ions heavier than Helium
have exactly the same speed and the same thermal widths of their velocity
distribution functions (von Steiger et al. 1995) for wind speeds above 400
km/s.

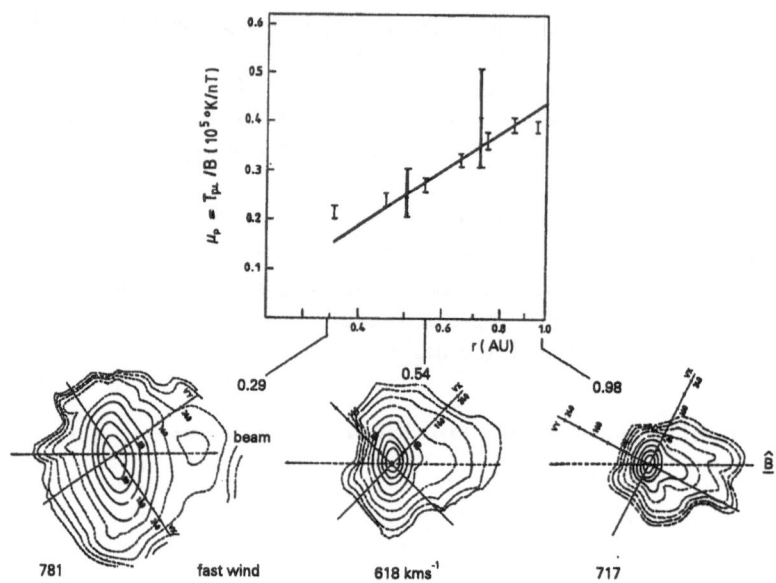

Fig. 4. Proton magnetic moment versus heliocentric distance (top) and selected velocity distribution functions measured in high-speed wind (Marsch 1991a). Solid isodensity contours correspond to 20% steps of the maximum and the last broken contour is at 0.1%. Note the large temperature anisotropy in the core and the proton beam along the magnetic field. The magnetic moment increases, thus indicating interplanetary heating. The solid line shows Tu's (1988) model results.

The electron distribution functions in velocity space are determined by the large-scale interplanetary magnetic field and electrostatic potential (related to the electron pressure gradient), by Coulomb collisions in the thermal energy range, and by various kinds of wave-particle interactions at all energies. Since electrons are subsonic, i.e. their mean thermal speed clearly exceeds the solar wind (ion) bulk speed, they may in a first approximation be considered as test particles that quickly explore the global setup of the heliospheric magnetic field. Its topology in the form of open field lines in coronal holes (CHs), of bottles and loops near the HCS, or flux ropes and plasmoids in connection with coronal mass ejections largely shapes the distributions. These can be decomposed into two components, a cold and almost isotropic collisional core and hot variably-skewed halo population and may be described by two convecting bi-Maxwellians as illustrated in Fig. 5. On open field lines the halo usually develops a high-energy extension with a very narrow pitch-angle distribution, the so-called strahl.

Aligned observations of the ICE and Ulysses spacecraft allow us to characterize the radial gradients of electron thermal parameters (Phillips et al.

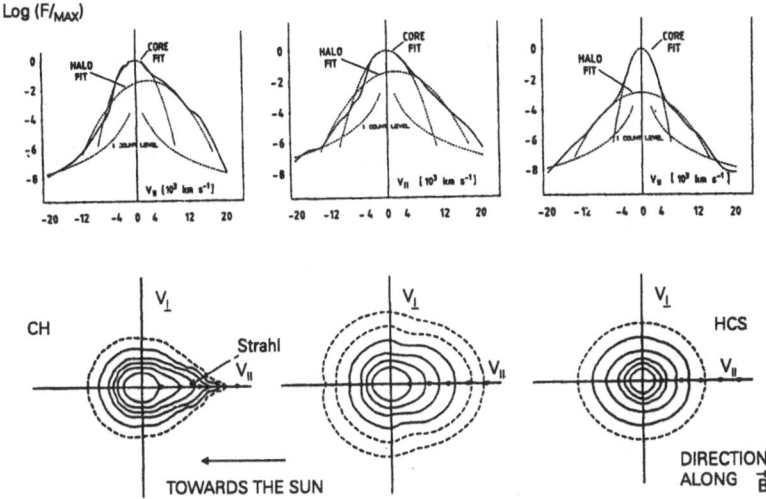

Fig. 5. Electron velocity distribution functions as energy spectra (top) and velocity space contours (bottom) for fast (left), intermediate and slow (right) wind (after Pilipp et al. 1987). Note the core-halo structure and the strahl of suprathermal electrons in fast solar wind.

1995). The core temperature was found to vary widely between isothermal and adiabatic, while the halo behaves more isothermally. The halo density falls off more steeply in dense plasma. McComas et al.(1992) have studied the halo electron parameters in the range from 1 to 4 AU and found that the halo represents always about 4% of the total distribution, indicating that the halo does not behave like being decoupled from the bulk electrons and may not consist of truely non-interactive test particles. The break point energy in the electron spectrum scales on average like 7 times the core temperature, a result predicted by Scudder and Olbert (1979a,b) for electrons mediated by Coulomb collisions.

3.3 Wave-particle interactions and microinstabilities

Significant progress has been made with the Ulysses electron measurements in the understanding of the electron heat flux regulation. It has been argued for a long time that whistler mode waves might be the primary agent for regulation of the electron heat flux. Feldman et al. (1976a,b) suggested a simple heat conduction law, in which the heat flux (q_e) is carried mainly by the halo electrons and scales essentially in proportion to the halo particle flux times the halo temperature. This proportionality has now experimentally been verified to a high degree by Scime et al. (1994), demonstrating that solar wind heat conduction has nothing to do with the local temperature gradient but with the thermal energy convected by halo electrons. The halo drift speed itself is closely tied to the Alfvén velocity or the local magnetic field, which is clear evidence for a wave regulation of the heat flux.

In an interaction with a wave a particle sees a stationary electric field in its rest frame of reference, if its velocity fulfils the condition for Landau resonance, or in the electromagnetic case for cyclotron resonance. Energy and momentum between particles and waves are exchanged as a result of this wave-particle interaction, and the velocity distributions are reshaped, until the free energy in the form of temperature anisotropy, beam drift or ion differential motion, and skewness or heat flux is reduced or removed. These processes are analytically described by quasilinear theory (e.g., Melrose 1986) and visualized by direct numerical simulation.

In summarizing the present state of our understanding of the role of microinstabilities in the solar wind (see again the reviews by Marsch 1991a,b), we may say that particle distributions are generally found to be marginally stable but prone to instability, and the salient unstable modes [and free energy sources] are: (1) ion acoustic wave [ion beam, electron heat flux]; (2) electromagnetic ion cyclotron wave [proton core temperature anisotropy]; (3) magnetosonic wave [proton beam, ion differential streaming]; (4) whistler, lower hybrid wave [core-halo drift, electron heat flux]. The quasilinear effects of these instabilities have not been explored in the solar wind context, let alone their nonlinear saturation and associated effects.

3.4   MHD Waves and turbulence

Ulysses observations have revealed new aspects of the nature of Alfvénic fluctuations at high latitudes. Firstly, their amplitudes or the field component variances are very large (Smith et al. 1995). Secondly, they are intimately linked with discontinuities (Tsurutani et al. 1997), which occur at rates of about 150 per day; thirdly, rotational discontinuities are an integral part of the Alfvén wave, such that phase glitches (of about 90°) often occur, which leads one to consider these directional fluctuations as phase-steepened non-linear Alfvén waves (Riley et al. 1995). This close co-existence of waves and structures supports the view that solar wind fluctuations consist of waves riding on structures or vice versa of multiple structures (discontinuities) embedded within waves. Observational support for this picture, which was developed from the Helios observations near 0.3 AU in the ecliptic plane, has been provided by Tu and Marsch (1994), who investigated the correlations at hourly scales between fluctuations of compressive variables like pressure, density, temperature and field magnitude under the assumption of weak turbulence, and found convincing evidence for the co-existence of Alfvén waves with fast-mode waves and pressure-balanced structures.

The radial evolution of the power spectrum of Alfvénic fluctuations is illustrated in Fig. 6 after Tu and Marsch (1995b), which shows the power spectra of Alfvén waves at 0.29 and 0.87 AU and the obvious steepening and decline in intensity with increasing distance from the Sun. The spectra are normalized to the WKB amplitude evolution (see section 5.6). Thus the dissipation becomes clearly visible. Wave energy is lost in the expanding wind to the particles and helps in maintaining non-adiabatic ion temperature profiles (see again Fig. 4).

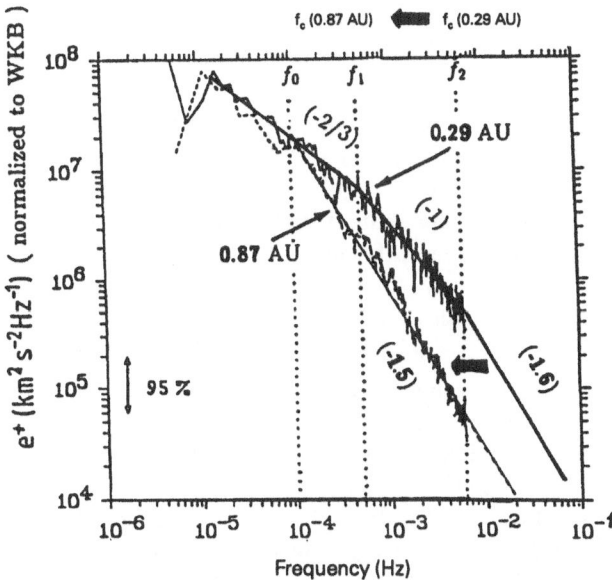

Fig. 6. The power spectra of Alfvénic fluctuations for two solar distances. Each spectrum is normalized to the WKB solution to emphasize turbulence evolution and dissipation, which become obvious in the difference between the two curves. Spectral slopes are given in brackets (Tu and Marsch 1995b).

4   Sources, gradients and boundary conditions

Among the major findings of SOHO is the observation that the fast and slow solar wind originates and accelerates in very different ways, which are intimately related with the global structures of the corona during the near-minimum period of the solar cycle. The slow wind starts accelerating only beyond 2 $R_\odot$ (see Fig. 3) and seems to peel slowly off the streamers' tops in an intermittent fashion (Sheeley et al. 1997), whereas evidence is accumulating that the fast wind accelerates rather rapidly, as compared with the canonical Alfvén-wave-driven high-speed wind (e.g. Leer et al. 1982 for an early review). Recent measurements with EISCAT seem to suggest that the full acceleration of the wind over the Sun's poles occurs within 10 $R_\odot$ (Grall et al. 1996).

The direct flow-velocity measurements clearly indicate strong near-Sun acceleration (Corti et al. 1997), in which the outflow speed increases from 50 km/s at 1.5 $R_\odot$ to 140 km/s at 2.3 $R_\odot$ in the Doppler-dimming data by UVCS obtained along a line of sight (LOS) cutting through a plume. According to the authors no significant difference in speed was obtained between plumes and darker lanes, although a signal of the "uncontaminated background" coronal hole proper could not be received.

The electron density and temperature in coronal holes are of paramount importance, since they not only characterize the collisional properties and the electron partial pressure in the coronal hole, but also determine the ionization state of Hydrogen, Helium and heavier elements and govern the radiative losses in the EUV and soft X-rays through collisional line excitation (see e.g. Mariska's book, 1992). Strictly speaking it is the detailed electron distribution function that matters; this is presently incapable of measurement close to the Sun, yet well known farther away, where nonthermal features prevail, such as the suprathermal tail, the strahl, and the core-halo structure (see Pilipp et al. 1987, Marsch 1991a, and Marsden 1995, for the details and Fig. 5).

Line ratio techniques have been used to determine the electron density and to establish with SOHO the electron density profiles for the first few 100″ above the limb in various regions. Wilhelm et al. (1997a) have used the Si VIII 1445.73 Å / 1440.51 Å line ratio, yielding a density that starts at 10^8 cm^{-3} at 20″ (1″ ≈ 715 km) and drops to about 6×10^7 cm^{-3} at an altitude of 300″ in the polar CH. This profile smoothly continues in the radial profile of n_e as estimated by the UVCS measurements of Oxygen O VI 1032 Å / 1037 Å line intensities and related emission measures (Corti et al. 1997). Given these densities, one can infer a corresponding flow speed of the plasma by exploiting the mass continuity equation and assuming a flow or flux tube geometry. This technique (e.g. applied by Fisher and Guhathakurta 1995) also implies that the wind is strongly accelerated close to the Sun.

One of the major surprises of SOHO is the fact that electrons are rather cold in CHs, evidence that comes from T_e estimates obtained from EUV emission line ratio. Wilhelm et al. (1997a) used the Mg IX 749.55 Å / 706.07 Å line ratio and concluded that T_e hardly ever reaches the canonical value of 1 MK, but remains below it and rapidly falls off with height in the CH. Similarly, David et al. (1997) derived, in a synergistic effort of the SUMER and CDS instruments using the line pair O VI 173 Å / 1032 Å, the electron temperature profiles in an equatorial streamer at the west limb and a coronal hole at the polar north limb. Again, their data shown in Fig. 7 demonstrates that electrons cool off rapidly and only have a T_e of about 0.5 MK at 1.3 $R_\odot$. These low electron temperatures are consistent with the Helios in-situ measurements almost a full solar cycle ago, which clearly indicated that interplanetary electrons are cold and their radial temperature gradients flat in fast flows associated with the CHs (see the review by Marsch, 1991a).

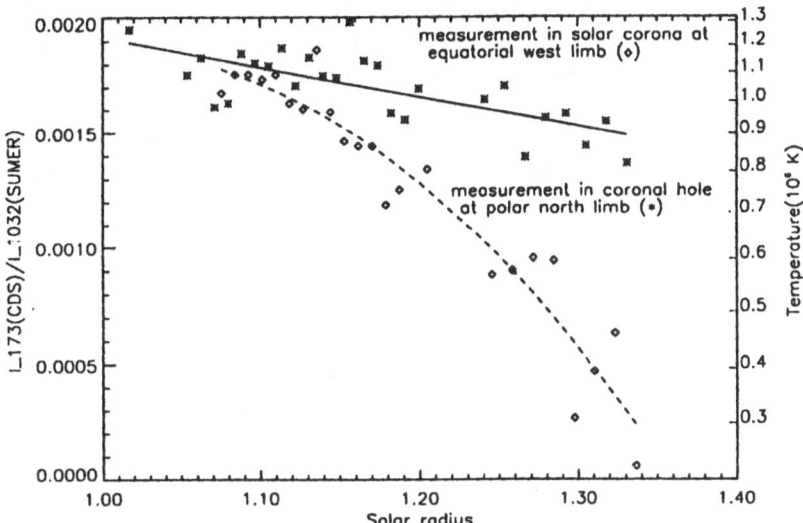

Fig. 7. Electron temperature gradients in the inner corona as obtained from EUV line-ratio diagnostics with the CDS/SUMER spectrometers on SOHO (David et al. 1997). Note that the electrons cool off rapidly in coronal holes.

In contrast to the electron observations, the protons, and more so the heavy ions, are rather hot in coronal holes. The ion temperatures are anisotropic as clearly indicated by the Oxygen line shapes derived from the UVCS instrument and presented in Cranmer et al. (1997) and by Kohl et al. (1997) in their report about first results from the UVCS instrument. For Oxygen $T_\perp$ is much larger than $T_\parallel$, a signature which has been familiar from the Helios proton measurements in perihelion near 60 $R_\odot$ and is interpreted as being the result of ion-cyclotron heating (see again Fig. 4). The equivalent velocity corresponding to the Doppler width of the O VI 1032 Å line is truely remarkable, amounting up to 600 km/s. If this is interpreted as a kinetic broadening, this would give coronal minor ion temperatures of more than 100 MK!

Where does the corona heating start? The SUMER instrument is able to measure the temperature gradient from the chromosphere through the TR into the lower corona. The widths of many EUV emission lines of different ions were found by Seely et al. (1997) to be much larger than the thermal width expected for a kinetic temperature equal to T_e or the line formation temperature. This trend is shown in Fig. 8. Arguments were put forward that the effective temperature in those lines is of a kinetic nature with no or weak contribution only of a "turbulent" velocity. After Seely et al. (1997) the kinetic temperature T_i for ions originating from such elements as Ne, Fe, Mg and Si would range between 2.5 and 6 MK for heights above the solar limb between 100″ and 200″. Each ion seems to live its own life and has a different temperature, perhaps resulting from the high-frequency-wave heating. This cyclotron heating happens naturally in a plasma, and the evidence for this in the interplanetary wind is strong and convincing (see again the solar wind reviews by Marsch 1991a,b and von Steiger et al. 1995). What remains elusive are clear signatures of outflow at the coronal base, if looked at on the disk. Chae et al. (1997) plotted the average Doppler shifts of various atoms and ions for the quiet Sun as a function of T_e and found mainly redshifts and velocities below 10 km/s.

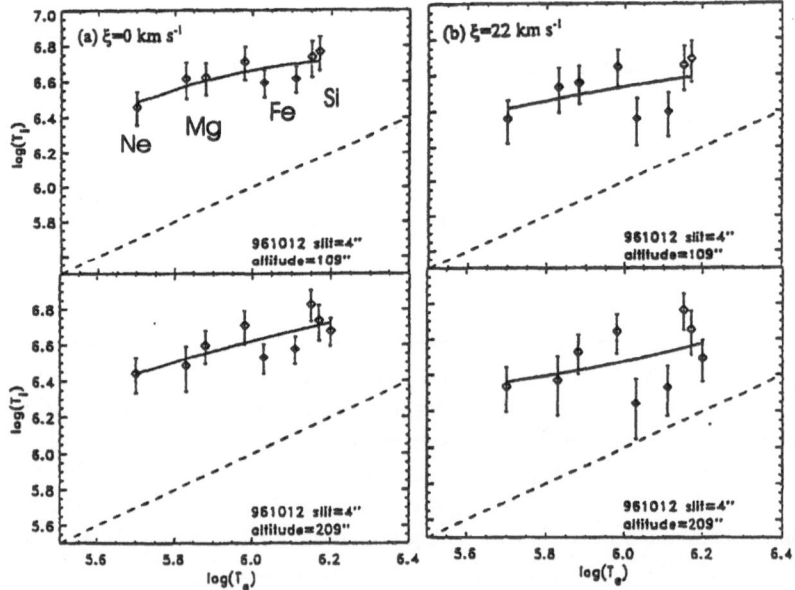

Fig. 8. Effective coronal heavy-ion temperatures plotted versus electron (ion formation) temperature for two heights in the corona. The dashed line corresponds to thermal equilibrium. Apparently, ions are much hotter than electrons. In the left boxes no turbulent broadening is assumed, whereas in the right panel an amplitude of $\xi = 22$ km/s was assumed (Seely et al. 1997).

5   Single-fluid (Parker) model

5.1   General remarks

In their modelling activities over the last years many authors have empha-
sized the point of view that an understanding of solar wind acceleration can
only be achieved if this problem is solved together with the coronal heating
problem. The status of models for coronal heating is discussed in the comprehensive Heidelberg proceedings (Ulmschneider et al. 1991). We stressed
in Section 4 the importance of properly defining solar wind boundary conditions, which are by no means obvious, given the complex observational
evidence for the highly dynamic nature of the transition region and lower
corona. Of course, one may not easily be prepared to abandon successful
conventional fluid models, usually implying stationary flows streaming along
flux tubes in simple solar magnetic field geometries.

5.2   Coronal expansion

In this introductory section we show that a hot corona cannot be confined
gravitationally. Let us simply assume that the heat flux is free of divergence.
Thus we have the integral

$$4\pi r^2 \kappa(T) \frac{dT}{dr} = const \tag{1}$$

Since $\kappa(T) = \kappa_0 T^{5/2}$, this equation can readily be integrated to give the
temperature profile

$$T(r) = T_\odot \left(\frac{R_\odot}{r} \right)^{2/7} \tag{2}$$

This can be used, with $p = 2nk_BT$, to obtain a solution for the pressure from
hydrostatic equilibrium, i.e.

$$\frac{dp}{dr} = -\frac{GM_\odot mn}{r^2} \tag{3}$$

is integrated by means of the given $T(r)$, which yields the pressure

$$p = p_\odot exp \left\{ \frac{7GM_\odot mn_\odot}{10p_\odot R_\odot} \left(\left(\frac{R_\odot}{r} \right)^{5/7} - 1 \right) \right\} \tag{4}$$

Apparently $p(\infty)$ is finite, and the problem that $p(\infty) \gg p$ in the interstellar
medium arises. As a consequence a hot, static corona cannot be confined!
The corona should be set in motion and start expanding (Parker 1958).

5.3   The isothermal solar wind

Following Parker's model we now take the plasma motion into account. Conservation of mass and momentum gives for spherical geometry the equations:

$$r^2 mnV = \dot{M} = const \tag{5}$$

$$mnV\frac{d}{dr}V = -\frac{dp}{dr} - \frac{GM_\odot mn}{r^2} \tag{6}$$

The total pressure is $p = 2nk_BT$ for a hydrogen plasma. Assume isothermal conditions, i.e. $T = T_\odot = const$. Then the sound speed c_s in the corona equals $(2k_BT_\odot/m)^{1/2}$ and is constant. The momentum equation, with the aid of the continuity equation, can thus be cast in the form

$$\left(\left(\frac{V}{c_s}\right)^2 - 1\right)\frac{dV}{V} = 2\left(1 - \frac{GM_\odot}{2c_s^2 r}\right)\frac{dr}{r} \tag{7}$$

Here $M_\odot$ is the mass of the Sun and G the gravitational constant, with the solar radius $R_\odot$. We introduce the Machnumber, $M = V/c_s$, and the critical point, $r_c = \frac{GM_\odot}{2c_s^2} = (\frac{V_\infty}{2c_s})^2 R_\odot$, in which V_∞ is the escape speed from the Sun. In these terms we can write the equation of motion as

$$\frac{dV}{dr} = \frac{2V}{r}\frac{1 - r_c/r}{M^2 - 1} \tag{8}$$

In order to obtain a flow with $\frac{dV}{dr} > 0$ everywhere, i.e. a continuous acceleration, the flow for $r < r_c$ must be subsonic with $M < 1$, and for $r > r_c$ it must be supersonic with $M > 1$. The only solution doing this is the critical one, passing through the critical point at $r = r_c$ where $M = 1$. This equation can readily be integrated and yields the $V(r)$ profile:

$$\left(\frac{V}{c_s}\right)^2 - ln\left(\frac{V}{c_s}\right)^2 = 4\left(ln\left(\frac{r}{r_c}\right) + \frac{r_c}{r}\right) + C \tag{9}$$

This formula implies for $V \gg c_s$ that $V \propto \sqrt{ln r}$ and $n \propto r^{-2}V^{-1}(r)$. Here C is an integration constant. This solution for an isothermal wind is the simplest solution giving a supersonic outflow of the plasma. The density $n(r) \to 0$ for $r \to \infty$, and similarly the pressure obeys $p(r) \to 0$ asymptotically, in compliance with the requirement that the wind expands practically into vacuum with zero interstellar pressure. Note that even fast flows are obtained for a high enough temperature ($T_\odot \approx 2 - 3$ MK), yet the initial acceleration is moderate. In order to obtain a fast wind at the canonical corona temperature of $1 - 2$ MK, one has to deposit additional momentum in the flow. This can be achieved with Alfvén waves, which are observed to come from the corona (see Section 3.4).

5.4   Basic single-fluid equations including Alfvén waves

For the sake of simplicity we consider again a steady, radially symmetric coronal expansion and wind with a mass density $\rho = mn$, flow speed V, pressure p, temperature T, and magnetic field B. Mass conservation is described by the continuity equation. In order to account for the nonradial expansion in coronal hole flows the area function of the flow tube is introduced which modifies the continuity equation, reading

$$\dot{M} = \rho V A = const \tag{10}$$

where the flow-tube area function is $A(r) = f(r)r^2$, and $f(r)$ models the non-spherical expansion. This nonradial expansion is often modelled after Kopp and Holzer (1976) by the function:

$$f(r) = \left(f_{max} e^{(r-R_1)\sigma} + f_1 \right) / \left(e^{(r-R_1)/\sigma} + 1 \right) \tag{11}$$

with free parameters f_{max}, R_1, σ to model the degree, location, and extent of the expansion region; f_1 is used to normalize so that $f(R_\odot) = 1$. The factor f_{max} may range between 1 and 10 (see also Banaszkiewicz et al. 1997).

For the radial evolution of the magnetic field Gauss law is used, implying that the field obeys the conservation equation

$$F_B = AB = const \tag{12}$$

As shown in Section 3.4, the fast streams are permeated by Alfvén waves. These waves have been shown to exert a wave pressure-gradient force on the background flow and can thus accelerate the flow (Jacques 1978; Leer et al. 1982). The momentum equation including the Alfvén wave pressure p_A is

$$V\frac{d}{dr}V = -\frac{1}{\rho}\frac{d}{dr}(p + p_A) - \frac{GM_\odot}{r^2} \tag{13}$$

The wave pressure is given by

$$p_A = \frac{1}{8\pi}\langle \delta B^2 \rangle = \frac{1}{2}\rho\langle \delta V^2 \rangle \tag{14}$$

i.e. by the mean value of the magnetic fluctuation energy of the waves.

5.5   Thermodynamics and energetics of the wind

There has been an increasing awareness among solar wind physicists that the problem of solar wind acceleration and coronal heating must be treated in concert (Hammer 1982a,b; Hollweg 1986; Withbroe 1988), and that the corona cannot simply be assumed as a given boundary when dealing with the wind acceleration but rather that both phenomena are intimately linked. For solar wind models this requires placing the boundary not at the bottom of a 10^6 K corona but even further down in the upper chromosphere or lower transition region (TR). The price to be paid by including these complications is that radiative losses in EUV and X-ray emission and conductive losses in terms of an inward directed heat flux must be accounted for and these need to be balanced by mechanical energy input, i.e. coronal heating must be considered.

The thermodynamics, i.e., the cooling of the plasma by radiation (loss function L) and conduction (heat flux F_c), and its heating by mechanical energy input at a volumetric rate H and by Alfvén wave dissipation at the rate Q_A, all these processes are described by the entropy or internal energy equation, which reads

$$\frac{\rho^\gamma}{\gamma - 1} V \frac{d}{dr}\left(\frac{p}{\rho^\gamma}\right) = -\frac{1}{A}\frac{d}{dr}(AF_c) + H - L + Q_A \qquad (15)$$

where γ is the adiabatic index, given by the ratio of specific heats and equal to 5/3 for a monoatomic isotropic gas. The classical expression for the electron heat flux in a collision dominated plasma is given by

$$F_c = -\kappa_e \frac{d}{dr} T_e \qquad (16)$$

with the thermal conductivity κ_e being proportional to $T_e^{5/2}$. Modified, but still unrealistic heat flux laws, have also been incorporated for the outer corona and distant wind because of our lack of understanding of heat conduction processes in the collisionless solar wind. The energy input has often ad hoc been modelled (e.g., Holzer and Axford 1970) as

$$H = H_\odot e^{-(r-R_\odot)/\lambda_m} \qquad (17)$$

where the damping length λ_m is an adjustable free parameter of the model. The heating rate H is determined by the energy flux required to maintain corona and wind. The associated mechanical energy flux F_m is thus defined by

$$F_m(r) = \frac{1}{A(r)} \int_r^\infty dr' A(r') H(r') \qquad (18)$$

and attains a typical value of $F_m(R_\odot) = 5 \times 10^5$ erg/(cm^2 s) in compliance with the overall energetics of the atmosphere. The wave dissipation has been modelled (Hollweg 1986), by using dimensional arguments and the idea of a turbulent cascade, as follows:

$$Q_A = \rho \langle \delta V^2 \rangle^{3/2}/\lambda_A \qquad (19)$$

where δV is the velocity fluctuation amplitude of the Alfvén waves and λ_A the dissipation or damping length, which is of the order of the turbulence correlation length. The scaling of this heating rate is derived from a Kolmogorov-type cascade dimensional argument.

Finally, we need to specify the radiative loss function (Mariska 1993), $L = n_e^2 \Lambda(T_e)$, which is proportional to the squared electron density and the function $\Lambda(T_e)$ which is determined by the collisionally excited, atomic radiation processes of the various highly-ionized minor elements in the corona. For a recent evaluation of Λ see the paper by Cook et al. (1989). Convenient parameterizations are provided by Rosner et al. (1978). Since we are considering a single-fluid model we may omit the index e for the electron temperature and density, which are then assumed equal to those of the protons.

We complete our compilation of the basic set of equations by quoting another global invariant of the atmosphere, describing the total energy budget. The divergence of the sum of the various energy fluxes must be zero:

$$-L+H = \frac{1}{A}\frac{d}{dr}\left(A(F_c + F_A) + \dot{M}\left(\frac{1}{2}V^2 + \frac{\gamma}{\gamma-1}\frac{p}{\rho} - \frac{1}{2}V_\infty^2\frac{R_\odot}{r}\right)\right) = \frac{1}{A}\frac{d}{dr}\mathcal{F} \tag{20}$$

Here $\mathcal{F}$ is the conserved energy flux in the distant solar wind. The terms in this equation from left to right give the radiative losses and mechanical heating and the divergence of the conductive, wave, kinetic enthalpy, and gravitational energy fluxes. The latter three, being proportional to the mass flux, only arise in an expanding atmosphere and become most important in the outer corona and wind. From this equation the asymptotic flow speed can readily be estimated by noting that at 1 AU all the solar wind energy essentially resides in the kinetic energy, whereas at the sonic critical point r_c the solar wind wave and particle thermal energy fluxes dominate, and where L and H can already be neglected, and therefore the total energy flux $\mathcal{F}$ is conserved. Thus we may solve for V_E at the Earth orbit and find:

$$V_E = \sqrt{2\mathcal{F}/\dot{M}} \tag{21}$$

an expression discussed thoroughly by Leer and Holzer (1980). To obtain a high terminal speed, V_E, one can either increase $\mathcal{F}$ or decrease $\dot{M}$.

5.6   Evolution equations for the wave spectrum and pressure

The radial evolution of the Alfvén waves is described by the wave energy exchange equation including dissipation (Jacques 1978), which reads

$$V\frac{d}{dr}p_A = Q_A + \frac{1}{A}\frac{d}{dr}(AF_A) \tag{22}$$

Here the Alfvén speed $V_A = B/\sqrt{4\pi\rho}$. The wave energy flux density may be written $F_A = p_A(3V + 2V_A)$. Note that Q_A is a source for internal plasma energy but a sink for wave energy. If dissipation can be neglected there is a simple integral expressing wave action conservation. By using the Alfvén Mach number $M_A = V/V_A = \rho V A\sqrt{4\pi\rho}/(\rho BA) = \dot{M}\sqrt{4\pi}/(F_B\sqrt{\rho}) = \sqrt{p_A/\rho}$, thus defining the mass density ρ_A at the Alfvén critical point r_A, we may cast this equation in the form

$$\frac{2}{M_A^2(1+M_A)}V\frac{d}{dr}\left(\rho_A M_A(1+M_A)^2\right) = -Q_A \tag{23}$$

This defines, for $Q_A = 0$, an adiabatic-type wave equation of state $\rho_A M_A(1+M_A)^2 = const$, often referred to as WKB solution, which allows us to infer adiabatic scaling properties of ρ_A with the plasma Mach number and density, reading $\rho_A \propto \rho^{3/2}$ for $M_A \gg 1$ and $\rho_A \propto \rho^{1/2}$ for $M_A \ll 1$. This is the wave analogue to the thermal adiabatic equation of state, which gives $p \propto \rho^{\gamma}$ with $\gamma = 5/3$. In this sense wave pressure can be considered as just another type of internal energy density of the wind, which is reshuffled between waves and particles when the dissipation term Q_A is switched on. One must consider complex spectral transfer equations if dissipation is to be adequately accounted for. This leads us to consider turbulence models of the solar wind.

Tu et al. (1984) and Tu (1988) modified the WKB theory in order to describe damping of interplanetary fluctuations and heating of the solar wind. The evolution of the fluctuations is described by the spectral transfer equation, which reads

$$\frac{1}{A}\frac{\partial}{\partial r}\left(A(3V+2V_A)P(f,r)\right) - V\frac{\partial}{\partial r}P(f,r) = -\frac{\partial}{\partial f}F(f,r) \tag{24}$$

Here $P(f,r)$ is the power spectrum and $F(f,r)$ is the spectral flux function describing the nonlinear couplings. Upon integration over the frequency domain $[f_L, f_H]$ we obtain the overall wave pressure defined by the integral

$$p_A = \frac{1}{8\pi}\int\limits_{f_L}^{f_H} df P(f,r) \tag{25}$$

and a direct interpretation of the dissipation rate as the difference of the spectral fluxes at the upper and lower boundary of the frequency domain: $-4\pi Q_A = F(f_L, r) - F(f_H, r)$ by comparison with (22).

This gives us a systematic interpretation of the meaning of Q_A, which requires a knowledge of the spectral fluxes at the energy-containing scales and dissipative scales. In a developed Kolmogorov cascade both fluxes match to produce a steady power-law spectrum with index $-5/3$, implying a flux function scaling after Tu (1988) like

$$F(f,r) = \frac{C_K}{(V+V_A)\sqrt{\rho}}f^{5/2}P(f,r)^{3/2} \tag{26}$$

This also means that $F(f_L, r) = F(f_H, r)$ and thus $Q_A = 0$. Here C_K is a Kolmogorov-type constant of order unity. If we make the assumption that $F(f_L, r) = 0$, which means the fluctuations are freely decaying and not replenished at low frequencies, then we find $Q_A = F(f_H, r)/4\pi$. The critical issue is of course the unknown spectra in the corona, for which plausible assumptions, guided by the interplanetary observations such as shown in Fig. 6, must be made in the models. The computational results of Tu et al. (1984) indicated that the cascade process gives significant dissipation for wave-fluctuation amplitudes as observed at 0.3 AU, provided reasonable values of $\langle \delta V \rangle$ at the Sun are assumed which are consistent with the EUV line widths, interpreted as being due to nonthermal Alfvén wave broadening.

5.7   Kinetic models of the wind

In the previous section simple fluid models of the solar wind including waves were reviewed. By virtue of its relative simplicity, the fluid picture has been favoured by most researchers for a theoretical description. However, from the very beginning of solar wind modeling the viewpoint was also put forward that the solar atmosphere may be conceived as a collisionless exosphere (e.g. Fahr and Shizgal 1983), in which case the boundary conditions at the base would essentially determine the structure of particle velocity distribution functions in the wind through application of Liouville's theorem. Scudder and Olbert (1976a,b) developed a theory of local and global processes affecting solar wind electron distributions, in particular the strahl or heat-carrying populations. Two distinct populations result from this theory, the cool core and hot halo. These species are identifiable in the Helios and even Ulysses data aquired between 1 and 4 AU (McComas et al. 1992). These authors also confirmed the model prediction that the break point energy between core and halo should typically be 7 times the core temperature. The roughly constant halo to core relative density of 0.04 is not explained and may be indicative of non-exospheric interactions of the halo electrons over scales of several AU.

Scudder (1992) has suggested that the corona itself could be visualized as an exoshere with the base in the TR, and that the corona might consist simply of those hot TR electrons capable of surmounting the gravitational and electrostatic potential barrier (velocity filtration), thus forming a hot corona without mechanical heating. However, as mentioned by Hollweg (1986), the following weak points of such a model must be kept in mind: Obtaining the coronal temperature inversion is only one part of the problem. The real issue is to maintain a 10^6 K-temperature corona by energy input against radiative losses (caused by collisional excitation!) and the losses in form of the solar wind and heat conduction. This problem can be handled in fluid models, whereas it remains to be demonstrated whether the exospheric approach can handle the energetics of corona and wind adequately.

6   Multi-fluid models

6.1   General remarks

The recent observational findings from Ulysses and SOHO in particluar, concerned with the coronal holes structure and the high speed solar wind, clearly indicate that we have to change our conventional perceptions and ideas about the plasma state of the corona as well as about how and where the solar wind is accelerated. In retrospect, it is not an exaggeration to conclude that we presently (in 1997) undergo a major change in "paradigm" of the coronal plasma physics, and of our understanding of coronal holes and fast polar solar wind as opposed to the slow equatorial solar wind. The particles become increasingly collisionless with height in the atmosphere, a notion which renders simple fluid models questionable. The multi-component plasma is thus far from local thermodynamic equilibrium and carries clear signatures of selective wave-particle interactions, as inferred from excessive line broadenings in heavy-ion EUV emission lines. The electrons are rather cool and the ions very hot, observations which make the single-fluid modelling efforts of the past obsolete. In coronal holes the terminal wind speed seems to be reached within 10 $R_\odot$, and the acceleration starts very close to Sun, with the "sonic" point being located near 2 to 3 $R_\odot$ according to modern two-fluid models.

6.2   The two-fluid equations

In order to account for temperature differences of electrons and protons we need to consider two seperate energy equations (Hartle and Sturrock 1968; Holzer and Axford 1970), in which there are different heating sources and collisional couplings between the two species. The radiative loss only appears in the electron equation, since it is the electrons that loose thermal energy by collisional EUV-line excitations and soft X-ray emissions. Classical collisional conduction is included for both species. The stationary energy equations read

$$V\frac{dT_p}{dr} + (\gamma-1)\frac{T_p}{A}\frac{d}{dr}(VA) = \frac{(\gamma-1)}{nk_BA}\frac{d}{dr}\left[A\kappa_p\frac{dT_p}{dr}\right]$$
$$- \nu_{ep}(T_p - T_e) + \frac{2Q_p}{3nk_B} \tag{27}$$

$$V\frac{dT_e}{dr} + (\gamma-1)\frac{T_e}{A}\frac{d}{dr}(VA) = \frac{(\gamma-1)}{nk_BA}\frac{d}{dr}\left[A\kappa_e\frac{dT_e}{dr}\right]$$
$$+ \nu_{ep}(T_p - T_e) + \frac{2Q_e}{3nk_B} - L\frac{(\gamma-1)}{nk_B} \tag{28}$$

If the temperature is anisotropic, we have to differentiate between the parallel and perpendicular degrees of freedom (e.g. Leer and Axford 1972):

$$\frac{dT_\parallel}{dr} = -\frac{2T_\perp}{V}\frac{dV}{dr} + \frac{2\nu_{pp}(T_\perp - T_\parallel)}{V} + \frac{Q_\parallel}{nVk_B} \tag{29}$$

$$\frac{1}{A}\frac{d(AT_\perp)}{dr} = -\frac{\nu_{pp}(T_\perp - T_\parallel)}{V} + \frac{Q_\perp}{nVk_B} \tag{30}$$

For $Q_{\parallel,\perp} = 0$ we retain the so-called double adiabatic equations of state. Here ν_{pp} is an effective collision frequency for the protons, quantifying the strength of the coupling between $T_\parallel$ and $T_\perp$. Marsch and Richter (1984) have determined empirical estimates of ν_{pp}, which turns out to be about 10^{-2} Ω_p, a size required to explain the in-situ-oberved temperature anisotropy. Detailed studies of fast solar wind models including proton temperature anisotropy have recently been carried out by Hu et al. (1997), who conclude that anisotropy has little influence on the dynamics yet imposes requirements on the unknown heating mechanism of the solar wind.

6.3 Hot protons in the inner corona

McKenzie et al. (1995) developed a theory for the high-speed solar wind with a simple dissipation length characterization of wave heating of the coronal plasma close to the Sun, with the dissipation length λ_m ranging between 0.25 and 0.5 $R_\odot$. The significant features of the solution based on the model equations of the previous sections are: rapid acceleration with a sonic point at about 2 $R_\odot$ and high maximum proton temperature, namley 8 to 10 MK. Such efficient near-Sun dissipation requires high-frequency Alfvén waves in the range 0.01 Hz to 1 kHz. The corresponding coronal proton speed and temperature are shown in Fig. 9. Note that the electrons are cool and the protons hot, thermal signatures consistent with the SOHO constraints discussed in Section 4. McKenzie et al. (1997) have extended these calculations further to include the proton temperature anisotropy and achieved results similar to the observed ones. Esser et al. (1997) have presented a parameter study on the effects of hot protons in the inner corona on the nascent solar wind. Some of their results and parameter profiles are shown in Fig. 10. Note that their density is in compliance with empirical constraints on n_e from coronagraph observations. If a high T_p is genuine then flow velocities of up to 800 km/s are readily obtained within 10 $R_\odot$ by a deposition of heat or momentum within a fraction of a solar radius above the Sun's surface.

Within the two-fluid model described above, Esser et al. (1987) had earlier carried out a parameter study of the effects of Alfvén waves in particular on the proton temperature profile. Esser et al. (1987) extended these studies to include heavy minor ions by adding their fluid equations to the basic model, with the intention of drawing possible inferences about the wind acceleration from minor ion EUV-spectral-line observations. Basically, the sloshing of ions by wave motion will spread their velocity distribution, an effect accounted for by giving them an effective temperature

$$T_{ieff} = T_i + \frac{m_i}{3k_B}\langle \delta V^2 \rangle \tag{31}$$

which includes the Alfvén wave amplitude and should show up in superthermal line broadenings, given by $\Delta\lambda = \lambda(2k_B T_{ieff}/m_i)^{1/2}/c$. Line-width measurements thus provide the most important direct diagnostic for coronal

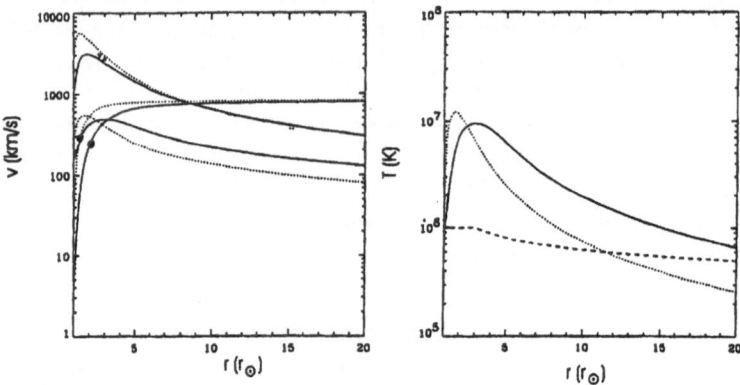

Fig. 9. Left: Variation of the solar wind bulk and thermal speed and Alfvén speed in the range from 1 to 20 $R_\odot$ for a damping length $\lambda_m = 0.25\ R_\odot$ (dotted lines) and $\lambda_m = 0.5\ R_\odot$ (full lines), with critical points marked by dots. Right: Variation of the proton (dashed line: electron) temperature in the same range (McKenzie et al. 1995).

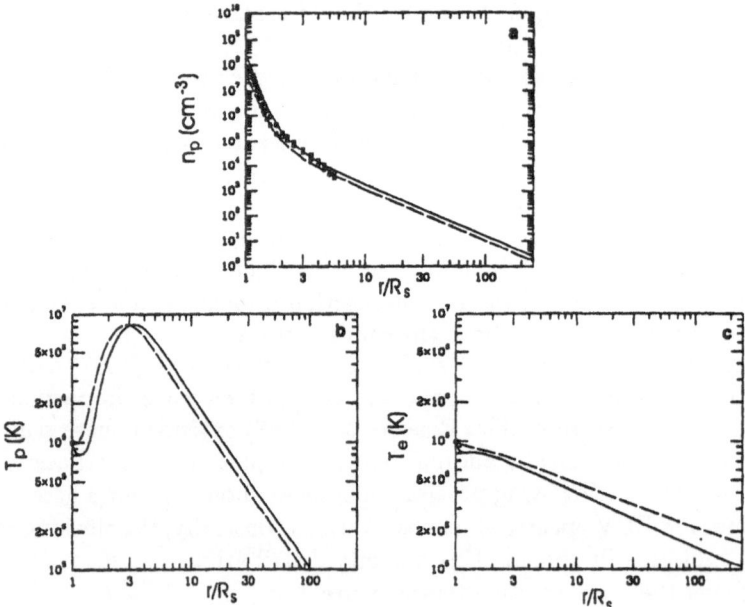

Fig. 10. Top: Electron densities derived from the SPARTAN polarization brightness measurements together with the model densities. Bottom left: Corresponding proton temperature versus distance from the Sun. Bottom right: Electron temperature for the same two models after Esser at al. (1997).

waves and turbulence. Such measurements have been extensively made by the SOHO EUV-spectrometers (see again Section 4). Li et al. (1997) studied in detail the influence of heavy ions on the fast wind a three-fluid model. They conclude that preferential heating of heavy ions, in particular of alpha particles, has an impact on the bulk flow which must be considered.

6.4   High-frequency-wave heating of the corona and acceleration of the solar wind

Yet none of the models discussed so far described the detailed physics of the heating mechanism. Therefore, Tu and Marsch (1997) elaborated the idea of Axford and McKenzie (1992), that high-frequency Alfvén wave dissipation is responsible for the heating, a process well established in the distant solar wind (Marsch et al. 1982), as shown in Figs. 4 and 6. The key model assumptions are: 1. proton-cyclotron damping of Alfvén waves, running into resonance in a rapidly declining magnetic field, gives fast heating close to the Sun; 2. The spectral evolution is slow, and therefore the spatial part of the power spectrum, $P(f, r)$, follows the WKB solution (obtained from (23) for $Q_A = 0$), i.e. when the spectrum is a separable function of r and f. Then from (23) and (25), with $f_H(r) = 0.1\ \Omega_p(r)$, the heating function (first derived by Tu 1987) is directly obtained by differentiation as

$$Q = -(V + V_A)\frac{P(f_H(r), r)}{4\pi}\frac{df_H(r)}{dr} \qquad (32)$$

The wave energy is thus eaten up by the particles at the local dissipation frequency, which declines with solar distance such that a larger and larger part of the upper range of the power spectrum is converted into heat, thus building up a large proton temperature. Input in the model is the unknown power spectrum for which assumptions have to be made, guided by what is observed in situ. The heating function can be given different spatial shapes depending on the detailed structure of the injected power spectrum. Without specifying any artificial heat source, the model yields that the electrons hardly reach a 1 MK temperature, whereas the protons attain temperatures of 2-3 MK and effective temperatures after (31) of up to 6-7 MK. Since proton heat conduction was included the temperature maximum was less than without conduction. The required wave amplitudes are compatible with constraints obtained from the SUMER observations (Wilhelm et al. 1997b; Seely et al. 1997). The dependence of the flow speed and location of the critical point upon the integrated wave amplitude in the range between 1 and 100 Hz is illustrated in Fig. 11.

The rapid acceleration is in all these new models achieved by the strong thermal pressure gradient, caused essentially by the steep temperature decline beyond the critical point. These models can account for both the heating of the corona and acceleration of the solar wind, because it is the high-frequency part of the spectrum which via dissipation directly heats the corona and builds up a strong thermal pressure, and it is the lower-frequency part which pushes additionally on the wind through the wave-pressure-gradient force acting beyond the Alfvén point near 10 $R_\odot$.

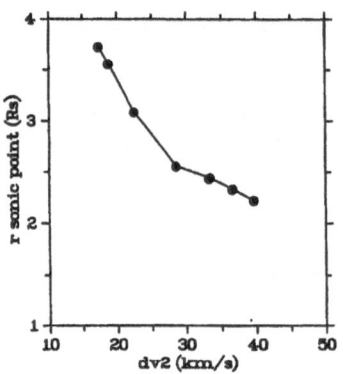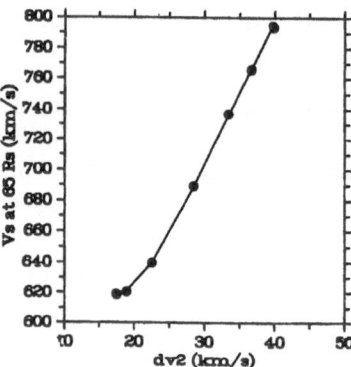

Fig. 11. Variation of the sonic point location (left) and solar wind velocity at 65 $R_\odot$ (right) as a function of the integrated wave amplitude at the coronal base for waves in the frequency range between 1 Hz and 100 Hz (Marsch and Tu 1997).

7   Modelling the Origin of the Solar Wind in the Chromospheric Network

7.1   Flows in Coronal Funnels Driven by High-Frequency Waves

Axford and McKenzie (1992) focused the attention on the chromospheric network (and the magnetic activity ocurring therein) as the region where the solar wind and the fluctuations it carries originate. Marsch and Tu (1997) developed a two-fluid model for the transition region, including for the first time upflow of the plasma in the magnetic funnels open to the overlying corona. A numerical study of the effects of high-frequency (up to 10 kHz) Alfvén waves on the heating of the TR region and acceleration of the nascent wind in the chromospheric network has been carried out. It was shown that waves with reasonable amplitudes can create via dissipation a substantial thermal pressure, which can drive locally supersonic flows in the rapidly expanding magnetic funnels. Fig. 12 shows the velocity and density profiles as a function of altitude in the solar atmosphere. The wave energy flux and spectrum are given at the lower boundary of the funnel and dissipate according to the heating function (32). A multiplicity of solution was obtained depending on the boundary conditions and the coronal pressure on top of the funnel. The calculations were time dependent and dynamic, as opposed to static TR models. It was also found that shocks may form in the funnel, and sometimes standing shocks were obtained, which could rather effectively heat the corona through a conversion of bulk kinetic into thermal energy.

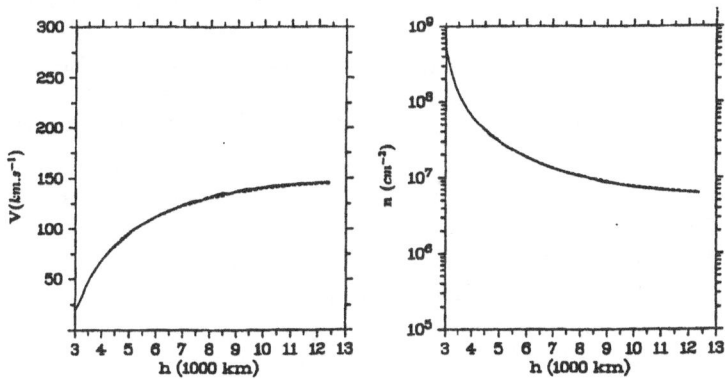

Fig. 12. Plasma flow velocity (left) and density (right) versus height in a coronal funnel of the solar transition region. The profiles correspond to a steady supersonic outflow. The parameters obtained at a height of 12000 km may serve as boundary values for a wind solution continuing these profiles into the outer corona (after Marsch and Tu 1997).

7.2   Ionization and fractionation in the chromosphere

In the solar wind community the attention of modellers has recently been drawn to the upper chromopshere as the location where the coronal and solar wind plasma is created in the first place, and also on the lower transition region where heavy ions in particular go with increasing temperature through multiple ionization stages, which are then frozen-in after recombination ceases in the dilute expanding wind (For an early comprehensive work of this subject see Hundhausen 1972 and Owocki and Scudder 1983). In this process the chemical composition of the solar atmosphere might change through fractionation associated with collisions, charge exchange and diffusion.

It has observationally been well established for a long time (e.g. Meyer 1996) that element abundances of solar wind and energetic particles are fractionated with respect to their first ionization potential (FIP effect) or time. Ulysses added to this the striking result that the FIP effect is considerably reduced in the fast wind emanating from the polar regions of the Sun (Geiss et al. 1995), implying different source conditions in the chromosphere and transition region below coronal holes in comparison with the equatorial streamer belt. The close link between the various atmospheric layers of the Sun and the solar wind is illustrated in Geiss et al. (1995), showing a superposed epoch plot of the systematic variations of the oxygen O^{7+}/O^{6+} freezing-in temperature, the Mg/O content and the wind speed in terms of the Helium velocity, which all vary while perfectly tracing each other and thus indicate a close relationship between the processes that set abundances and accelerate the wind.

Various scenarios have been developed to explain the fractionation. For the most recent review see Henoux and Somov (1992). The only models that make quantitative predictions are by von Steiger and Geiss (1989) and Marsch et al. (1995) and Peter (1996a,b), who suggested that velocity-dependent fractionation of elements can occur through a diffusion process in the upper chromosphere where a large fraction of the gas is still neutral. The modelled region is located between the supergranulation cells in the vertical magnetic fields of the coronal funnels. This atmospheric layer is assumed to be simply stratified and one-dimensional. The flows at chromospheric levels are subsonic such that the advective terms can be neglected. For sake of simplicity a unique and constant temperature was assumed. No ambipolar effects, $(T_e = 0)$, were included and gravity neglected. Only photo-ionisation was considered. Improved versions of this model, incorporating all the previously neglected terms, were numerically solved in the thesis work of Peter (1996b). The ratios of the densities obtained for various pairs of chemical elements after diffusion through a few ionization-diffusion lengths reproduces the fractionation pattern observed in slow wind remarkably well, as is shown in Fig. 13. The results of Peter (1996a) give the classical enhancement by a factor of about 4 for the heavy low-FIP elements in the slow wind and of about 2 in fast streams in good agreement with observations.

8   Conclusions

In the past years considerable progress has been made in characterising the coronal plasma state and in determining the radial profiles of key parameters in the polar coronal holes. This success is largely due to the unprecedented and concerted effort of all the SOHO coronal instruments. The new empirical constraints have to be accepted and digested by the theorists concerned with modelling the fast wind. Certainly, the single fluid descriptions are outdated and multi-fluid models are a future must. It became also obvious that an thourough understanding of the microstate of the coronal plasma as diagnosed by means of EUV-line shapes and in-situ-measured ion compositions will require a new approach, going beyond the MHD paradigm and classical transport but involving instead the tools and concepts familiar from space plasmas in the heliosphere and planetary magnetospheres.

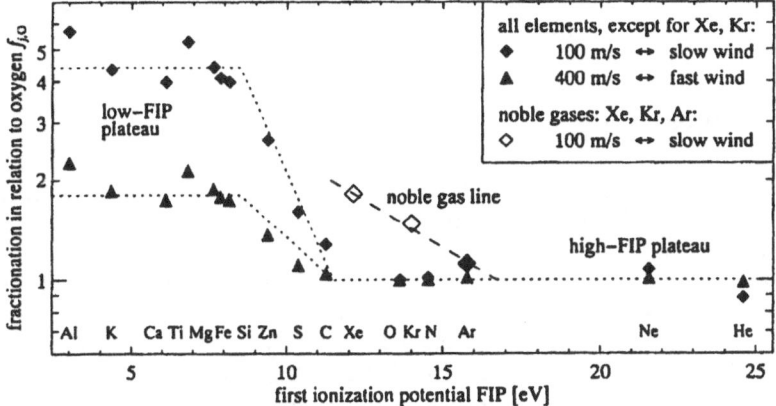

Fig. 13. Calculated fractionation for different hydrogen velocities in the chromosphere and for different elements. The dotted lines indicate the measured low- and high-FIP plateaus (Peter 1996a,b).

In these collisionless systems plasma waves and particles are intimately linked through instabilities and wave-particles interactions, processes that ask for a kinetic description on the level of distribution functions and for new transport schemes and coefficients quite different from the widely used isotropic conductivity and viscosity, invoked e.g. for the dissipation in reconnection and microflares and for the heating in current sheets in the TR and corona and in loops. Similarly, it is time to abandon parameterized and unspecified heating functions so common in coronal hole (and loop alike) models, and to face the difficult problem of identifying the treats of the real heating processes in the data and of developing the associated microphysics, which in my opinion will involve waves at frequencies much higher than considered in conventional coronal heating models. This is already the state of the art in solar radio astronomy (see e.g. the book of Benz, 1993). Similarly, kinetic physics will open new avenues for research on the coronal heating and solar wind acceleration problem.

We conclude with listing some important unsolved problems and prospects for future research. The origin and acceleration of the solar wind, in the form of fast streams from CHs, slow wind associated with the HCS, and transient flows in coronal mass ejections, remains a fundamental problem to be solved. Sources and boundaries of the nascent wind in the magnetically structured solar atmosphere should be identified clearly. The primordial ionization and element fractionation together with the mass supply to the wind and the abundance variations need to be understood. The role of MHD and plasma waves in heating the corona and accelerating the ions has to be clarified.

Development of a consistent theory of the non-equilibrium thermodynamics and transport of the solar wind based on wave-particle interactions stands as a future challenge for theoreticians.

References

Axford, W.I., McKenzie, J.F. (1992): *The origin of high speed solar wind streams* (in *Solar Wind Seven*, ed. E. Marsch and R. Schwenn, Pergamon Press, Oxford, England), 1-4

Banaszkiewicz, M., Axford, W.I., McKenzie, J.F. (1997): *An analytic solar magnetic field model* (Geophys. Res. Lett.), in press

Barnes, A. (1992): *Acceleration of the solar wind* (Rev. Geophys. **30**), 43

Barnes, A., Gazis, P.R., Phillips, J.L. (1995): *Constraints on solar wind acceleration mechanisms from Ulysses plasma observations: The first polar pass* (Geophys. Res. Lett. **22**), 3309-3311

Benz, A.O. (1993): *Plasma Astrophysics, Kinetic Processes in Solar and Stellar Coronae* (Kluwer Academic Publishers, Dordrecht, The Netherlands)

Bertaux, J.-L., Quemerais, E., Lallement, R. (1996): *Observations of a sky Lyman α groove related to enhanced solar wind mass flux in the neutral sheet* (Geophys. Res. Lett. **23**), 3675-3678

Burlaga, L.F. (1990): *Magnetic clouds* (in *"Physics of the Inner Heliosphere"*, Vol. 1, R. Schwenn and E. Marsch (eds.), Springer Verlag, Berlin, Heidelberg, New York)

Chae, J., Yun, H.S., Poland, A.I. (1997): *Temperature dependence of UV line average Doppler shifts in the quiet Sun* (Astrophys. J.), submitted

Cook, J.W., Cheng, C.C., Jacobs, V.L., Antiochos, S.K. (1989): *Effect of coronal elemental abundances on the radiative loss function* (Astrophys. J. **338**), 1176-1183

Corti, G., Poletto, G., Romoli, M., Michels, J., Kohl, J., Noci, G. (1997): *Physical parameters in plume and interplume regions from UVCS observations* (Proceedings of "The Fifth SOHO Workshop", ESA-SP), in press

Cranmer, S.R., Field, G.B., Kohl, L. (1997): *The impact of UVCS/SOHO observations on models of ion-cyclotron resonance heating of the solar corona* (Proceedings of the Cambridge Cool Stars Meeting), in press

David , Gabriel, A.H., Bely-Debau, F. (1997): *Temperature structure in coronal holes* (Proceedings of "The Fifth SOHO Workshop", ESA-SP), in press

De Forest, C.E., Scherrer, P.H., Bogart, R.S., Bush, R.I., Duvall, T., Hoeksema, J.T., Kosovichev, A.G., Schou, J., Tarbell, T., Title, A. (1997): *Recent results from MDI: A status report* (Proceedings of "The Fifth SOHO Workshop", ESA-SP), in press

De Forest, C.E., Gurman, J.B. (1997): *Quasi-periodic compressive waves in polar plumes* (Proceedings of "The Fifth SOHO Workshop", ESA-SP), in press

Esser, R., Holzer, T.E., Leer, E. (1987): *Drawing inferences about solar wind acceleration from coronal minor ion observations* (J. Geophys. Res. **92**), 13377-13389

Esser, R., Habbal, S.R., Coles, W.A., Hollweg, J.V. (1997): *Hot protons in the inner corona and their effect on the flow properties of the solar wind* (J. Geophys. Res. **102**), 7063-7074

Fahr, H.J., Shizgal, B. (1983): *Modern exospheric theories and their observational relevance* (Rev. Geophys., Space Phys. **21**) 75-124

Feldman, W.C., Asbridge, J.R., Bame, S.J., Gary, S.P., Montgomery, M.D. (1976a): *Electron parameter correlations in high-speed streams and heat flux instabilities* (J. Geophys. Res. **81**), 2377-2382

Feldman, W.C., Asbridge, J.R., Bame, S.J., Gary, S.P., Montgomery, M.D., Zink, S.M. (1976b): *Evidence for the regulation of solar wind heat flux at 1 AU* (J. Geophys. Res. **81**), 5207-5211

Fisher, R., Guhathakurta, M. (1995): *Physical properties of polar coronal rays and holes as observed with the Spartan 201-01 coronagraph* (Astrophys. J. **447**), L139-L142

Fleck, B., Domingo, V., Poland, A. (1995): *The SOHO Mission* (Solar Physics **126**, No. 1-2, Kluwer Academic Publishers, Dordrecht, The Netherlands)

Geiss, J., Gloeckler, G., von Steiger, R. (1995): *Origin of the solar wind from composition data* (Space Science Reviews **72**) 49-60

Grall, R.R., Coles, W.A., Klinglesmith, M.T., Breen, A.R., Williams, P.J.S., Markkanen, J., Esser, R. (1996): *Rapid acceleration of the polar solar wind* (Letters of Nature **379**), 429-432

Habbal, S.R., Woo, R., Fineschi, S., Kohl, J.L. (1997): *Transition from fast to slow solar wind in the inner corona* (Proceedings of "The Fifth SOHO Workshop", ESA-SP), in press

Hammer, R. (1982a): *Energy balance of stellar corona: I. Methods and examples* (Astrophys. J. **259**), 767-778

Hammer, R. (1982b): *Energy balance of stellar corona: II. Effect of coronal heating* (Astrophys. J. **259**), 779-791

Hartle, R.E., Sturrock, P.A. (1968): *Two-fluid model of the solar wind* (Astrophys. J. **151**), 1155-1170

Henoux, J.C., Somov, B.V. (1992): *First ionization potential fractionation* (Proceedings of the First SOHO Workshop, **ESA-SP 348**), 325-330

Hoeksema, J.T. (1995): *The large-scale structure of the heliospheric current sheet during the Ulysses epoch* (Space Science Reviews **72**), 137-148

Hollweg, J.V. (1986): *Transition region, corona, and solar wind in coronal holes* (J. Geophys. Res. **91**), 4111-4125

Holzer, T.E., Axford, W.I. (1970): *The theory of stellar winds and related flows* (Ann. Rev. Astron. Astrophys. **8**), 31-61

Hu, Y.Q., Esser, R., Habbal, S.R. (1997): *A fast solar wind model with anisotropic proton temperature* (J. Geophys. Res. **102**), 14,661-14,676

Hundhausen, A.J. (1972): *Coronal Expansion and Solar Wind* (Springer-Verlag, New York)

Isenberg, P.A. (1991): *The solar wind* (in *Geomagnetism*, Vol. 4, edited by J.A. Jacobs, Academic Press), 1-85

Jacques, S.A. (1978): *Solar wind models with Alfvén waves* (Astrophys. J. **226**), 632-649

Kohl, J.L., Noci, G., Anonucci, E., Tondello, G., Huber, M.C.E., Gardner, L.D., Nicolosi, P., Strachan, L., Fineschi, S., Raymond, J.C., Romoli, M., Spadaro, D., Panasyuk, A., Siegmund, O.H.W., Benna, C., Ciaravella, A., Cranmer, S.R., Giordano, S., Karovska, M., Martin, R., Michels, J., Modigliani, A., Naletto,

G., Pernechele, C., Poletto, G., Smith, P.L. (1997): *First results from the SOHO ultraviolett coronagraph spectrometer* (Solar Physics), in press

Kopp, R.A., Holzer, T.E. (1976): *Dynamics of coronal hole regions, I. Steady polytropic flows with multiple critical points* (Solar Phys. **49**), 43-56

Koutchmy, S., Zirker, J.B., Steinolfson, R.S., Zhugzda, J.D. (1991): *Coronal activity* (in *Solar Interior and Atmosphere*, A.N. Cox, W.C. Livingston, M.S. Matthews (Eds.), The University of Arizona Press, Tucson, USA), 1044-1086

Lamy, P., Llebaria, A., Koutchmy, S., Reynet, P., Molodensky, M., Howard, R., Schwenn, R., Simnett, G. (1997): *Characterization of Polar Plumes from LASCO-C2 Images in Early 1996* (Proceedings of "The Fifth SOHO Workshop", ESA-SP), in press

Leer, E., Axford, W.I. (1972): *A two fluid model with anisotropic proton temperature* (Solar Phys. **23**), 238-250

Leer, E., Holzer, T.E. (1980): *Energy addition to the solar wind* (J. Geophys. Res. **85**), 4681-4688

Leer, E., Holzer, T.E., Flå, T. (1982): *Acceleration of the solar wind* (Space Sci. Rev. **33**), 161-200

Li, X., Esser, R., Habbal, S.R. (1997): *Influence of heavy ions on the high-speed solar wind* (J. Geophys. Res. **102**), 17,419-17,432

Mariska, J.T. (1992): *The Solar Transition Region* (Cambridge Astrophysics Series **23**, Cambridge, University Press)

Marsch, E., Goertz, C.K., Richter, K. (1982): *Wave heating and acceleration of solar wind ions by cyclotron resonance* (J. Geophys. Res. **87**), 5030-5044

Marsch, E., Richter, A.K. (1984): *Helios observational constraints on solar wind expansion* (J. Geophys. Res. **89**), 6599-6612

Marsch, E. (1991a): *Kinetic physics of the solar wind plasma* (in *Physics of the Inner Heliosphere*, Vol. II, Eds. R. Schwenn und E. Marsch, Springer-Verlag, Heidelberg), 45-133

Marsch, E. (1991b): *MHD Turbulence in the Solar Wind* (in *Physics of the Inner Heliosphere*, Vol. II, Eds. R. Schwenn and E. Marsch, Springer Verlag, Heidelberg), 159-241

Marsch, E. (1994): *Theoretical models for the solar wind* (Adv. Space Res. **14**), (4)103-(4)121

Marsch, E., von Steiger, R., Bochsler, P. (1995): *Element fractionation by diffusion in the solar chromosphere* (Astron. Astrophys. **301**), 261-276

Marsch, E., Tu, C.-Y. (1997): *The effects of high-frequency Alfvén waves on coronal heating and solar wind acceleration* (Astron. Astrophys. **319**), L17-L20

Marsch, E., Tu, C.-Y. (1997): *Solar wind and chromospheric network* (Solar Physics **0**), 1-20

Marsden, R.G. (1985): *The Sun and the Heliosphere in Three Dimensions* (D. Reidel Publishing Company, Dordrecht, The Netherlands)

Marsden, R.G. (1995): *The High Latitude Heliosphere* (Space Science Reviews **72**, Kluwer Academic Publishers, Dordrecht, The Netherlands)

McComas, D.J., Bame, S.J., Feldman, W.C., Gosling, J.T., Phillips, J.L. (1992): *Solar wind halo electrons from 1-4 AU* (Geophys. Res. Lett. **19**), 1291-1294

McKenzie, J.F. (1991): *Solar corona and wind* (J. Geomag. Geolectr. **43**, Suppl.), 45-58

McKenzie, J.F., Banaszkiewicz, M., Axford, W.I. (1995): *Acceleration of the high speed solar wind* (Astron. Astrophys. **303**), L45-L48

McKenzie, J.F., Banaszkiewicz, M., Axford, W.I. (1997): *The fast solar wind* (Geophys. Res. Lett.), in press

Melrose, D.B. (1986): *Instabilities in Space and Laboratory Plasmas* (Cambridge University Press, Cambridge)

Meyer, J.-P. (1996): *Abundance anomalies in the solar outer atmosphere* (ASP Conf. Series, Astr. Soc. of the Pacific (S.S. Holt and G. Sonneborn, Eds.))

Neugebauer, M. (1981): *Observations of solar wind helium* (Fundam. Cosmic Phys. **7**), 131-199

Owocki, S.P., Scudder, J.D. (1983): *The effect of a non-Maxwellian electron distribution on oxygen and iron ionization balances in the solar corona* (Astrophys. J. **270**), 758-768

Parker, E.N. (1958): *Dynamics of the interplanetary gas and magnetic fields* (Astrophys. J. **128**), 664-684

Parker, E.N. (1965): *Dynamical theory of the solar wind* (Space Sci. Rev. **4**), 666-708

Peter, H. (1996a): *Velocity-dependent fractionation in the solar chromosphere* (Astron. Astrophys. **312**), L37-L40

Peter, H. (1996b): *Mehrflüssigkeitsmodelle der unteren Sonnenatmosphäre und Schlussfolgerungen für den Sonnenwind* (Ph.thesis, Univ. Göttingen, Report MPAE-W-100-97-06)

Peter, H., Marsch, E. (1997): *Ionization layer of hydrogen in the solar chromosphere and the solar wind mass flux* (Proceedings of "The Fifth SOHO Workshop", ESA-SP), in press

Pilipp, W.G., Miggenrieder, H., Montgomery, M.D., Mühlhäuser, K.-H., Rosenbauer, H., Schwenn, R. (1987): *Characteristics of electron velocity distribution functions in the solar wind derived from the Helios plasma experiment* (J. Geophys. Res. **92**), 1075-1092

Phillips, J.L., Feldman, W.C., Gosling, J.T., Scime, E.E. (1995): *Solar wind plasma electron parameters based on aligned observations by ICE and Ulysses* (Adv. Space Res. **16**), (9)95-(9)100

Riley, P., Sonett, C.P., Balogh, A., Forsyth, R.J., Scime, E.E., Feldman, W.C. (1995): *Alfvénic fluctuations in the solar wind: A case study using Ulysses measurements* (Space Sci. Rev. **72**), 197-200

Rosner, R., Tucker, W.H., Vaiana, G.S. (1978): *Dynamics of the quiescent solar corona* (Astrophys. J. **220**), 643-665

Schwenn, R. (1983): *Direct correlations between coronal transients and interplanetary disturbances* (Space Sci. Rev. **34**), 85-99

Schwenn, R. (1990): *Large scale structure of the interplanetary medium* (in *Physics of the Inner Heliosphere*, Vol. 1, R. Schwenn and E. Marsch (eds.), Springer Verlag, Berlin, Heidelberg, New York), 99-181

Schwenn, R., Marsch, E., Eds. (1990, 1991): *Physics of the Inner Heliosphere* (Springer Verlag, Berlin, Heidelberg, New York)

Schwenn, R., Inhester, B., Plunkett, S.P., Epple, A., Podlipnik, B., Bedford, D.K., Bout, M.V., Brueckner, G.E., Dere, K.P., Eyles, C.J., Howard, R.A., Koomen, M.J., Korendyke, C.M., Lamy, P.L., Llebaria, A., Michels, D.J., Moses, J.D., Moulton, N.E., Paswaters, S.E., Simnett, G.M., Socker, D.G., St. Cyr, O.C.,

Tappin, S.J., Wang, D. (1997): *First view of the extended green line emission corona at solar activity minimum using the LASCO-C1 coronograph on SOHO* (Solar Physics), in press

Scime, E.E., Bame, S.J., Feldman, W.C., Gary, S.P., Phillips, J.L. (1994): *Regulation of the solar wind electron heat flux from 1 to 5 AU: Ulysses observations* (J. Geophys. Res. **99**), 23,401-23,410

Scudder, J.D., Olbert, S. (1979): *A theory of local and global processes which affect solar wind electrons, 1. The origin of typical 1 AU velocity distribution functions – Steady state theory* (J. Geophys. Res. **84**), 2755-2772

Scudder, J.D., Olbert, S. (1979): *A theory of local and global processes which affect solar wind electrons, 2. Experimental support* (J. Geophys. Res. **84**), 6603-6620

Scudder, J.D. (1992): *Why all stars should posses circumstellar temperature inversions* (The Astrophys. J. **398**), 319-349

Seely, J.F., Feldman, U., Schühle, U., Wilhelm, K., Curdt, W., Lemaire, P. (1997): *Turbulent velocities and ion temperatures in the solar corona obtained from sumer line widths* (The Astrophys. J. **484**), L87-L90

Sheeley, N.R.,Jr., Wang, Y.-M., Hawley, S.H., Brueckner, G.E., Dere, K.P., Howard, R.A., Koomen, M.J., Korendyke, C.M., Michels, D.J., Paswaters, S.E., Socker, D.G., St. Cyr, O.C., Wang, D., Lamy, P.L., Llebaria, A., Schwenn, R., Simnett, G.M., Plunkett, S., Biesecker, D.A. (1997): *Measurements of flow speeds in the corona between 2 and 30 $R_\odot$* (Astrophys. J. **484**), 472-478

Smith, E.J., Neugebauer, M., Balogh, A., Bame, S.J., Lepping, R.P., Tsurutani, B.T. (1995): *Ulysses observations of latitude gradients in the heliospheric magnetic field: Radial component and variances* (Space Science Review 72), 165-170

Thieme, K.M., Marsch, E., Schwenn, R. (1990): *Spatial structures in high-speed streams as signatures of fine structures in coronal holes* (Annales Geophysicae 8, 11), 713-724

Tsurutani, B.T., Ho, C.M., Arballo, J.K., Lakhina, G.S., Glassmeier, K.-H., Neubauer, F.M. (1997): *Nonlinear electromagnetic waves and spherical arc-polarized waves in space plasmas* (Plasma Phys. Control Fusion **39**), A237-A250

Tu, C.-Y., Pu, Z.-Y., Wei, F.-S. (1984): *The power spectrum of interplanetary Alfvénic fluctuations: Derivation of the governing equations and its solution* (J. Geophys. Res. **89**), 9695-9702

Tu, C.-Y. (1987): *A solar wind model with the power spectrum of Alfvénic fluctuations* (Solar Phys. **109**), 149-186

Tu, C.-Y. (1988): *The damping of interplanetary Alfvénic fluctuations and the heating of the solar wind* (J. Geophys. Res. **93**), 7-20

Tu, C.-Y., Marsch, E. (1994): *On the nature of compressive fluctuations in the solar wind* (J. Geophys. Res. **99**), 21,481-21,509

Tu, C.-Y., Marsch, E. (1995a): *MHD structures, waves and turbulence in the solar wind: Observations and theories* (Space Science Reviews **73**), 1-210

Tu, C.-Y., Marsch, E. (1995b): *Comment on "Evolution of energy-containing turbulent eddies in the solar wind" by W.H. Matthaeus, S. Oughton, D.H. Pontius Jr., and Y. Zhou* (J. Geophys. Res. **100**), 12,323-12,328

Tu, C.-Y., Marsch, E. (1997): *Two-fluid model for heating of the solar corona and acceleration of the solar wind by high-frequency Alfvén waves* (Solar Physics **171**), 363-391

Ulmschneider, R., Priest, E.R., Rosner, R., Eds. (1991): *Mechanisms of Chromo-spheric and Coronal Heating* (Springer Verlag Heidelberg)

von Steiger, R., Geiss, J. (1989): *Supply of fractionated gases to the corona* (Astron. Astrophys. **225**), 222-238

von Steiger, R. Geiss, J., Gloeckler, G., Galvin, A.B. (1995): *Kinetic properties of heavy ions in the solar wind from SWICS/Ulysses* (Space Science Review **72**), 71-76

Wilhelm, K., Lemaire, P., Curdt, W., Schühle, U., Marsch, E., Poland, A.I., Jordan, S.D., Thomas, R.J., Hassler, D.M., Huber, M.C.E., Vial, J.-C., Kühne, M., Siegmund, O.H.W., Gabriel, A., Timothy, J.G., Grewing, M., Feldman, U., Hollandt, J., Brekke, P. (1997a): *First results of the SUMER telescope and spectrometer on SOHO* (Solar Physics **170**), 75

Wilhelm, K., Marsch, E., Dwivedi, B., Hassler, M., Lemaire, P., Huber, M.C.E. (1997b) *The corona above polar coronal holes as seen by SUMER on SOHO* (Astrophysical J.), submitted

Winterhalter, D., Gosling, J.T., Habbal, S.R., Kurth, W.S., Neugebauer, M. (Eds.) (1995) *Solar Wind Eight* (AIP Conference Proceedings **382**)

Withbroe, G.L. (1988): *The temperature structure, mass and energy flow in the corona and inner solar wind* (Astrophys. J. **325**), 442-467

Woch, J., Axford, W.I., Mall, U., Wilken, B., Livi, S., Geiss, J., Gloeckler, G., Forsyth, R.J. (1997): *SWICS/Ulysses observations: The three-dimensional struc-ture of the heliosphere during solar minimum* (Geophys. Res. Lett.), in press

Woo, R., Habbal, S.R. (1997): *Extension of coronal structure into interplanetary space* (Geophys. Res. Lett. **24**), 1159-1162

Plasma Diagnostics for the Solar Atmosphere

Helen E. Mason

Department of Applied Mathematics and Theoretical Physics, Silver Street, CAMBRIDGE CB3 9EW, UK

Abstract. Superb observations are now being obtained with the ESA/NASA Solar Heliospheric Observatory (SOHO). In these lectures, I discuss the spectroscopic diagnostic techniques used to analyse ultraviolet and X-ray spectra. Intensities of the spectral emission lines can be used to determine the electron density and temperature structure, element abundances and dynamic nature of different features in the solar transition region and corona. To ensure that these techniques are accurate it is necessary to model all the important atomic processes with the best available atomic data calculations. I discuss the analysis of SOHO spectra and look towards other astrophysical UV and X-ray observations.

1 Introduction

In studying the solar atmosphere we are searching for the fundamental mechanisms for heating the corona and accelerating the solar wind. We want to understand the role of magnetic fields and the dynamic nature of the atmosphere. We are confronted by motions and inhomogeneities on a smaller and smaller scale as the spatial, spectral and temporal resolution of our instruments increases. Recent images and movies obtained with SOHO and YOHKOH illustrate well the inhomogeneous and dynamic nature of the solar atmosphere.

2 SOHO Mission

The SOHO mission, which was launched in December 1995, includes four EUV/UV instruments for studying the solar atmosphere. The Coronal Diagnostic Spectrometer (CDS) (160-800 Å) is being used to study the relationship of the transition region and coronal emission - the CDS is designed to determine the physical parameters of the solar atmosphere. The Solar Ultraviolet Measurement of Emitted Radiation (SUMER) instrument (500-1600 Å) can resolve the fine structures in the chromosphere and transition region, flows and wave motions. The EUV Imaging Telescope (EIT) has provided superb images and movies of the solar atmosphere in four spectral bands centered on the lines He II 304 Å, Fe IX 171 and 174 Å, Fe XII 195 Å and Fe XV 284 Å, covering the temperature range $4 \ 10^4$ K to $3 \ 10^6$ K. The Ultraviolet Coronagraph Spectrometer (UVCS) can observe the solar corona from its base out to 10 R_0. The UVCS spectroscopic diagnostics are based on the measurement of the intensities and spectral line profiles of resonantly and Thompson scattered H I Lyman-α, the collisionally excited and resonantly scattered Li-like resonance lines of O VI (1032 and 1037 Å); it can also detect other coronal lines such as Si XII (499 and 521 Å) and Fe XII (1242 Å). The Large Angle Spectroscopic Coronagraph (LASCO) provides images of the solar corona from 1.1 to 30 R_0 in several wavelength bands in the visible 5300-6000 Å. UVCS and LASCO have provided fascinating observations of streamers, coronal mass ejections and many other large scale dynamic phenomena.

The SOHO observations have dispelled any notion we might have had of a *quiet* Sun. Even during solar minimum, the solar atmosphere is far from tranquil; it is characterised by transient and dynamic features. In the CDS observations, the O V emission at 10^5 K show EUV brightenings in intensity which Harrison (1997a) has christened "blinkers". The O V intensities increase by a factor of 2-4 on timescales of 5 mins, over an area of 20-40". These blinkers seem to be located at network junctions, with a frequency such that approximately 3000 are in progress on the solar surface at any one time. Harrison estimates the energy content to be approximately 10^4 erg/cm^2/sec, which is more than two orders of magnitude less than the radiation losses of the upper chromosphere and corona. They could be related to nano-flares (Parker, 1988) but their energy content is only 10^{-6} of an average flare.

3   Coronal Diagnostic Spectrometer

In these lectures, I shall concentrate mainly on the results obtained with the CDS, since EIT and SUMER are covered by other lecturers. The CDS instrument consists of a Wolter II grazing incidence telescope feeding two spectrometers. The Normal Incidence Spectrometer (NIS) covers two wavelength ranges (307-379 and 513-633 Å) using a microchannel plate/CCD combination detector. The Grazing Incidence Spectrometer (GIS) has four microchannel plate plus spiral anode detectors which cover the wavelength ranges 151-221, 256-341, 393-492 and 659-785 Å. The CDS principal investigator is Dr R.A. Harrison at the Rutherford Appleton Laboratory (RAL). The home page at RAL (http://solg2.bnsc.rl.ac.uk) provides an excellent overview of the CDS observations in the form of an atlas.

The CDS allows us to take slices through the solar atmosphere at different temperatures between 2 10^4 - 3 10^6 K. Simultaneous rasters, up to 4' x 4' in size, can be obtained at different wavelengths, with a high spatial resolution (approximately 3") (see Figure 1). The NIS channels cover ions from a wide temperature range (2 10^4 - 6 10^6 K), including the low temperature emission from He I at 584Å (2 10^4 K); transition region emission from O V 630 Å (2.5 10^5 K); the Mg IX 368 Å line at coronal temperatures (10^6 K) and the Fe XVI 335 and 361 Å lines observed in active regions (2 10^6 K).

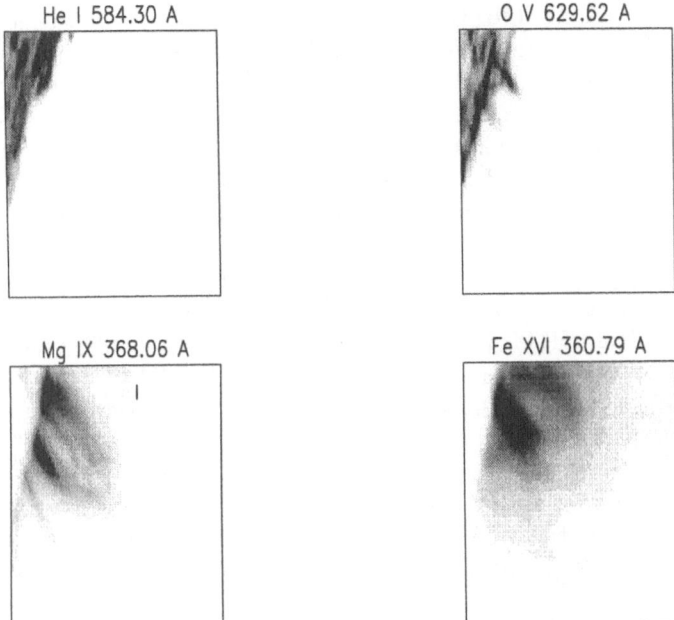

SOHO/CDS NIS Raster, 6−Sep−1996 06:24:33

LARGEBP2 −− Active Region on Limb −− s4594r00.fits
Center = (1011",−334"), Size = 244"x240"

Fig. 1. CDS-NIS rasters of an active region, showing emission from He I, O V, Mg IX and Fe XVI. Taken from Mason and Pike (1997)

Carrington rotation maps have been compiled from the CDS synoptic observations taken in He I, O V and Mg IX (Harrison, 1997b). These show the evolution of coronal holes and active regions spanning several solar rotations.

A key feature of the CDS instrument is the capability of obtaining simultaneous rasters and spectral information. This allows the determination of temperature and density estimates from diagnostic line ratios; emission measure distributions from sequences of ions and elemental abundance variations; flow patterns from Doppler shifts. Details of the CDS instrument and science objectives can be found in Harrison *et al* (1995). First results from CDS are published in Harrison *et al* (1997). Spectroscopic diagnostics using CDS are discussed in Mason *et al* (1997).

The spectroscopic diagnostic techniques depend on an accurate knowledge of the relevant atomic physics processes, including radiative decay rates, electron collisional excitation, ionisation and recombination. The electron excitation rates required for the analysis of CDS spectra were the subject of an Atomic Data Assessment Workshop (sponsored by CDS and SUMER) held in March 1992 (Lang, 1994). These atomic data are being continuously improved. The Iron Project, an international collaboration of atomic physicists, aims to calculate new electron excitation rates for all the iron ions (Hummer *et al*, 1993). Several atomic databases and analysis packages have been developed and adapted for application to the SOHO data. The Atomic Data and Analysis Structure (ADAS) package has been adapted for analysis of CDS observations (Summers *et al*, 1996, Brookes *et al*, 1997). ADAS solves the generalised collisional-radiative model for equilibrium and non-equilibrium conditions. The CHIANTI atomic database (Dere *et al*, 1997) and associated software are being extensively used to analyse SOHO observations. In Figure 2 a sample active region spectrum from NIS is shown to be in excellent agreement with a CHIANTI synthetic spectrum.

A useful review of spectroscopic diagnostics for the solar atmosphere was written by Gabriel and Mason (1982). A more recent review for solar and stellar plasmas in the VUV (100-2000 Å) was published by Mason and Monsignori Fossi (1994). A detailed comparison of the CHIANTI synthetic spectrum with the SERTS active region spectrum was carried out by Young *et al* (1997). An initial survey of CDS active region observations has been published by Fludra *et al* (1997).

4   The Atomic Processes

We are concerned with atomic processes in a hot ($T > 2\ 10^4$ K) and low electron number density ($N_e < 10^{12}$ cm^{-3}) plasma. We assume that the spectral lines are optically thin, which is valid for the outer atmosphere of the Sun and stars. The notation used is such that Fe XIV (or Fe^{+13}) is the element iron with thirteen electrons stripped off. Taking Fe XIV as an example, the ground configuration $3s^23p$ has two levels - $^2P_{0.5}$ and $^2P_{1.5}$ - the transition between these two levels gives rise to the coronal green line at 5303 Å, which is observed by LASCO. The transitions between the excited configurations $3s3p^2$ and $3s^23p$ are at around 300-400 Å (seen by CDS-NIS) and those between $3s^23d$ and $3s^23p$ are at shorter wavelengths around 200 Å.

4.1   Emission Lines

An ion in an excited state can spontaneously emit radiation: the process is called *bound-bound emission*:

$$X_j^{+m} \implies X_i^{+m} + h\nu \qquad (1)$$

where an atom X of charge state m, in a bound state j emits a photon of energy $\Delta E_{i,j}$ $(= h\nu_{i,j} = \frac{hc}{\lambda_{i,j}})$ (ergs) to arrive at a lower energy state i. The emissivity (power per unit volume) is given by:

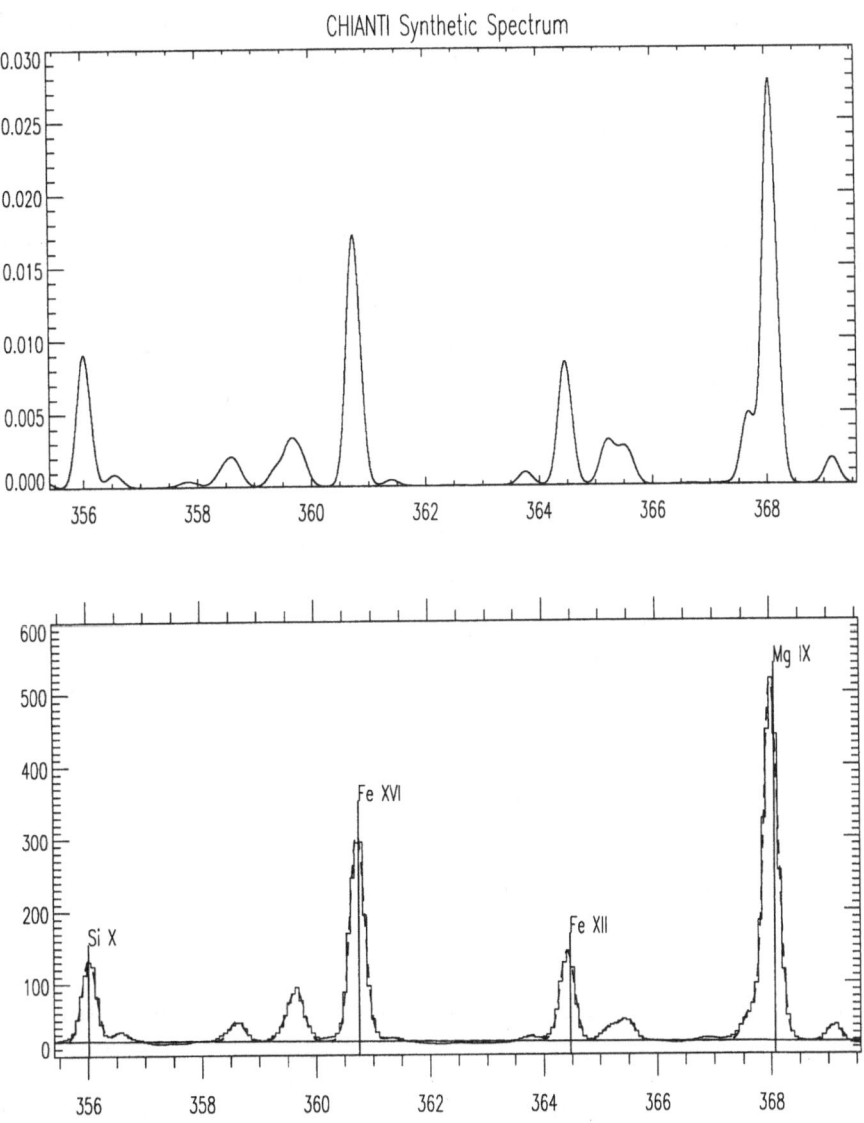

Fig. 2. A sample CDS-NIS active region spectrum (355-370 Å) is shown in the lower plot - the dashed line is a multi-Gaussian line fit with background. The upper plot is a standard active region spectrum from CHIANTI. Taken from Mason and Pike (1997).

$$P(\lambda_{i,j}) = N_j(X^{+m})A_{j,i}\Delta E_{i,j} \qquad erg\ cm^{-3}sec^{-1} \qquad (2)$$

where $A_{j,i}$ (sec^{-1}) is the Einstein spontaneous emission coefficient; $N_j(X^{+m})$ is the number density of the level j of the ion, which can be expressed as:

$$N_j(X^{+m}) = \frac{N_j(X^{+m})}{N(X^{+m})}\frac{N(X^{+m})}{N(X)}\frac{N(X)}{N(H)}\frac{N(H)}{N_e}N_e \qquad cm^{-3} \qquad (3)$$

where $\frac{Nj(X^{+m})}{N(X^{+m})}$ is the population of level j relative to the total number density, $N(X^{+m})$, of ion X^{+m} and is a function of the electron temperature and density; $\frac{N(X^{+m})}{N(X)}$ is the ionization ratio of ion X^{+m} which is predominantly a function of temperature (Arnaud and Rothenflug, 1985, Arnaud and Raymond, 1992); $\frac{N(X)}{N(H)}$ is the element abundance relative to Hydrogen which varies in different astrophysical plasmas and also in different solar features; $\frac{N(H)}{N_e}$ is the Hydrogen abundance relative to electron density (~ 0.8); N_e is the electron number density (cm^{-3}).

The intensity of a spectral line from a column of plasma, volume of emission V, of cross section area **A**, is then given by:

$$I(\lambda_{i,j}) = \frac{1}{4\pi\mathbf{A}}\int_V P(\lambda_{i,j})\,dV \qquad erg\ sec^{-1}st^{-1}cm^{-2} \qquad (4)$$

In low density plasmas the collisional excitation processes are generally faster than ionization and recombination timescales, therefore the collisional excitation is dominant over ionization and recombination in producing excited states. The number density population of level j must be calculated by solving the statistical equilibrium equations for a number of low lying levels and including all the important collisional and radiative excitation and de-excitation mechanisms.

4.2   The Coronal Model Approximation

There are some cases (e.g.: optically allowed, electric dipole transitions) for which the assumption is made that the population of the upper level of transition j occurs mainly via collisional excitation from the ground state g and that the radiative decay overwhelms any other depopulation process. This is called the *coronal model* approximation. The statistical equilibrium equations can be solved as a two-level system for each transition:

$$N_g(X^{+m})N_e C^e_{g,j} = N_j A_{j,g} \qquad (5)$$

where the electron collisional excitation rate coefficient is $C^e_{g,j}$ (cm^3 sec^{-1}). If $A_{j,g} \gg N_e C^e_{g,j}$, then the population of the upper level j is negligible in comparison with that of the ground level g, so $[\frac{N_g(X^{+m})}{N(X^{+m})} \simeq 1]$. For a typical

UV transition in Fe XIV, at coronal densities and temperatures, we find that $A_{j,g}$ is approximately 10^{10} sec^{-1}, whereas $N_e C_{g,j}^e$ is around unity.

The emitted power in the line for the coronal model (per unit volume and time) is given by:

$$P(\lambda_{g,j}) = \frac{N(X^{+m})}{N(X)} \frac{N(X)}{N(H)} \frac{N(H)}{N_e} C_{g,j}^e \Delta E_{g,j} N_e^2 \qquad (6)$$

and the spectral line intensity is proportional to N_e^2

4.3   Electron Collisional Excitation and De-excitation

The collisional excitation rate coefficient for a Maxwellian electron velocity distribution with a temperature $T_e(K)$, is given by:

$$C_{i,j}^e = \frac{8.63 \times 10^{-6}}{T_e^{1/2}} \frac{\Upsilon_{i,j}(T_e)}{\omega_i} \, exp\left(\frac{-\Delta E_{i,j}}{kT_e}\right) \qquad (7)$$

where ω_i is the statistical weight of level i, k is the Boltzmann constant and $\Upsilon_{i,j}$ is the thermally-averaged collision strength:

$$\Upsilon_{i,j}(T_e) = \int_0^\infty \Omega_{i,j} \, exp\left(-\frac{E_j}{kT_e}\right) d\left(\frac{E_j}{kT_e}\right) \qquad (8)$$

Here the collision strength (Ω) is a symmetric, dimensionless quantity, which is related to the electron excitation cross-section; E_j is the energy of the scattered electron relative to the final energy state of the ion. The electron de-excitation rates are obtained by the application of the principle of the detailed balance.

The solution of the electron-ion scattering problem is complex and takes a great deal of computing resources. The accuracy of a particular calculation depends on two main factors. The first is the representation which is used for the target wavefunctions, the second is the type of scattering approximation chosen. The target must take account of configuration interaction and allow for intermediate coupling for the higher stages of ionization. The main approximations used for electron-ion scattering are *Distorted Wave* (DW), *Coulomb Bethe* (CBe) and the more elaborate *Close-Coupling* (CC) approximation. The DW approximation neglects the coupling of the channels (target + scattering electron). Since the scattering electron sees a central field potential, the DW approximation is only valid for systems which are a few times ionized. For high partial wave values of the incoming electron, the CBe approximation is valid, when it is assumed that the scattering electron does not penetrate the target. In the CC approximation, the scattering electron sees individual target electrons, the channels are coupled and a set of integro-differential equations are solved. The CC approximation is the most accurate (better than 5%) but it is also the most expensive in terms of computing resources. New CC calculations are being calculated as part of the Iron Project (http://www.am.qub.ac.uk/projects/iron) for the many coronal ions, including Fe XIV (Storey *et al*, 1996) and Fe XII (Binello *et al*, 1997). Fe XII is a particularly useful ion for SOHO, since different spectral lines from Fe XII are observed by CDS, SUMER, UVCS and EIT.

4.4   Proton Collisional Excitation and De-excitation

The proton collisional excitation and de-excitation rates should also be considered. They become comparable with electron collisional processes only for transitions where $\Delta E_{i,j} \ll kT_e$. This happens, for instance, for transitions between fine structure levels at high temperatures, for example the Fe XIV transition in the ground configuration : $3s^2 3p$ ($^2P_{0.5}$ - $^2P_{1.5}$).

4.5   Ionization Balance

The degree of ionization of an element is obtained by equating the ionization and recombination rates that relate successive stage of ionization.

$$N^{+m}(q_{col} + q_{au} + q_{ct}) = N^{+m+1}(\alpha_r + \alpha_d + \alpha_{ct}) \tag{9}$$

The dominant processes in optically thin plasmas are *collisional ionization - direct ionization from the inner and outer shells* (q_{col}) and excitation followed by *autoionization* (q_{au}); *radiative recombination* (α_r) and *dielectronic recombination* (α_d); charge transfer (α_{ct}, q_{ct}).
The process of radiative recombination is:

$$X_n^{+m+1} + e \Rightarrow X_{n'}^{+m} + h\nu \tag{10}$$

with n and n' the quantum state of the ions. The inverse process is photoionization, which is a dominant process for many low density astrophysical plasmas, but not for the transition region and corona of stars.
The process of dielectronic recombination is:

$$X_n^{+m+1} + e \Rightarrow (X_{n'}^{+m})^{**} \Rightarrow X_{n''}^{+m} + h\nu \tag{11}$$

an electron is captured by an ion with charge $+m + 1$ to form a doubly excited state $()^{**}$ of an ion X with charge $+m$. This ion can then either *autoionize* back again or undergo a spontaneous radiative transition of the inner excited electron to a state below the first ionization limit. Dielectronic recombination is the dominant recombination mechanism at high temperatures, at least a factor of 20 higher than radiative recombination.
The inverse process to *dielectronic recombination* is *autoionization*:

$$(X_n^{+m})^{**} \Rightarrow X_{n'}^{+m+1} + e' \tag{12}$$

The symbol $()^{**}$ again indicates a doubly excited state of the ion X_n^{+m}.
Direct collisional ionization from the inner and outer shells of the ground configuration can be expressed as:

$$X_n^{+m} + e \Rightarrow X_{n'}^{+m+1} + e + e'. \tag{13}$$

5   CHIANTI

The CHIANTI atomic database and analysis software (Dere *et al*, 1997) is the result of an international collaboration (USA/Italy/UK) to provide a comprehensive dataset for ions of astrophysical interest seen in the wavelength region $\lambda > 50$ Å. There are three basic sets of files - the observed and theoretical energy levels, the wavelengths and radiative data for each transition in an ion and the electron collisional data for each transition. The latter are fitted using the Burgess and Tully (1992) approach. CHIANTI also includes a set of IDL routines to solve the statistical equilibrium equations; to provide theoretical line intensity ratios; to calculate CDS synthetic spectra and to determine the differential emission measure. CHIANTI can be accessed via the web site at NRL (http://wwwsolar.nrl.navy.mil/chianti.html), Cambridge or Arcetri. It has been incorporated into the CDS science analysis software.

6   Plasma Diagnostics

6.1   Emission Measure Analysis

The emitted power for an allowed transition in the simple coronal model, can be re-expressed in the form:

$$P(\lambda_{g,j}) = \frac{N(X)}{N(H)} \frac{N(H)}{N_e} G(T, \lambda_{g,j}) \Delta E_{g,j} N_e^2 \qquad erg \ cm^{-3} sec^{-1} \qquad (14)$$

Generally the function $G(T, \lambda_{g,j})$, called the *contribution function*, is strongly peaked in temperature. It is given by:

$$G(T, \lambda_{g,j}) = \frac{N(X^{+m})}{N(X)} C_{g,j}^e \qquad (15)$$

The emission measure is a measure of the amount of plasma emitting the observed radiation as a function of temperature and it is the primary characteristic which any theoretical model should predict. Following Pottasch (1963), one can assume that each spectral line is emitted from a uniform volume V, over a temperature range ΔT (usually about 0.3 on a $Log_{10}T_e$ scale) around the temperature T_{max} corresponding to the peak value of its contribution function, $G(T_{max})$. Taking a constant value for the contribution function $< G(T) >$ over ΔT, the intensity can be expressed as:

$$I_{\lambda_{g,j}} = \frac{1}{4\pi A} \frac{N(X)}{N(H)} \frac{N(H)}{N_e} \Delta E_{g,j} < EM >< G(T) > \qquad (16)$$

where $< G(T) > \Delta T = \int_{\Delta T} G(T, \lambda_{g,j}) dT$. and the volume emission measure $< EM >$ is defined as:

$$< EM > = \int_V N_e^2 dV \qquad (17)$$

This method, with some refinements, can be used to derive the isothermal $< EM >$ as a function of temperature using lines emitted at different values of T_{max}. The average electron density $< N_e^2 >=< EM > /V$ can be crudely deduced assuming that the spectral line is emitted over a homogeneous, isothermal volume estimated from images at that temperature.

More sophisticated inversion methods involve re-writing the intensity flux as:

$$I(\lambda_{ij}) = \frac{1}{4\pi} \frac{N(H)}{N_e} \frac{N(X)}{N(H)} \Delta E_{i,j} \int_T G(T)\phi(T)dT \; ergs \; cm^{-2} \; s^{-1} \; st^{-1} \qquad (18)$$

where the $\phi(T)$ is differential emission measure (DEM) function, for a plane parallel atmosphere. The DEM relates to the amount of material in the temperature interval T and $T + dT$ and the temperature gradient along the line of sight. The DEM gives information about the structure of the atmosphere, which can be related to the energy transfer mechanisms - eg conductive flux. Often, it is more convenient to express the DEM on a $Log_{10}T_e$ scale. To derive the DEM from the observations one must solve a set of integral equations. The problem, given a set of observed spectral intensities, the values of the abundances and the $G(T)$ functions, is to invert the system of integral equations. The inversion problem itself is not simple and requires some assumptions about the nature of the solution. A series of workshops was sponsored by CDS in 1990/91 to study differential emission measure techniques (Harrison and Thompson, 1992). It was found that most codes eventually gave consistent results, but that the DEM derived depends rather critically on the methods used to constrain the solution and also on the errors in the observed intensities and atomic data.

The differential emission measures derived by G. Del Zanna from the CDS-GIS spectra for a coronal hole, the quiet Sun and an active region are given in Figure 3. In the quiet Sun and active region spectrum, the Fe XV and Fe XVI lines are observed and can be used to determine the drop in the DEM at high temperatures. We note that the DEM falls rapidly in the coronal hole spectrum. This is constrained by the lack of any Fe XIII and Fe XIV lines. Bromage et al (1997) and Del Zanna and Bromage (1997) have recently carried out a careful analysis of the electron temperature and density structure for the coronal hole. Using the CHIANTI atomic database, their preliminary differential emission measure shows that the coronal hole had a maximum temperature of $8 \; 10^5$ K and an electron density of $2.5 \; 10^8$ cm^{-3} from a Si IX diagnostic line ratio [342/350]. This electron density is about a factor of two less than the value they derived for the surrounding quiet Sun.

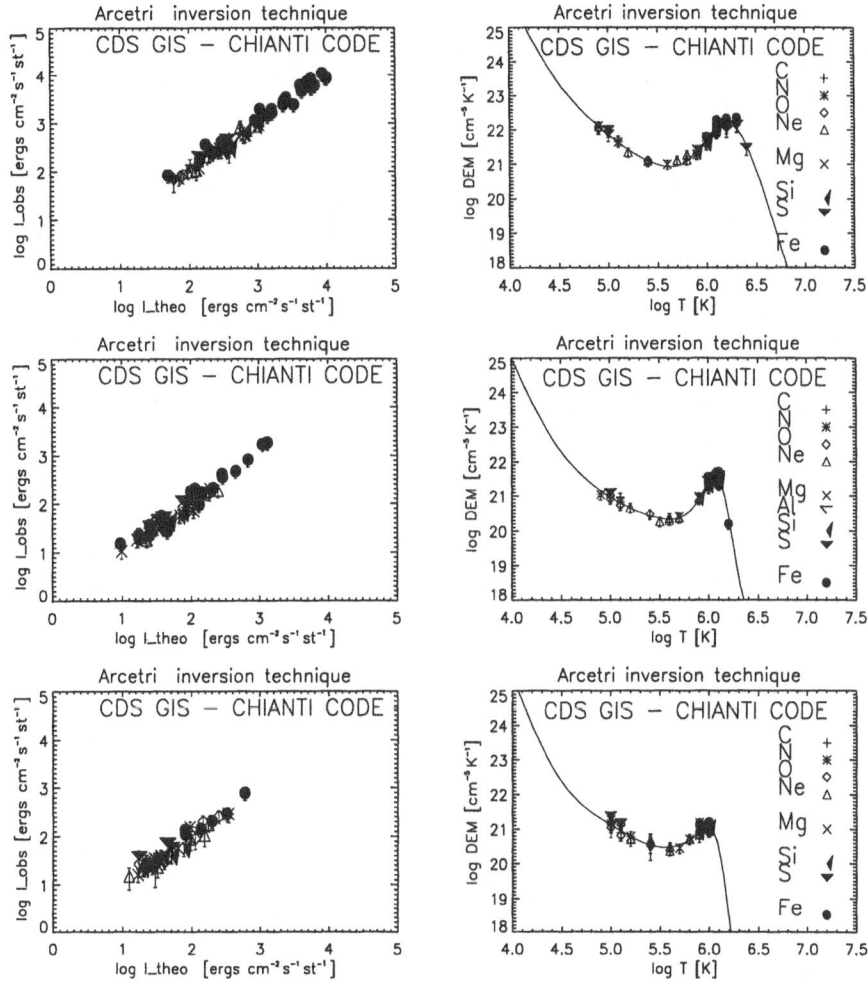

Fig. 3. The derived DEM's (right hand plots) and comparison of observed ($LogI_{obs}$) and theoretical ($LogI_{theo}$) intensities (left hand plots) for an active region, the quiet Sun and a coronal hole (top to bottom). Taken from Mason *et al* (1997).

6.2   Electron Density Diagnostics

The electron pressure ($N_e T_e$) is an important parameter in any theoretical model for the plasma. Experience from solar observations is that the plasma often exists in the form of unresolved filamentary structures, even down to the best spatial resolution which has yet been obtained. At one extreme is the solar transition region, where only a very small fraction of the observed emitting volume is actually filled with plasma. The situation is a little better in the corona, but even here the filamentary nature of the emission is evident.

The determination of electron density from spectral line intensity ratios from the same ion, makes no assumption about the size of the emitting volume or the element abundance value. It therefore provides a powerful and important diagnostic for the solar plasma.

Spectral lines may be grouped into different categories according to the behaviour of the upper level population: allowed lines collisionally excited from the ground level; forbidden or intersystem lines originating from a *metastable level* - m; allowed lines excited from a metastable level.

For simplicity we again consider a two level model. For forbidden and intersystem transitions the radiative decay rate is generally very small ($A_{m,g} \simeq 10^0 - 10^2$ sec^{-1}) collisional de-excitation then becomes an important depopulating mechanism ($A_{m,g} \simeq N_e C^e_{m,g}$) and may even be the dominant mechanism; moreover the population of the metastable level becomes comparable with the population of the ground level and we have:

$$N_m = \frac{N_g N_e C^e_{g,m}}{N_e C^e_{m,g} + A_{m,g}} \tag{19}$$

For small electron densities, $N_e \to 0$, $A_{m,g} \gg N_e C^e_{m,g}$, then the intensity has the same dependence on the density as an allowed line (I^A):

$$I_{m,g} \simeq N_e^2 \tag{20}$$

For very large values of electron density, $N_e \to \infty$, the collisional depopulation dominates, $N_e C^e_{m,g} >> A_{m,g}$; the metastable level is in Boltzman equilibrium with the ground level:

$$\frac{N_m}{N_g} = \frac{C^e_{g,m}}{C^e_{m,g}} = \frac{\omega_m}{\omega_g} \, exp \left(\frac{-\Delta E_{g,m}}{kT} \right) \tag{21}$$

The line intensity has the form:

$$I_{m,g} \simeq N_e \tag{22}$$

For intermediate values of electron density $A_{m,g} \simeq N_e C^e_{m,g}$; the population of the metastable level is significant and the intensity varies as:

$$I_{m,g} \simeq N_e^\beta \qquad 1 < \beta < 2 \tag{23}$$

The intensity ratio of a forbidden to an allowed transition (I^F/I^A) for different spectral lines from the same ion can be used to determine an average electron density for the emitting volume. This value is independent of the elemental abundance, ionisation ratio and any assumptions about the size of that volume. Alternatively the intensity ratio of two forbidden lines from the same ion could be used ($I^F/I^{F'}$). A good example of this type of diagnostic ratio occurs in the transition region ion O IV. The $2s^2 2p \, ^2P_J$ - $2s2p^2 \, ^4P'_J$, which fall at around 1400 Å have been used to determine the electron density in the transition region. The value obtained from these line ratios was around

10^{11} cm^{-3}, at least two orders of magnitude higher than the estimated from emission measure analyses - indicating that the transition region is composed of filamentary structures (Dere *et al*, 1987), well below the spatial resolution of solar instruments.

If the population of metastable level (m) is comparable with the ground level (g), then other excited levels (k) can be populated from this metastable level as well as from the ground level and the dependence of the intensity on electron density becomes:

$$I_{k,m} \simeq N_e^\beta \qquad 2 < \beta < 3 \qquad (24)$$

Fe XIV is formed at coronal temperatures (1.8 10^6 K), it is a good example of the latter diagnostic case. The metastable level is actually the upper level in the ground configuration ($3s^2 3p\ ^2P_{1.5}$). The transition probability to the ground level is only 60 sec^{-1}. For low electron densities (10^8 cm^{-3}), almost all the population for Fe XIV is in the ground level, but as electron density is increased (10^{10} cm^{-3}), the upper level begins to have a significant population aswell. The spectral line at 334.2 Å is excited from the ground level($3s^2 3p\ ^2P_{0.5}$), whereas the line at 353.8 Å is excited from the upper level in the ground configuration ($3s^2 3p\ ^2P_{1.5}$). The intensity ratio of these two lines (Figure 4) varies with electron density, reflecting the level population changes in the ground configuration. For active region observations, the ratio of [353.8/334.2] gives electron densities varying from around 10^9 cm^{-3} in the quiet part to around $10^{10.5}$ cm^{-3} in the heart of the active region (Mason et al, 1997). Care must be taken when using this Fe XIV diagnostic ratio in very hot (flaring) regions, since the 353.8 Å line is blended with an Ar XVI line (formed at $\log_{10}$T(K)$\sim$ 6.6) in the CDS data.

The CDS wavelength range is designed to cover several electron density diagnostic lines. In particular, the 307-379 Å band of NIS contains several excellent density diagnostics, covering mainly coronal temperatures ($1 - 2\ 10^6$ K), see Mason *et al* (1997).

Si X, formed at around 1.3 10^6 K, provides a useful electron density diagnostic with lines at 347.4 Å and 356.0 Å, which are prominent feature of most NIS spectra. Si X has a ground configuration $2s^2 2p$ with levels $^2P_{0.5}$ and $^2P_{1.5}$ analagous to Fe XIV. The line at 347.4 Å is excited from the ground level whereas the line at 356.0 Å is predominantly excited from the upper level in the ground configuration. The ratio [356.0/347.4] reflects the change in population between the levels in the ground configuration. Around active regions the lines have similar intensities, suggesting densities close to 10^9 cm^{-3}, however in the hotter parts of active regions the 356.0 Å line can be found to be around two times more intense than 347.4 Å line. This is close to the high density limit of Si X, implying a density $> 10^{10}$cm^{-3}. By contrast, above the limb we can find the Si X ratio approaching the low density limit. Mason and Pike (1997) have been studying the temperature and density structure of a limb active region. From a comparison of Mg IX

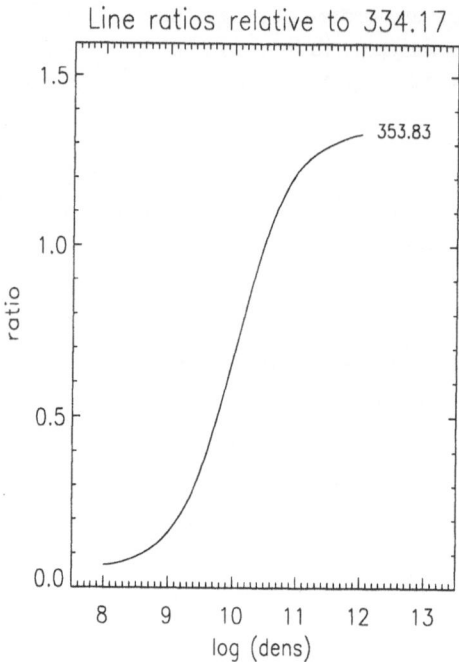

Fig. 4. The Fe XIV density sensitive [334.2/353.8] intensity ratio. Taken from Mason *et al* (1997).

(10^6 K) and Fe XVI (3 10^6 K), they find that the hottest part of the active region is low lying. They also obtain an electron density map from the Si X intensity ratio [356/347] (Figure 5). It is evident that the most dense part of the active region is also low lying. It seems that all the action in this active region is taking place low down, rather than in the high coronal loops. This could possibly be associated with emergent flux.

There are also some useful density diagnostics within the GIS bandpasses. For example, the Fe XIII intensity ratio [203.8/202.0] is seen to vary from around 1 to 4, suggesting densities of between 10^9 and 10^{11} cm^{-3}. This diagnostic ratio has been studied in both solar and stellar spectra.

6.3   Electron Temperature Diagnostics

The simplest but crudest method of deducing the plasma temperature is to assume ionization equilibrium. Since many ions are formed over the same range of temperature, line ratios can be plotted as a function of temperature of the emitting isothermal plasma.

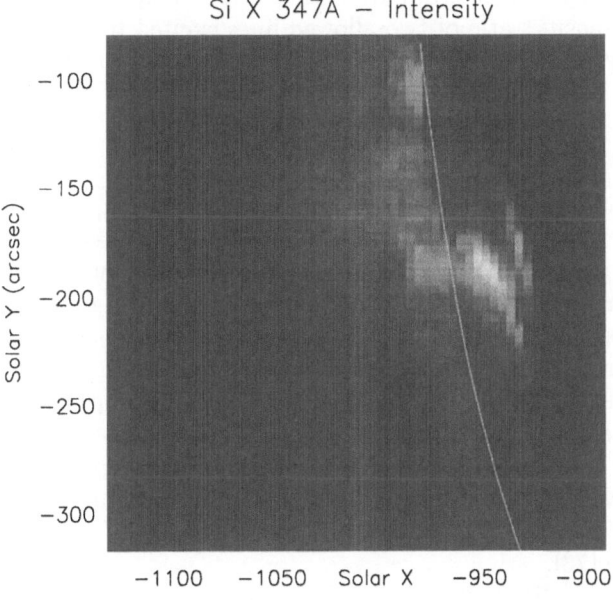

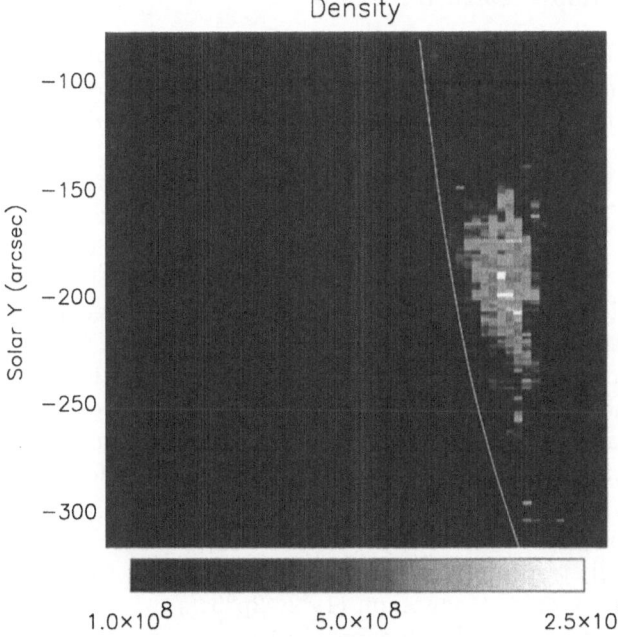

Fig. 5. Si X intensity (upper raster) and an electron density map derived from Si X line ratios (lower raster, with the scale in cm^{-3} underneath). Taken from Mason and Pike (1997).

A more accurate determination of electron temperature can be obtained from the intensity ratio of two allowed lines excited from the ground level g but with significantly different excitation energy. The ratio is given by:

$$\frac{I_{g,j}}{I_{g,k}} = \frac{\Delta E_{g,j} \Upsilon_{g,j}}{\Delta E_{g,k} \Upsilon_{g,k}} \, exp \left(\frac{\Delta E_{g,k} - \Delta E_{g,j}}{kT_e} \right) = F(T_e) \qquad (25)$$

The ratio is sensitive to the change in electron temperature if $(\Delta E_{g,k} - \Delta E_{g,j}) \geq kT_e$ assuming that the lines are emitted by the same isothermal volume with the same electron density. Such spectral lines are far apart in wavelength and it may be necessary to use lines from different instruments. This gives rise to major uncertainties in the derived temperature due to the relative calibration of the intensities.

It was suggested that the Li-like ions [(2s-2p)/(2s-3p)] would be useful as temperature diagnostics. However their $G(T)$ functions are very broad and it is not clear that the lines are formed in the same temperature region. Lines with narrower $G(T)$ funtions would be better. Good temperature diagnostics in EUV-UV spectral range are the ratios of O V [629/172], O V [1218/629], O VI [1032/173].

6.4   Abundance Determination

There has been a great deal of discussion and controversy recently about variations in the elemental abundances in the solar atmosphere (Meyer, 1993). It appears that the element abundances in the solar wind differ from those in the photosphere. In fact, it seems that the coronal abundances vary in different solar features. The behaviour depends on the value for the First Ionization Potential (FIP) - which is the energy required to ionize the neutral atom. The ions with FIP greater than 10 eV appear to behave differently from those with FIP's less than 10 eV. It is believed that this could reflect the ionization and acceleration processes for the solar corona, low down in the solar atmosphere.

One approach to determining element abundances is to use the detailed shape of the $< EM >$ or DEM distribution for ions from the same element and apply an iterative procedure to normalising the curves for different elements. Another procedure is to use the intensity ratios for individual spectral lines which have very similar $G(T)$ functions (Widing and Feldman, 1992), but different FIP's. For example neutral Neon, with a FIP of 21.6 eV, has a closed shell structure $(1s^2 2s^2 2p^6)$ which is difficult to ionize, whereas it is much easier to ionize Magnesium $(1s^2 2s^2 2p^6 3s^2)$, with a FIP of 7.6 eV.

Short lived explosive brightenings in active regions are frequently observed by CDS in transition region emission (Mason et al, 1997). Besides O V, strong increases are seen in other transition region lines such as Mg V and Mg VI. Young and Mason (1997) have carried out a detailed investigation of bright transition region emission in a newly emerging flux region. In particular they determine the relative elemental abundance of Mg to Ne using specific spectral line ratios. They study the emission lines from two brightenings. The first - A, appears to be associated with a low lying loop and is very intense in many transition region ions. It displays a low Mg/Ne abundance of 0.26 ± 0.05 (consistent with the photospheric value) and a high electron density of 10^{11} cm^{-3}, from the O IV line ratios [626/553]. The other brightening - B, lies at the base of larger coronal loops, which extend high in the active region. These loops show a larger Mg/Ne abundance of 2.4 ± 0.3 and a lower electron density of $10^{9.2}$ cm^{-3}, from Mg VII line ratios, [319/368]. These two regions exhibit almost an order of magnitude difference in the Mg/Ne abundance ratio, but are only 1' apart. The plasma with photospheric abundance and a high electron density seems to come from the low lying loop structures, which could be emerging flux.

6.5   Spectral Line Profiles

Line shifts and broadenings give information about the dynamic nature of the solar and stellar atmospheres. The transition region spectra from the solar atmosphere are characterised by broadened line profiles. The nature of this excess broadening puts constraints on possible heating processes. Systematic redshifts in transition region lines have been observed in both solar spectra and stellar spectra of late type stars. On the Sun, outflows of coronal material have been correlated with coronal holes, a probable source of the solar wind. The excess broadening of coronal lines above the limb provides information on wave propagation in the solar wind. Initial results from CDS on spectral line broadening and shifts are discussed in Brekke *et al* (1997a, 1997b).

If the optically thin spectral lines can be fitted with gaussian profiles, the intensity per unit wavelength I_λ is defined as:

$$I_\lambda = \frac{I}{\sqrt{2\pi}\sigma}exp[-(\lambda - \lambda_0)^2/2\sigma^2] \qquad (26)$$

where $I = \int I_\lambda d\lambda$ is the integrated intensity and σ is the gaussian width given by:

$$\sigma^2 = \frac{\lambda^2}{2c^2}(\frac{2kT}{M} + \xi^2) + \sigma_I^2 \qquad (27)$$

for a Maxwellian velocity distribution of temperature T, usually assumed to be the temperature corresponding to peak abundance of the ion. Here M is the ion mass, σ_I is the gaussian instrumental width and ξ is the most probable non-thermal velocity *ntv*. Dere and Mason (1993) studied the ntv's for a range of ions from transition region to coronal temperatures, using the HRTS observations.

Macrospicules often occur in the polar coronal holes. Fortuitously, one was observed by CDS on the southern limb, during a synoptic scan in He I

(584 Å, 2 10^4K), O V (629 Å, 2.5 10^5K) and Mg IX (368 Å, 10^6K). This data was analysed by Pike and Harrison (1997). The macrospicule had an extent of more than 40" (31,000 km) and a width of 20" (14,000 km) and was bright in Mg IX as well as O V and He I. They found strong evidence for acceleration and outflows along the western footpoint, which, depending on the orientation of the macrospicule, could indicate velocities up to 580 km/sec, consistent with solar wind flows.

6.6 Summary of CDS Spectroscopic Diagnostics

CDS is a complex instrument to operate and observing sequences must be designed to address specific scientific objectives. Because of telemetry limitations, there has to be a trade off between spectral and spatial coverage. In order to obtain quantative results for physical parameters, it is necessary to develop specialist analysis software, incorporating large atomic databases, such as CHIANTI. It is also important to provide high accuracy atomic data for the solar ions, such as that provided by the Iron Project. Accurate intensity and wavelength calibrations are required (Landi *et al*, 1997).

Many new and exciting results from SOHO are being published - including papers from the Fifth SOHO Workshop held in Oslo and in the special issues of Solar Physics. The study of transition region brightenings in the quiet Sun and active regions will help us to fathom the heating mechanism for the solar atmosphere; the study of coronal holes, macrospicules etc. will contribute towards our knowledge of solar wind sources. The determination of electron density and temperature distributions for coronal holes, the quiet Sun and active regions will enable us to define the important energy transfer processes. In these lectures, I have concentrated on work which uses the spectroscopic diagnostic potential of CDS. There are many different topics being studied with CDS data - such as prominences, coronal mass ejections, streamers.

7   Solar Flares and X-ray Observations

SOHO is a quiet Sun mission, launched at solar minimum, but other solar satellites, such as the Solar Maximim Mission, SMM, and YOHKOH were designed to study the active phase of the Sun. During solar flares, the plasma temperature can exceed 10^7 K and the X-ray emission becomes very intense. Iron can be stripped of almost all its electrons. Take for example Fe XXI which is abundant at 10^7 K and contains density diagnostic line ratios. The ground configuration is $2s^2 2p^2$ and the transition ($^3P_1$ - $^3P_0$), which lies at 1354.1 Å, has been extensively studied in solar spectra, for example with the SMM-UVSP instrument (Mason *et al*, 1986). The transitions between the ground configuration and the first excited configuration $2s^2 2p^2$ - $2s2p^3$ give rise to lines at around 100 Å. These flare lines were only recorded once in solar spectra by OSO-5, but are now being extensively studied in stellar spectra from EUVE. The transitions $2s^2 2p^2$ - $2s^2 2p3d$ fall in the X-ray wavelength range at around 12 Å. These were also observed by OSO-5, and later by SMM. A more recent analysis of the $2s^2 2p^2$ - $2s^2 2p4d$ transitions (Phillips *et al*, 1996) indicates a very high electron density for solar flares, possibly in excess of $10^{12} \, \text{cm}^{-3}$. The inner shell transitions (when a 1s electron is excited) from Fe XXI and other highly ionized iron ions fall at around 2 Å. There are a many useful diagnostic satellite lines, which I do not have time to discuss in these lectures. They have been covered in other reviews (cf Gabriel and Mason, 1982) The next few years should provide some exciting X-ray observations of solar (SOLARB) and astrophysical (AXAF and XMM) plasmas.

8   Acknowledgements

I acknowledge the financial support of PPARC. SOHO is an ESA/NASA project. I thank all the members of the CDS team, but in particular Richard Harrison and Dave Pike for their support and encouragement.

References

Arnaud, M and Raymond, J. (1992): *Astrophys. J.*, **398**, 394

Arnaud, M. and Rothenflug, R. (1985): *Astron. Astrophys. Suppl. Ser.*, **60**, 425

Binello, A.M., Mason, H.E. and Storey, P.J. (1997): *Astron. Astrophys.*, in press.

Brekke, P., Kjeldseth-Moe, O., Brynildsen, N., Maltby, P., Haugan, S.V.H., Harrison, R.A., Thompson, W.T. and Pike, C.D. (1997a): *Solar Physics*, **170**, 163

Brekke, P., Kjeldseth-Moe, O. and Harrison, R.A. (1997b): *Solar Physics*, in press

Bromage, B.J.I., Del Zanna, G., DeForest, C., Thompson, B. and Clegg, J.R. (1997): *The Fifth SOHO Workshop, Oslo, June 1997*, ESA SP, in press

Brookes, D.H., Fischbacher, G., Fludra, A., Harrison, R.A., Innes, D., Landi, E., Landini, M., Lang, J., Lanzafamme, A., Loch, S., McWhirter, R.W.P., Summers, H.P. and Thompson, W.T. (1997a,b): *The Fifth SOHO Workshop, Oslo, June 1997*, ESA SP, in press

Burgess, A. and Tully, J.A. (1992): *Astron. Astrophys.*, **254**, 436

Del Zanna, G. and Bromage, B.J.I. (1997): *The Fifth SOHO Workshop, Oslo, June 1997*, ESA SP, in press

Dere, K.P., Bartoe, J.-D., Brueckner, G.E., Cook, J.W. and Socker, D.G. (1987) *Science*, **238**, 1267

Dere, K.P., Landi, E., Mason, H.E., Monsignori Fossi, B.C. and Young, P.R. (1997): *Astron. Astrophys. Suppl. Ser.*, in press

Dere, K.P. and Mason, H.E. (1993): *Sol. Phys.*, **144**, 217

Fludra, A., Brekke, P., Harrison, R.A., Mason, H.E., Pike, C.D., Thompson, W.T., and Young, P.R. (1997): *Solar Physics*, in press

Gabriel, A.H. and Mason, H.E. (1982), *Solar Physics*, **Ch 10** in *Applied Atomic Collision Physics, Vol 1*, eds. H.S.W. Massey, B. Benderson and E.W. McDaniel (Academic Press)

Harrison, R.A. (1997a): *Solar Physics*, in press

Harrison, R.A. (1997b): *The Fifth SOHO Workshop, Oslo, June 1997*, ESA SP, in press

Harrison, R.A., Fludra, A., Pike, C.D., Payne, J., Thompson, W.T., Poland, A.I., Breeveld, E.R., Breeveld, A.A., Culhane, J.L., Kjeldseth-Moe, O. and Aschenbach, B. (1997) *Solar Physics*, **170**, 123

Harrison, R.A., Sawyer, E.C., Carter, M.K., Cruise, A.M., Culter, R.M., Fludra, A., Hayes, R.W., Kent, B.J., Lang, J., Parker, D.J., Payne, J., Pike, C.D., Peskett, S.C., Richards, A.G., Culhane, J.L., Norman, K., Breeveld, A.A., Breeveld, E.R., Al Janabi, K.F., McCalden, A.J., Parkinson, J.H., Self, D.G., Thomas, P.D., Poland, A.I., Thomas, R.J., Thompson, W.T., Kjeldseth-Moe, O., Brekke, P., Karud, J., Maltby, P., Aschenbach, B., Bauninger, H., Kuhne, M., Hollandt, J., Siegmund, O.H.W., Huber, M.C.E., Gabriel, A.H., Mason, h.E. and Bromage, B.J.I. (1995): *Solar Physics*, **162**, 233

Harrison, R.A. and Thompson, A.M. (eds.) (1992): *RAL-91-092*

Hummer, D.G., Berrington, K.A., Eissner, W., Pradhhan, A.K., Saraph, H.E. and Tully, J.A. (1993): *Astron. Astrophys.*, **279**, 298

Landi, E., Landini, M., Pike, C.D. and Mason, H.E. (1997): *Solar Physics*, in press

Lang, J. (ed.) (1994): *A.D.N.D.T.*, **57**

Mason, H.E. and Monsignori Fossi, B.C. (1994): *The Astron. Astrophys. Rev.*, **6**, 123

Mason, H.E. and Pike, C.D. (1997): *The Tenth Cambridge Workshop on Cool Stars, Stellar Systems and the Sun*, held at Cambridge, USA July 15-19, 1997, in preparation

Mason, H.E., Shine, R.A., Gurman, J.B. and Harrison, R.A. (1986): *Astrophys. J.*, **309**, 435

Mason, H.E., Young, P.R., Pike, C.D., Harrison, R.A., Fludra, A., Bromage, B.J.I., and Del Zanna G. (1997): *Solar Physics*, **170**, 143

Meyer, J.-P. (1993): *Adv. Space Res.*, **13(9)**, 377

Parker, E.N. (1988): *Astrophys. J.*, **330**, 474

Phillips, K.J.H., Bhatia, A.K., Mason, H.E. and Zarro, D.M. (1996): *Astrophys. J.*, **466**, 549

Pike, C.D. and Harrison, R.A. (1997):*Solar Physics*, in press

Pottasch, S.R. (1963): *Astrophys. J.*, **137**, 945

Storey, P.J. Mason, H.E. and Saraph, H.E. (1996): *Astron. Astrophys.*, **309**, 677

Summers H.P., Brooks D.H., Hammond T.J., Lanzafamme A.C. and Lang J. (1996): *RAL Technical report RAL-TR-96-017*, March 1996

Widing, K.G. and Feldman, U (1992): *Proc of the Solar Wind Seven Conference, Goslar, Germany, 16-20 Sept. 1991*, eds. E. Marsch, R. Schwenn, p405

Young P.R., Landi E., and Thomas R.J. (1997): *Astron. Astrophys.*, in press

Young P.R. and Mason, H.E. (1997): *Solar Physics*, in press

Radiative Transfer
and Radiation Hydrodynamics

Mats Carlsson

Institute of Theoretical Astrophysics, P.O. Box 1029 Blindern, N-0315 Oslo, Norway

Abstract. Radiation plays an important role, firstly in determining the structure of stars through the dominant role radiation plays in the energy balance (in some objects also in the momentum balance), secondly because we diagnose astrophysical plasma through the emitted electromagnetic radiation.

These lectures discuss the diagnostic use of optically thick spectral lines and continua with special emphasis on the Sun. Modern methods to solve the equations of radiative transfer and statistical equilibrium are outlined. We stress the importance of solving self-consistently for the coupling between radiation and hydrodynamics for understanding the dynamic outer atmosphere of the Sun and review some results from such radiation hydrodynamics simulations.

1 Introduction

To infer the physical conditions of stars and other astrophysical objects we are almost exclusively restricted to an analysis of the photons that escape from the medium. Before the photons reach us they have interacted with the matter so that the information about local conditions in the atmosphere has been convolved both in space and in time. The diagnostic deciphering of the information content in the radiation is made complicated by the fact that most plasmas both emit and absorb photons at the same time. In addition, photons can travel large distances thus coupling distant parts of the plasma. It is therefore necessary to study the processes of emission, absorption and radiative transfer in order to make proper inferences from observations of radiation.

Radiation also plays an important role in determining the structure of stars through the role of radiation in the energy balance and for some objects also in the momentum balance. Velocity fields in the stellar atmosphere Doppler shift the narrow absorption profiles of spectral lines thus affecting the coupling between the radiation and the local thermodynamic properties of the plasma. Radiative losses damp the temperature increase in the compression phase of a hydrodynamic disturbance. There is thus a strong coupling between the hydrodynamics and the radiation in a dynamic atmosphere and we have to treat this coupling self-consistently to gain an understanding of dynamic phenomena in stellar atmospheres.

The outline of these lectures is as follows: In Section 2 we discuss basic concepts and the basic equation of radiative transfer, in Section 3 we discuss the concept of height of formation in the context of the diagnostic use of spectral information. In the absence of Local Thermodynamic Equilibrium (LTE) the condition of statistical equilibrium can be used to determine the population densities. These equations are set up in Section 4, a general method of solving non-linear equations is outlined in Section 5 and the particular application to the equations of statistical equilibrium and radiative transfer is given in Section 6. Finally, the equations of radiative transfer and population rate equations are combined with the hydrodynamic equations into the equations of radiation hydrodynamics in Section 7 with special emphasis on recent results for the Solar chromosphere.

2   Equation of Radiative Transfer

We assume a general familiarity with basic radiative transfer theory and will just give an incomplete summary in order to define notation and highlight important aspects. For an excellent and more complete discussion we refer to Mihalas (1978). Furthermore, we restrict the discussion to the case of a one-dimensional plane-parallel atmosphere (all quantities are constant on a plane at a given height in the atmosphere).

Along a ray we look at how the monochromatic specific intensity $I_\nu(s)$ (normally only called the intensity) changes:

$$dI_\nu(s) = [\eta_\nu(s) - \chi_\nu(s)I_\nu(s)]\, ds \tag{1}$$

where $\eta_\nu(s)$ is called the emissivity, $\chi_\nu(s)$ the opacity (sometimes called extinction or absorption) and ds is the geometrical distance along the ray. Note that the number of photons removed from the ray is proportional to the opacity but also to the intensity itself. The direction dependence enters mainly through the dependence of ds on direction for a given height difference. Both η_ν and χ_ν are often direction *independent* in the absence of velocity fields. However, the opacity can be very direction dependent in the presence of velocity fields.

If we introduce the geometrical height, z, along the normal increasing outwards, the angle between the ray and the normal of the atmosphere, θ, and the directional cosine, $\mu = \cos\theta$, we get one standard form of the equation of radiative transfer:

$$\mu\frac{dI_{\nu\mu}(z)}{dz} = \eta_\nu(z) - \chi_\nu(z)I_{\nu\mu}(z) \tag{2}$$

The right hand side give the source (η_ν) and sink ($\chi_\nu I_{\nu\mu}$) of photons. Since all quantities depend on height the positional coordinate (z) is implicit and normally dropped from the equations.

The emissivity tends to be roughly proportional to the opacity and it is therefore useful to use the ratio of the two instead of the emissivity itself. This ratio is called the source function, $S_\nu \equiv \eta_\nu/\chi_\nu$. Furthermore, it is useful to use an "optical distance" instead of the geometrical distance. For the radiative transfer a long geometrical distance with small opacity is equivalent to a short geometrical distance with large opacity. The useful quantity is therefore the product of the two. We define the optical depth, τ_ν, from

$$d\tau_\nu = -\chi_\nu dz \qquad (3)$$

Note that the optical depth increases downwards along the normal, the geometrical height has opposite direction and we therefore get a minus sign in the equation above.

With these definitions we get the one-dimensional equation of radiative transfer in its standard form:

$$\mu \frac{dI_{\nu\mu}}{d\tau_\nu} = I_{\nu\mu} - S_\nu \qquad (4)$$

Note that the source term (S_ν) is now negative and the sink term ($I_{\nu\mu}$) is positive. This is because the optical depth scale has a direction opposite of the normal. For a ray directed towards us (out of the atmosphere) we get a positive value of μ but a negative $d\tau_\nu$.

Since this is a first order ordinary differential equation we need one boundary condition. Normally we know, or can approximate, the intensity incident on each boundary. This means that the boundary condition is given at one boundary for all positive values of μ and at the other boundary for negative values of μ. This fact complicates the solution of the equation. If the atmosphere has an upper boundary but the lower boundary is the interior of the sun we call this a semi-infinite atmosphere and normally use the boundary conditions $I_\nu^-(0) = 0$, $I_\nu^+(\tau_{\max}) = S_\nu(\tau_{\max})$ where superscripts denote the sign of μ. We then get the following solution:

$$I_{\nu\mu}(\tau_\nu) = \begin{cases} \frac{1}{\mu} \int_{\tau_\nu}^{\infty} S_\nu(t) e^{-(t-\tau_\nu)/\mu} dt & \text{if } \mu > 0 \\ \frac{1}{-\mu} \int_{0}^{\tau_\nu} S_\nu(t) e^{-(\tau_\nu - t)/(-\mu)} dt & \text{if } \mu < 0 \end{cases} \qquad (5)$$

Note the use of $(-\mu)$ for $\mu < 0$ to get a positive quantity.

The interpretation of the solution is that we get a contribution to the outgoing intensity from a given depth, t, over a given small distance along the ray, dt/μ (where the factor $1/\mu$ accounts for the projection of the distance dt along the normal to the distance along the ray) which is equal to the source function there exponentially attenuated with the optical distance from t to the point where we are evaluating the intensity, τ_ν. This optical distance equals the optical distance along the normal $(t - \tau_\nu)$ projected to the direction of the ray (the factor $1/\mu$ in the exponential). These contributions are summed up (the integral) from all depths below the evaluation point. The interpretation of the solution for the incoming radiation is equivalent.

This solution is called the *formal solution* because the source function has to be known (in addition to the opacities giving the optical depth scale). In the solar outer atmosphere the source function depends on the intensity itself such that a direct integration as in (5) is not possible. We then have to use an iterative method to find the solution. In all such iterative methods we need to evaluate the formal solution in each iterative step. It is thus important to find efficient methods for the formal solution even when we do *not* know the source function a priori.

One standard method for the evaluation of the formal solution is the Feautrier method (Feautrier 1964). The method is efficient, stable, accurate and gives the intensity at all points along the ray in one step. For more accurate versions of the Feautrier method see Kunasz & Hummer (1974), Auer (1976) and Rybicki & Hummer (1991).

The opacity can be split into an absorption part and a scattering part. Absorption can consist of excitation of an atom by absorption of a photon followed by a collisional de-excitation. The net effect is then the destruction of a photon converting the energy to local thermal energy. If the de-excitation instead is radiative we end up with the photon just changing direction and no coupling to the local conditions. Such a process is called a scattering event. The distinction between absorption and scattering becomes more problematic if atoms with more than two levels are considered but the two-level case is still instructive. Using this distinction and assuming isotropic scattering we get an emissivity consisting of a thermal part and a scattering part:

$$\eta_\nu = \kappa_\nu B_\nu + \sigma_\nu J_\nu \qquad (6)$$

where κ_ν is absorption, σ_ν scattering, $\chi_\nu = \kappa_\nu + \sigma_\nu$, B_ν the Planck function and J_ν the mean intensity:

$$J_\nu(z) = \frac{1}{2} \int_{-1}^{1} I_{\nu\mu}(z)d\mu \qquad (7)$$

The source function can then be written as

$$S_\nu = \frac{\kappa_\nu}{\kappa_\nu + \sigma_\nu} B_\nu + \frac{\sigma_\nu}{\kappa_\nu + \sigma_\nu} J_\nu \qquad (8)$$

It is the scattering part that introduces a global coupling in the source function through the dependency on the mean intensity.

3   Height of Formation

3.1   Contribution Functions

For the diagnostic use it is of interest to find the depth of formation of the observed quantity. From equation (5) it is obvious that the intensity is formed over a range of depths. It is useful to define a *contribution function to the intensity*, C_I, with the property:

$$I_{\nu\mu}(0) = \int C_I(x)dx \qquad (9)$$

Note that two quantities enter into the definition of the contribution function; the contribution *to* a given observable quantity (in this case the intensity) *on* a given depth-scale (here denoted with x). The integration is performed over the whole atmosphere. Since many different integrands give the same integrated quantity we need additional constraints in our definition, see Magain (1986) for a discussion. Using the integrand from the formal solution is an appropriate choice. If we use monochromatic optical depth as the depth variable we get from equation (5):

$$C_I(\tau_\nu) = \frac{1}{\mu} S_\nu(\tau_\nu) e^{-\tau_\nu/\mu} \qquad (10)$$

The monochromatic optical depth-scale is not very useful when comparing the formation depth at different wavelengths. To transform the contribution function to another depth-scale we use:

$$C_I(x)dx = C_I(\tau_\nu)d\tau_\nu \qquad (11)$$

Some useful contribution functions are the contribution function to the intensity on a geometric height-scale (z), a standard optical depth-scale (τ_0, where χ_0 is the opacity at a standard wavelength, usually 500 nm) and the logarithm of the standard optical depth ($\lg\tau_0$):

$$C_I(z) = \frac{1}{\mu} S_\nu(\tau_\nu) e^{-\tau_\nu/\mu} \chi_\nu \qquad (12)$$

$$(13)$$

$$C_I(\lg\tau_0) = \frac{\ln 10}{\mu} S_\nu(\tau_\nu) e^{-\tau_\nu/\mu} \frac{\chi_\nu}{\chi_0} \tau_0 \qquad (14)$$

Note that these contribution functions give information on where the *intensity* is formed. If we are interested in where the *absorption* is formed we *cannot use the contribution function to the intensity* but need to consider the contribution function to the *relative absorption* defined as:

$$R_\nu \equiv \frac{I_c - I_{\nu\mu}}{I_c} \qquad (15)$$

where I_c is the continuum intensity. Note that R_ν is zero for the continuum and is negative for an emission line.

To deduce the contribution function for the relative absorption we need to write the transfer equation for the relative absorption, find the formal solution and identify the integrand as the contribution function; all steps analogous to how the contribution function to the intensity was found. See Magain (1986) for details.

The relative absorption is typically formed higher than the intensity. In extreme cases the absorption may be formed at a very different place than the intensity (*e.g.* telluric absorption lines where the intensity is formed in the Sun and the absorption in the Earth's atmosphere).

Typical contribution functions to the intensity on a geometric height-scale (note the specification of both the quantity (intensity) and the depth-scale) are shown in Fig. 1.

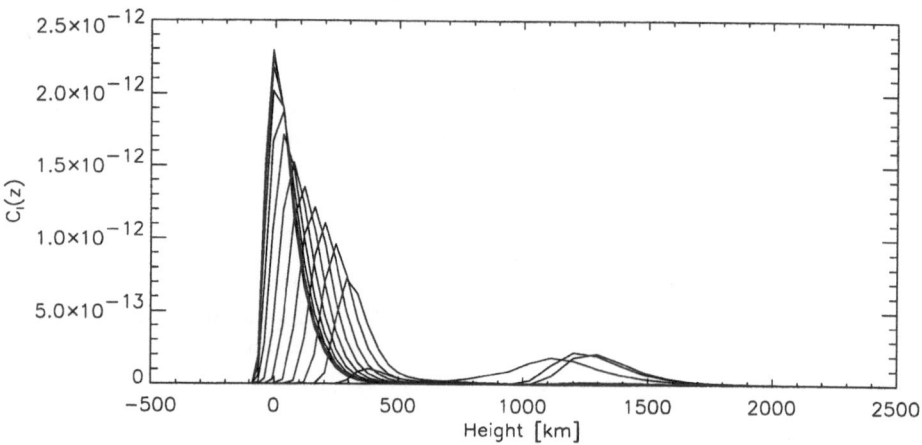

Fig. 1. Typical contribution functions to the intensity on a height scale at different frequencies from continuum (lowest height) to the line center for a Solar atmosphere. Note a typical width of about 200 km.

3.2   Response Functions

We are often interested in deducing the bulk velocity from the Doppler-shift of a line or the temperature change from the change of intensity. In these cases we should use the *response function* instead of a contribution function. The response function gives how a given observed quantity changes (*e.g.* Doppler-shift of line-center) when we change a physical parameter in the atmosphere (*e.g.* velocity) as a function of depth. There are thus three defining quantities in a response function. From this definition we get:

$$\Delta V = \int R_{\Delta V, v}(x) v(x) dx \qquad (16)$$

where ΔV is the Doppler-shift of line-center and $v(x)$ is the vertical velocity. The Doppler-shift response function to bulk-velocity is often similar to the contribution function to the relative absorption. The width of the response function makes it impossible to measure velocity variations that have shorter typical length-scales. Specifically, it is not possible to detect high frequency waves (with a wavelength shorter than the width of the response function) in a stellar atmosphere. Waves with a wavelength not much longer than the width of the response function will give Doppler-shift amplitudes that are substantially smaller than the velocity amplitude of the wave. This is illustrated in Fig. 2.

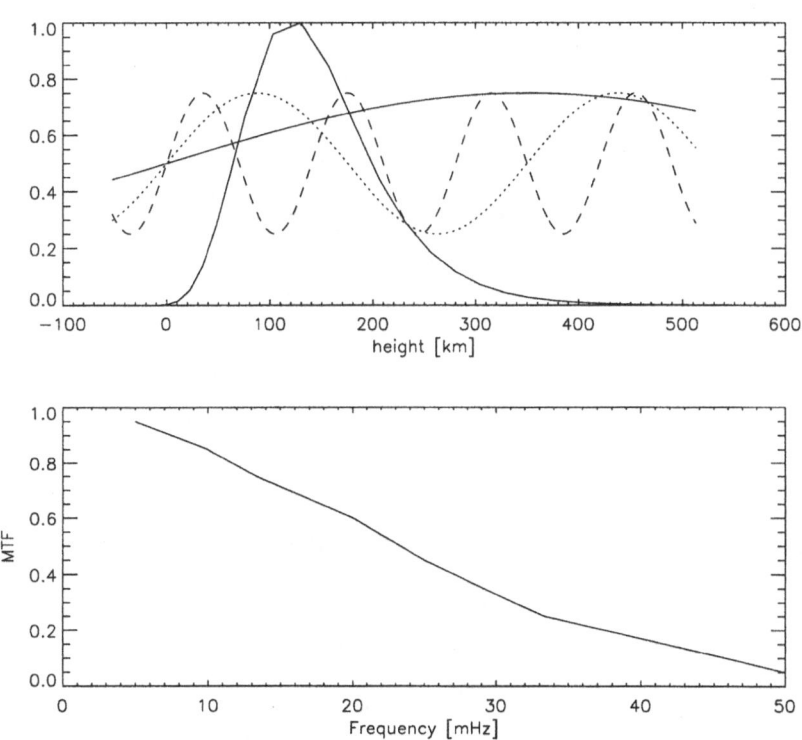

Fig. 2. *Upper panel:* Response function to line-center Doppler-shift for a typical spectral line in the solar atmosphere (*solid*) and three sinusoidal velocity fields with frequency 5 mHz (*solid*), 20 mHz (*dotted*) and 50 mHz (*dashed*). *Lower panel:* Amplitude of Doppler-shift compared to amplitude of the sinusoidal wave as function of frequency. Waves with a frequency of 23 mHz give a Doppler-shift of only half the amplitude of the wave due to the width of the response function.

4 Statistical Equilibrium

We here give a slightly simplified, phenomenological, picture of what is involved in radiative transfer problems in order to give a feel for what it is all about. For a more rigorous treatment the reader is again referred to standard textbooks like Mihalas (1978).

In order to calculate the specific intensity from the formal solution (Equation 5) we need to know the source function, S_ν and the monochromatic optical depth, τ_ν. The latter can be calculated if we know the monochromatic opacity, χ_ν. The opacity for a given process is equal to the cross-section for the process at a given frequency multiplied by the number density of the particle. The total opacity at a given frequency is the sum of the opacities of the processes that can take place at that frequency. To calculate the opacity we thus need cross-sections which, at least in principle, are available from atomic physics and we need population densities.

If we assume Local Thermodynamic Equilibrium (LTE) the population density of a given energy level of an atom is given by the Saha and Boltzmann equations. The ratio of the population density to the mass density is then given by the abundance of the element and the *local* values of the electron density and the temperature. With a given temperature structure as a function of height we can calculate the mass density and the electron density from the hydrostatic equilibrium and charge conservation equations (see *e.g.*, Gray 1992). If we in addition assume that the source function is given by the Planck function we can calculate all we need for the formal solution. Radiative transfer under the assumption of LTE is thus trivial in principle. In practice the calculation of a spectrum from a stellar atmosphere with given temperature structure is still non-trivial due to the need for atomic data (line strengths, broadening parameters) for millions of lines.

If we cannot assume LTE the problem becomes much more complex. The source functions will no longer be given by the Planck function and the population densities will not be given by the Saha and Boltzmann equations. In a *static* atmosphere we get the necessary extra equations from demanding that the population density of a given energy level shall be constant in time. We end up with the equations of statistical equilibrium:

$$n_i \sum_{j \neq i}^{n_l} P_{ij} - \sum_{j \neq i}^{n_l} n_j P_{ji} = 0 \tag{17}$$

where n_i is the population density of energy level i, P_{ij} is the transition probability of a transition from energy level i to energy level j and n_l is the number of energy levels considered. The first term thus gives the number of transitions per unit time from energy level i to all other energy levels and the second term gives the number of transitions into energy level i. Statistical equilibrium requires these rates to be equal. We have one such equation for each energy level considered. This set is linerarly dependent and we have to replace one of the equations with the condition of particle conservation,

$$\sum_{j=1}^{n_l} n_j = n_{tot} \tag{18}$$

where n_{tot} is the total number of particles of that element.

To solve this coupled set of equations we only need to know the transition probabilities since the total number of particles of a given element is given by the abundance and the mass density. The problem lies in the fact that the transition probabilities in general depend on the radiation field. We have

$$P_{ij} = R_{ij} + C_{ij} \tag{19}$$

with C_{ij} being the collisional transition probability (in principle known from atomic physics) and R_{ij} the radiative transition probability. For transitions between lower bound state i and upper bound state j we have:

$$\begin{aligned} R_{ji} &= A_{ji} + B_{ji}\bar{J}_{ij} \\ R_{ij} &= B_{ij}\bar{J}_{ij} \end{aligned} \tag{20}$$

with A_{ji}, B_{ij} and B_{ji} the Einstein coefficients for spontaneous emission, absorption and stimulated emission, respectively. All these are given by atomic physics. $\bar{J}_{ij}$ is the absorption profile ($\phi_{\nu\mu}$) weighted integrated mean intensity:

$$\bar{J}_{ij} = \frac{1}{2} \int_{-1}^{1} \int_{0}^{\infty} \phi_{\nu\mu} I_{\nu\mu} d\nu d\mu \tag{21}$$

For transitions between two bound states the opacity and emissivity are

$$\chi_{\nu\mu} = \chi_{\nu c} + \frac{h\nu_{ij}}{4\pi} B_{ij}\phi_{\nu\mu}\left(n_i - \frac{g_i}{g_j}n_j\right) \tag{22}$$

$$\eta_{\nu\mu} = \eta_{\nu c} + \frac{h\nu_{ij}}{4\pi}\phi_{\nu\mu}A_{ji}n_j \tag{23}$$

where subscript c indicates background processes that do not depend on the population densities. g_i and g_j are the statistical weights for the lower and upper levels, respectively. The line opacity is related to the upward radiative rate. The last term in the line opacity comes from the treatment of stimulated emission as negative absorption.

The radiative rates introduce a *global* coupling since the intensity depends on conditions throughout the atmosphere, see Equation 5. In order to find the population densities we have to know the intensities and in order to find the intensitites we need the population densities. The equations of statistical equilibrium and radiative transfer constitute a coupled set of non-local, non-linear equations that have to be solved simultaneously. A strategy for solving sets of non-linear equations is by linearization. The general technique is outlined in Section 5 with the particular application to the coupled set of statistical equilibrium and radiative transfer equations given in Section 6.

5   Linearization

A classical method to solve a non-linear equation or a set of non-linear equations is the Newton-Raphson method. We may formulate this iterative method as a perturbation method and will illustrate the procedure by first applying it to a simple non-linear equation:

$$x^2 = 2 \tag{24}$$

At the n'th stage in an iterative method we do not have the exact solution x but instead an approximation to the solution, $x^{(n)}$. When this approximation is inserted in the equation we do not get the right hand side exactly but have an error, $E^{(n)}$:

$$(x^{(n)})^2 = 2 + E^{(n)} \tag{25}$$

We look for an addition to the current approximation that will fulfill the equation:

$$(x^{(n)} + \delta x)^2 = 2 \tag{26}$$

Expanding the parenthesis we get

$$(x^{(n)})^2 + 2x^{(n)}\delta x + \delta x^2 = 2 \tag{27}$$

The essence of linearization consists of neglecting non-linear terms in the perturbation (the term δx^2) and defining an approximate correction, $\delta x^{(n)}$, from the resulting equation:

$$(x^{(n)})^2 + 2x^{(n)}\delta x^{(n)} = 2 \tag{28}$$

Subtracting Equation 25 results in an equation for the approximate correction:

$$2x^{(n)}\delta x^{(n)} = -E^{(n)} \tag{29}$$

The iterative procedure thus starts by calculating the error from the current estimate of the solution (Equation 25), continues by calculating the approximate correction (Equation 29) and adding that correction to get the next estimate of the solution:

$$E^{(n)} = (x^{(n)})^2 - 2 \tag{30}$$

$$\delta x^{(n)} = -\frac{E^{(n)}}{2x^{(n)}} \qquad (31)$$

$$x^{(n+1)} = x^{(n)} + \delta x^{(n)} \qquad (32)$$

To start the iteration we need a starting approximation. Starting with $x^{(1)} = 1$ we get the following values:

n	$x^{(n)}$	$\lg|x - \sqrt{2}|$
1	1	-0.4
2	1.5	-1.1
3	1.416	-2.6
4	1.414216	-5.7
5	1.414213562	-11.8

The convergence is rapid with about double the number of significant figures per iteration — the convergence is quadratic. One may also note that the equation has two solutions but we get convergence only to one; which one depends on the starting approximation.

Linearization is a very general and powerful technique. Instead of one equation we may have a system of equations (like the statistical equilibrium equations). We may very easily add more constraints or equations to the system (like an energy equation instead of assuming a given temperature stratification). In the linearization step (defining the approximate correction, Equations 27–28) we may introduce additional approximations (like an approximate treatment of the radiative transfer). However, if additional approximations are introduced we lose the property of quadratic convergence.

The approximations introduced do *not* affect the accuracy of the method, only the convergence rate (or lack thereof). The accuracy is only set by the accuracy with which we can calculate the error from a given approximation. In the next section we will show how these methods can be used to solve the coupled equations of statistical equilibrium (SE) and radiative transfer (RT).

6   Linearization of the SE and RT Equations

Following the steps of the previous section we introduce a current approximation in the statistical equilibrium equations (Equation 17):

$$n_i^{(n)} \sum_{j \neq i}^{n_l} P_{ij}^{(n)} - \sum_{j \neq i}^{n_l} n_j^{(n)} P_{ji}^{(n)} = E_i^{(n)} \qquad (33)$$

We perturb the population densities and all variables that depend on the population densities:

$$n_i^{(n+1)} = n_i^{(n)} + \delta n_i^{(n)} \qquad (34)$$

$$P_{ij}^{(n+1)} = P_{ij}^{(n)} + \delta P_{ij}^{(n)} \tag{35}$$

and require

$$n_i^{(n+1)} \sum_{j \neq i}^{n_l} P_{ij}^{(n+1)} - \sum_{j \neq i}^{n_l} n_j^{(n+1)} P_{ji}^{(n+1)} = 0 \tag{36}$$

Expansion, linearization and subtraction gives

$$\delta n_i^{(n)} \sum_{j \neq i}^{n_l} P_{ij}^{(n)} + n_i^{(n)} \sum_{j \neq i}^{n_l} \delta P_{ij}^{(n)} - \sum_{j \neq i}^{n_l} \delta n_j^{(n)} P_{ji}^{(n)} \tag{37}$$

$$- \sum_{j \neq i}^{n_l} n_j^{(n)} \delta P_{ji}^{(n)} = -E_i^{(n)}$$

with the perturbed rates given by Einstein coefficients and perturbed intensities. For an upward bound-bound rate we get

$$\delta P_{ij}^{(n)} = B_{ij} \delta \bar{J}_{ij}^{(n)} = B_{ij} \frac{1}{2} \int_{-1}^{1} \int_{0}^{\infty} \phi_{\nu\mu} \delta I_{\nu\mu}^{(n)} d\nu d\mu \tag{38}$$

since the collisional rates and profile functions do not depend on the population densities and we assume the electron density and temperature to be given.

We still have to write the perturbed intensities in terms of perturbations in population densities and it is here we need the equations of radiative transfer.

We therefore take the transfer equation

$$\mu \frac{dI_{\nu\mu}^{(n)}}{dz} = -\chi_{\nu\mu}^{(n)} I_{\nu\mu}^{(n)} + \eta_{\nu\mu}^{(n)} \tag{39}$$

add perturbations, linearize and subtract the transfer equation to obtain

$$\mu \frac{d}{dz} \delta I_{\nu\mu}^{(n)} = -\chi_{\nu\mu}^{(n)} \delta I_{\nu\mu}^{(n)} - I_{\nu\mu}^{(n)} \delta \chi_{\nu\mu}^{(n)} + \delta \eta_{\nu\mu}^{(n)} \tag{40}$$

We introduce an equivalent source function perturbation from

$$\delta S_{\nu\mu}^{(n)} = \delta \eta_{\nu\mu}^{(n)} / \chi_{\nu\mu}^{(n)} - I_{\nu\mu}^{(n)} \delta \chi_{\nu\mu}^{(n)} / \chi_{\nu\mu}^{(n)} \tag{41}$$

and define a monochromatic optical depth along the ray from

$$d\tau_{\nu\mu}^{(n)} = -\chi_{\nu\mu}^{(n)} dz / \mu \tag{42}$$

We can then write a transfer equation for the *perturbations*:

$$\frac{d}{d\tau_{\nu\mu}^{(n)}} \delta I_{\nu\mu}^{(n)} = \delta I_{\nu\mu}^{(n)} - \delta S_{\nu\mu}^{(n)} \tag{43}$$

Through the formal solution of the transfer equation the intensity perturbation can be written as an integral over depth of the equivalent source function perturbation. This perturbation can be expressed in terms of the opacity and emissivity perturbations (using Equations 22–23) and since these depend on the population density perturbations we have achieved our goal of writing everything in terms of population density perturbations. We have a coupled set of equations for the population density corrections; one equation for each atomic energy level for each depth in the atmosphere. We thus have a matrix equation of order $n_l n_\tau$ where n_τ is the number of depth-points. This matrix is in principle full because the intensity perturbation at one depth depends on the population density perturbations at all other depths through the transfer equation. Calculating the matrix is time-consuming because we need to calculate on the order of $(n_l n_\tau)^2$ exponentials in the formal solution. Solving the matrix equation is also time consuming since the coefficient matrix is full.

In order to have a more efficient method we should find additional approximations that contain the basic physics (in order to keep a decent convergence rate) but lead to savings in the time it takes to construct the matrix and also in the solution time.

One such approximation was found by Scharmer (1981) who extended the Eddington-Barbier relation by replacing the formal solution integral with a one-point quadrature formula. The intensity at one point is thus approximated by the source function in another point multiplied by some weight,

$$I_{\nu\mu}(\tau_{\nu\mu}) = \omega S_{\nu\mu}(\tau'_{\nu\mu}) \tag{44}$$

By demanding that the formula shall be exact for a linear source function Scharmer arrived at the following quadrature points and weights:

$$\begin{cases} \tau'_{\nu\mu} = \tau_{\nu\mu} + 1, & \omega = 1, & \mu > 0 \\ \tau'_{\nu\mu} = \frac{\tau_{\nu\mu}}{1 - e^{-\tau_{\nu\mu}}} - 1, & \omega = 1 - e^{-\tau_{\nu\mu}}, & \mu < 0 \end{cases} \tag{45}$$

For the outgoing intensity these relations say that the intensity is equal to the source function one unit of optical depth further down along the ray. This one-point quadrature formula is exact for a linear source function but is also a rather good approximation when the source function is not a linear function of optical depth because of the rather narrow contribution function to the intensity. Since the approximation is only used in the calculation of the approximate correction and *not* in the calculation of the error it will only affect the convergence rate and not the accuracy of the result.

Experience has shown that the Scharmer operator is close to the optimal choice for the convergence rate. For the outgoing intensity we need no exponentials at all and for the incoming intensity we need only $n_l n_\tau$ exponentials thus reducing the work needed to construct the matrix by a large factor. The resulting matrix is also close to a banded structure with large portions identically zero. The quadrature point for the incoming intensity approaches

$\tau_{\nu\mu}/2$ for small values of $\tau_{\nu\mu}$. With seven grid-points per decade of optical depth this corresponds to just two points higher up in the atmosphere. We thus get only two sub-diagonals with non-zero elements below the diagonal of the matrix. This fact leads to large savings in the solution time.

The main draw-back of the Scharmer operator is closely related to the advantages. The operator takes explicitly into account the non-local nature of the radiative transfer and this leads to good convergence properties. However, the generalization to multi-dimensional radiative transfer is almost impossible when the treatment of the radiative transfer is non-local. The methods of choice for multi-dimensional radiative transfer problems are therefore the local methods developed by Olson et al. (1986) and Rybicki & Hummer (1991). See Hubeny (1992) for a review. A schematic illustration of the different radiative transfer operators is shown in Fig. 3

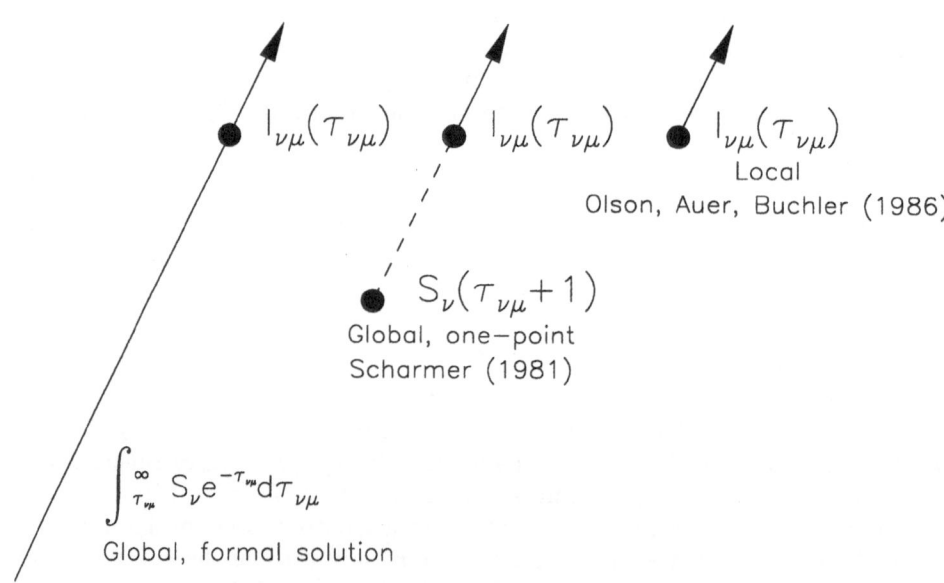

Fig. 3. Schematic illustration of different radiative transfer operators. The formal solution writes the intensity at optical depth $\tau_{\nu\mu}$ along the ray as an integral over depth and is thus global. The Scharmer operator replaces the integral over depth with a one-point quadrature formula. Since the quadrature point is non-local the operator is still global. The OAB operator is completely local.

7 Radiation Hydrodynamics

The previous discussion on how to solve for the outgoing intensity assumed that the atmosphere was given, in particular the variation of electron density and temperature throughout the atmosphere. This is called the forward solution of the *diagnostic* problem. However, the *structure* of the atmosphere, including the bulk velocity, density, electron density and temperature, is affected by the radiation. To solve for these four additional quantities in a one-dimensional plane-parallel atmosphere we need four additional equations. These are the equations of conservation of momentum, mass, charge and energy. The momentum equation may have a significant contribution from radiation pressure but in the solar case this radiative term can be neglected. The energy equation will be dominated by the radiative terms since the energy transport in stellar atmospheres often is dominated by radiation. The other dominant term in the energy equation may be energy transport by convection. In outer parts of the atmosphere we may have large terms in the energy balance from waves and shocks.

For a static atmosphere we neglect all variations with time and set the bulk velocity to zero. The above four equations then reduce to three: hydrostatic equilibrium, conservation of charge and energy. If we in addition assume LTE we have reduced the radiation hydrodynamics problem to the standard "classical" model atmosphere problem where we assume

- one dimensional geometry (either plane parallel or spherical)
- hydrostatic equilibrium (no systematic velocities and no change in time)
- convective-radiative equilibrium
- Local Thermodynamic Equilibrium (LTE)

Convective energy transport is in the classical models described with a mixing-length phenomenological formalism (*e.g.*, Mihalas 1978).

The assumption of LTE means that the distribution of population densities over energy states can be calculated from the Boltzmann and Saha equations using *local* values of the temperature and electron density restricting the effects of the potentially global coupling from the radiation. This is an enormous simplification of the problem. There is often no *a priori* justification for the use of this assumption other than the reduction of computational effort.

The above assumptions lead to only three parameters describing the stellar atmosphere: the effective temperature, the acceleration of gravity and the abundances of the elements. These few parameters all have a clear physical meaning. The fact that such classical stellar atmosphere models can reproduce observed spectra to some detail is one of the major success stories in modern astrophysics. Additional fudge parameters are, however, often needed to cover-up for the inaccuracy of the assumptions potentially masking important aspects of the physics. Such fudge parameters include the so-called microturbulence (giving additional line broadening) and macroturbulence.

In the outer parts of stellar atmospheres it is clear that LTE is a bad approximation. It is possible to replace this assumption with the statistical equilibrium equations and still solve the static model atmosphere problem. See Auer & Mihalas (1969, 1970, 1972), Mihalas & Auer (1970), Werner (1986, 1987, 1989), Anderson (1985, 1989, 1991), Dreizler & Werner (1992, 1993) and Hubeny & Lanz (1995) for examples.

For other trends in stellar atmosphere modelling see Carlsson (1995).

In the next sections we look at some specific aspects of the radiation hydrodynamics problem. We set the scene by deriving the simple condition of radiative equilibrium in Section 7.1, discuss some numerical problems and how they can be overcome in Sections 7.2–7.3, outline how the equations can be solved in Section 7.4 and discuss results in Section 7.5.

7.1   Radiative Equilibrium

If the only mode of energy transport is through radiation we get the condition of radiative equilibrium as our energy equation in a static atmosphere. Integrating the RT equation (2) over angle we get

$$\frac{dF_\nu}{dz} = 2\pi \int_{-1}^{1} \chi_{\nu\mu}(S_{\nu\mu} - I_{\nu\mu})d\mu \tag{46}$$

where F_ν is the monochromatic radiation flux density. Integrating over frequency we get the total flux divergence:

$$\frac{dF}{dz} = 2\pi \int_{0}^{\infty} \int_{-1}^{1} \chi_{\nu\mu}(S_{\nu\mu} - I_{\nu\mu})d\mu d\nu = 0 \tag{47}$$

with the last equality being the condition of radiative equilibrium.

This energy equation only specifies that the total radiation flux density does not change with height but does not specify the value of the flux density. We therefore have to specify the flux density at least at one height in addition to using the above equation:

$$F = \sigma T_{\text{eff}}^4 \tag{48}$$

with σ being the Stefan-Boltzmann constant and T_{eff} the effective temperature of the star.

If the opacity and source functions are isotropic (not unreasonable for stationary atmospheres) we can perform the integration over angle in (47):

$$\frac{dF}{dz} = 4\pi \int_{0}^{\infty} \chi_\nu(S_\nu - J_\nu)d\nu = 0 \tag{49}$$

7.2   Preconditioning

One problem in implementing the condition of radiative equilibrium in a code is that the driving term is the *difference* between the mean intensity and the source function. At large optical depths these quantities are almost equal and the difference may be lost in the numerical truncation errors. If possible one should then *precondition* the equations by analytically taking out the large terms that almost cancel, see Rybicki (1972) and Scharmer & Carlsson (1985) for examples.

7.3   Treatment of Millions of Spectral Lines

In an ordinary star there are millions of spectral lines such that a standard discretization, adequatly sampling the variations in opacity and mean intensity, would require an enormous number of frequency points. Two basic methods exist to overcome this problem.

In **Opacity Sampling** (OS) a large number of points are distributed over the spectrum. Each individual spectral line is not properly sampled but if the number of points is large enough the *statistical* properties are properly sampled and the flux divergence is adequately represented. The calculated *spectrum* is, however, undersampled and can not be compared directly to an observed spectrum. For cool stars on the order of 10^4 frequency points are needed.

In the **Opacity Distribution Function** (ODF) approach the opacities are redistributed within narrow wavelength bins (typically 50 Å) to create a smoother function of opacity as a function of wavelength. This smoother function can be sampled with fewer frequency points than the original opacity distribution, see Fig. 4. On the order of 500 points may be needed, a factor of 20 fewer than with the opacity sampling method. One disadvantage is that high opacities are lined up in depth and this may be unphysical because different spectral lines may dominate the opacity at different depths, mainly because of differences in temperature and therefore ionization balance. Another disadvantage is that the calculation of the opacity distribution functions is timeconsuming and has to be redone if the abundances are changed.

In LTE one may even sample ODFs over the whole spectrum if the variation of the Planck function is taken into account. In this way one may approximately calculate the radiative flux divergence with only a handful of frequency points. This is the approch chosen to treat the radiative energy transport in the 3D radiation-hydrodynamic simulations by Nordlund & Stein, see Nordlund (1982) for details.

7.4   Complete Linearization

Treating the forward diagnostic problem we ended up with one equation for each energy level considered (see Section 6). The full radiation hydrodynamics problem is not much more complicated from a methodological point of

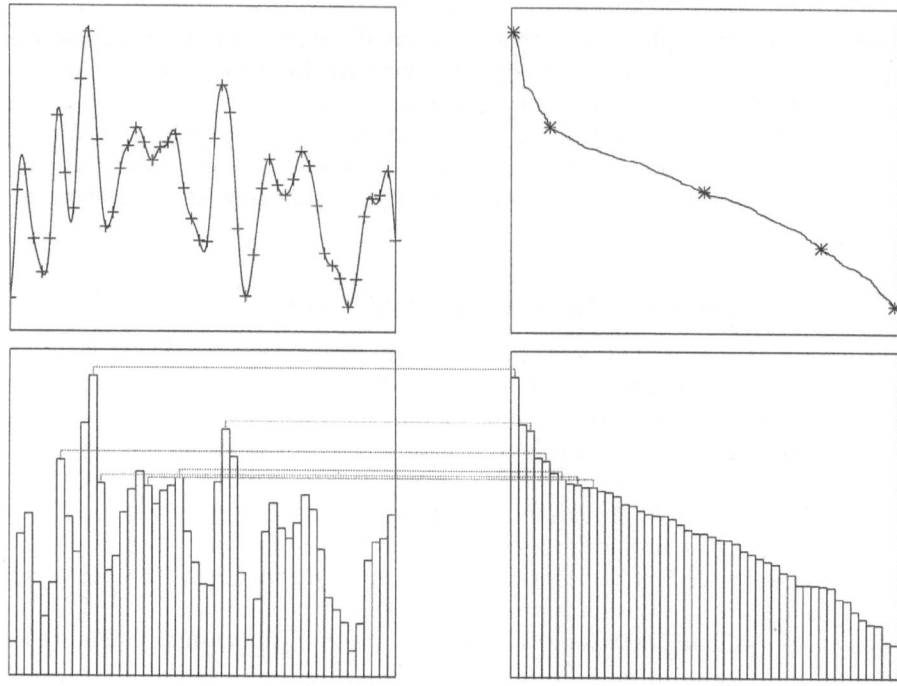

Fig. 4. Opacity distribution function method. *Left panels:* The original opacity as a function of wavelength within a narrow band. *Right panels:* The same opacities but now sorted with all the high opacities at one end of the wavelength band. The lines between the bottom panels show how every second of the ten largest opacities have been moved in the sorting process. Only a handful of points is needed to well approximate the right hand distribution while the original distribution demands a large number of points (*upper panels*).

view. We need four extra equations for the four extra variables bulk velocity, density, electron density and temperature: conservation of momentum, mass, charge and energy. Taking into account the time-variation means the statistical equilibrium equations are replaced by the rate equations. They state that the net rate into an energy level no longer is zero but is equal to the time derivative plus an advection term giving the net inflow into a volume element:

$$\frac{\partial n_i}{\partial t} + \frac{\partial}{\partial z}(n_i v) = \sum_{j \neq i}^{n_l} n_j P_{ji} - n_i \sum_{j \neq i}^{n_l} P_{ij} \qquad (50)$$

The method of linearization can still be used. The difference in the rate equations is that we get many more terms. The δP_{ij} terms in Equation (37) now get contributions also from δC_{ij} since the collisional rates depend on both electron density and temperature. We also get terms from the profile functions. From the $\delta\chi_{\nu\mu}$ term we used to have only δn_i and δn_j terms but we now need the whole expression:

$$\delta\chi_{\nu\mu} = \sum_x \frac{\partial\chi_{\nu\mu}}{\partial x}\delta x \qquad (51)$$

with x being all the variables we solve for ($\{n_i\}_1^{n_l}, v, \rho, n_e, T$). All perturbations thus have to be expanded into perturbations of the variables we solve for. In the diagnostic problem the radiative transfer equation was incorporated into the statistical equilibrium equations through the use of Scharmer's operator. We got expressions of an equivalent source function perturbation $\delta S_{\nu\mu}$ that could be written in terms of δn_i and δn_j through the expressions for $\chi_{\nu\mu}$ and $\eta_{\nu\mu}$. In the radiation hydrodynamics problem we get the same expressions but with perturbations in all variables we solve for through the new expressions for $\delta\chi_{\nu\mu}$ and $\delta\eta_{\nu\mu}$. In the energy equation we treat the radiative transfer in the same way as in the rate equations.

In the diagnostic problem we could treat one atomic species at a time but here we need to include all the energy levels of all the elements that contribute to the energy balance or the electron density. With a six-level model for hydrogen and a six-level model for ionized calcium (taking into account the most important energy levels for the Solar chromosphere) we end up with 10 rate equations, 2 particle conservation equations and 4 hydrodynamic equations per depth-point. With 100 depth-points we total 1600 equations in 1600 unknowns. This set has to be solved for each time-step. To enable the treatment of shocks it is an advantage to formulate the equations on an adaptive depth-grid. This means that the grid is neither fixed relative to the solar center (Eulerian grid) nor relative to the moving fluid (Lagrangian grid) but something in between. The grid locations are solved for together with the other variables and we add one equation that describes where we want the grid-points to be. One may choose to set the grid density to be proportional to gradients in the variables. See Dorfi & Drury (1987) for details.

7.5   Dynamics

The classical assumption of hydrostatic equilibrium means there are no systematic velocity fields in the atmosphere. The balance between the pressure gradient and the gravity leads to an exponentially decreasing density with height. In the absence of damping, any acoustic waves excited are therefore bound to rapidly increase in amplitude and form shocks in the outer atmosphere; weak shock theory predicts this to happen at about 1Mm above the visible solar surface (Stein & Schwartz 1972). Any classical model atmosphere will therefore fail to describe the outer atmosphere — dynamic phenomena will quickly dominate and it will be *meaningless* to use mean quantities to describe the atmosphere.

There is a large literature on acoustic waves in stellar atmospheres, see reviews by Narain & Ulmschneider (1990, 1996) for references.

To properly describe the behaviour of waves in the chromosphere it is necessary with a fully consistent coupling between non-LTE radiation and the hydrodynamics. Such a consistent treatment was achieved only rather recently (Carlsson & Stein 1992, 1994, 1995, 1997a, 1997b) The equations were formulated along the lines given in the previous section. The starting atmosphere was taken from a static convective-radiative equilibrium model thus having no chromospheric temperature rise. As bottom boundary condition was taken a velocity field deduced from observations of the Doppler shift of an iron line formed around 280 km height (Lites et al. 1993). We will here only summarize some of the conclusions. For details, the reader is referred to the original papers.

It was found that statistical equilibrium at the instantaneous values of the hydrodynamic variables was a *bad* approximation. At times of compression in a wave the instantaneous values would give increased hydrogen ionization with the energy increase absorbed by an increase of the hydrogen ionization energy with only a small temperature increase as a consequence. Because of the long timescales for hydrogen ionization and recombination this is not a realistic picture. Hydrogen does not have time to ionize in the compression phase above about 500 km height and the energy therefore goes into increased temperature instead of into hydrogen ionization energy. This leads to a much sharper temperature increase over shock fronts than would be the case with infinitely fast ionization/recombination rates. See Carlsson & Stein (1992) for details.

Another conclusion from the simulation work is that our traditional picture of stellar chromospheres has to be radically rethought. Carlsson & Stein (1994, 1995) find that the temperature rise exhibited in semiempirical models of the non-magnetic solar chromosphere is mainly a result of non-linear averaging of a shock dominated atmosphere. It is here important to stress that *any* dynamic atmosphere will lead to an *overestimate* of the mean temperature when mean intensities in the blue are used as diagnostics. The simulations show a mean gas temperature that does *not* increase with height giving mean intensities that give a semi-empirical temperature rise (Fig. 5). The simulations do not claim to contain the full picture of the Solar chromosphere; *e.g.,* they lack the effect of an incident radiation field from the corona and high frequency waves from the photosphere. Both of these effects will favour a slow temperature increase. The essential conclusions are, however:

 − a mean temperature structure can *not* be deduced from mean intensities in the blue part of the spectrum.
 − in a dynamic chromosphere the mean temperature is not only difficult to deduce but is also a *meaningless* and even *misleading* quantity. The energy balance can only be deduced from a dynamical model and not from any such mean model.

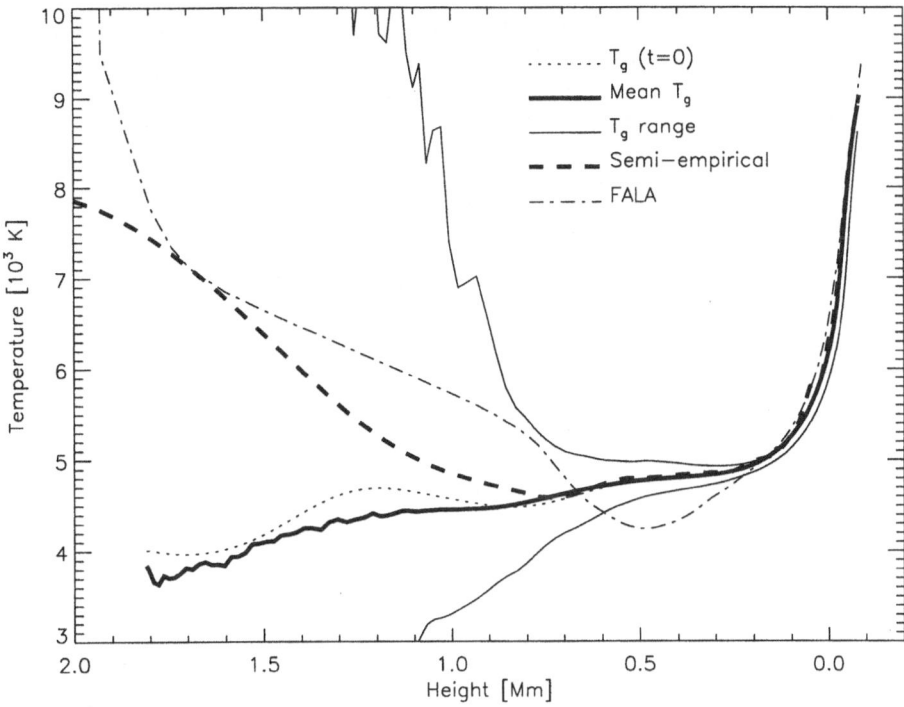

Fig. 5. Time average of the temperature in a dynamical simulation (Carlsson & Stein 1995), the range of temperatures in the simulation, the semiempirical model that gives the best fit to the time average of the intensity as a function of wavelength calculated from the dynamical simulation, the starting model for the dynamical simulation and the semiempirical model FALA (Fontenla et al. 1993). The maximum temperatures are only reached in narrow shock spikes of short duration. The semiempirical model giving the same intensities as the dynamical simulation shows a chromospheric temperature rise while the mean temperature in the simulation does not.

The diagnostics of the Solar chromosphere is made difficult by the lack of good observational diagnostics. In the optical region the continuum opacity is much too small and we thus need a large line opacity. This leaves lines originating from the ground state of the dominant ionization stage of abundant elements. Most of these resonance lines lie in the UV region of the spectrum with the notable exception of the two resonance lines from singly ionized calcium, CaII. These Fraunhofer H and K lines show very strong absorption with a slight emission as a chromospheric signature near line center. The emission is in the form of two emission peaks with absorption in between. In non-magnetic regions outside the chromospheric network the emission is assymmetric with almost never a red emission peak and the violet peak (named H_{2V} for the H-line and K_{2V} for the K-line) only prominent at times with a period of about three minutes. It has turned out that this assymmetry and the temporal behaviour is difficult to reproduce in theoretical models. Another way of seeing this difficulty is that the feature is a good diagnostic of the atmospheric conditions.

The simulations mentioned above do, however, reproduce the CaII H_{2V} behaviour; even in some detail, see Fig. 6.

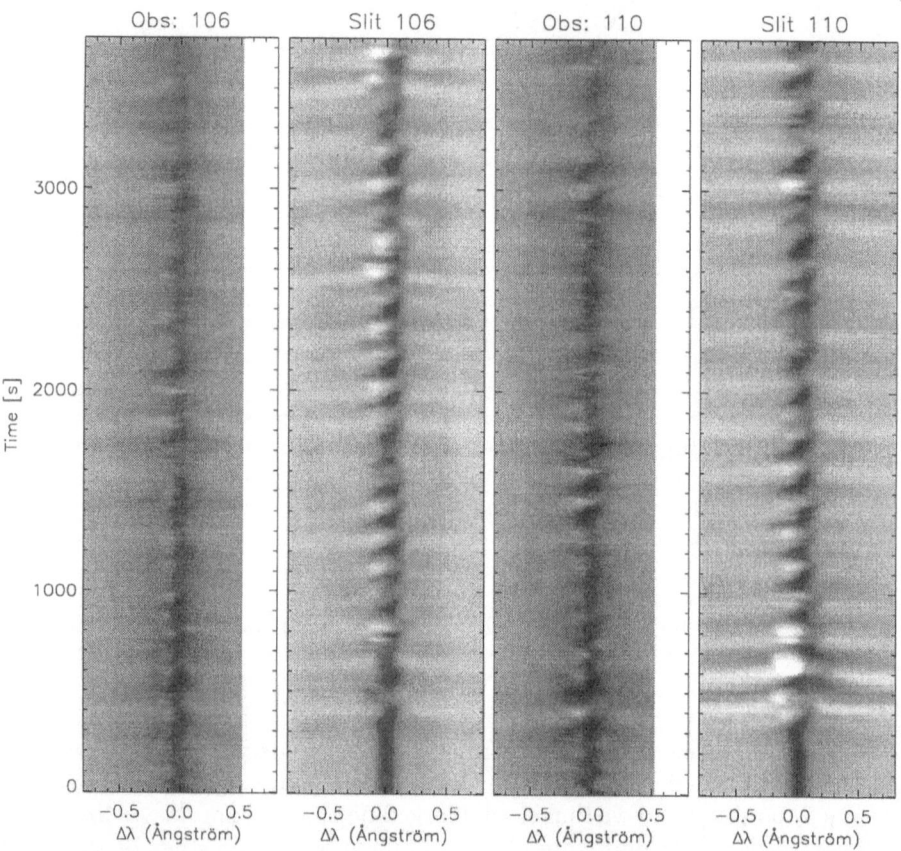

Fig. 6. The computed CaII H line intensity as a function of wavelength and time compared with observations for slit positions 106 and 110. To mimic effects of scattered light and seeing the simulations have been convolved with a Gaussian with FWHM of 5 km/s, scattered light of 1% of the continuum intensity has been added and the sequence have been shifted in time with a random function sharply peaked at no shift. There is close correspondence between the simulated and observed grains after the initial startup period. From Carlsson & Stein (1997b).

The formation of the Ca II H_{2V} grains is explained in detail in Carlsson & Stein (1997a).

In the UV part of the spectrum we have access to more spectral features formed in the Solar chromosphere. All continua shortward of 152 nm are formed above 500 km height. With the SUMER spectrograph on-board the SOHO spacecraft this wavelength region is accessible. First results show that the non-magnetic chromosphere is indeed very dynamic. Continuum intensity variations are similar to the predictions from simulations (see Carlsson et al. 1997) while line emission variations are not. Emission lines from neutral elements show emission all the time while they sometimes disappear in the simulations. These lines are formed higher than the Ca II resonance lines and one explanation for the discrepancy may be that these layers are influenced by hotter material in magnetic flux-tubes that cross areas that lower down only have weak magnetic fields.

To further our understanding of the dynamics of the outer layers of the Solar atmosphere we need detailed observations coupled with detailed radiation-hydrodynamics simulations. We may then get enough constraints to construct a meaningful *dynamic* picture of the solar atmosphere. Anything short of a consistent dynamic picture has to be treated with utmost caution.

Acknowledgements

This work was supported by a grant from the Norwegian Research Council The computations were supported by a grant from the Norwegian Research Council, tungregneutvalget.

References

Anderson, L. 1991, in L. Crivellari, I. Hubeny, D. G. Hummer (eds.), Stellar Atmospheres: Beyond Classical Models, NATO ASI Series C-341, KLuwer, Dordrecht, p. 29

Anderson, L. S. 1985, ApJ, 298, 848

Anderson, L. S. 1989, ApJ, 339, 558

Auer, L. H. 1976, J. Quant. Spectrosc. Radiat. Transfer, 16, 931

Auer, L. H., Mihalas, D. 1969, ApJ, 158, 641

Auer, L. H., Mihalas, D. 1970, MNRAS, 149, 65

Auer, L. H., Mihalas, D. 1972, ApJS, 24, 193

Carlsson, M. 1995, in P. Lilje, P. Maltby (eds.), Frontiers of Astrophysics, Proc. Rosseland Centenary Symposium, Institute of Theoretical Astrophysics, Oslo, p. 181

Carlsson, M., Judge, P. G., Wilhelm, K. 1997, ApJ, 486, L63

Carlsson, M., Stein, R. F. 1992, ApJ, 397, L59

Carlsson, M., Stein, R. F. 1994, in M. Carlsson (ed.), Proc. Mini-Workshop on Chromospheric Dynamics, Institute of Theoretical Astrophysics, Oslo, p. 47

Carlsson, M., Stein, R. F. 1995, ApJ, 440, L29

Carlsson, M., Stein, R. F. 1997a, ApJ, 481, 500

Carlsson, M., Stein, R. F. 1997b, in C. E. Alissandrakis, G. Simnett, L. Vlahos (eds.), Proc. Eighth European Solar Physics Meeting, Lecture Notes in Physics, Springer, p. 159

Dorfi, E. A., Drury, L. O. 1987, J. Comput. Phys., 69, 175

Dreizler, S., Werner, K. 1992, in U. Heber, C. S. Jeffery (eds.), The Atmospheres of Early-Type Stars, Lecture Notes in Physics **401**, Springer, 436

Dreizler, S., Werner, K. 1993, A&A, 278, 199

Feautrier, P. 1964, C.R. Acad. Sci. Paris, 258, 3189

Fontenla, J. M., Avrett, E. H., Loeser, R. 1993, ApJ, 406, 319

Gray, D. F. 1992, The Observation and Analysis of Stellar Photospheres, Cambridge University Press, Cambridge UK (second edition)

Hubeny, I. 1992, in U. Heber, C. S. Jeffery (eds.), The Atmospheres of Early-Type Stars, Lecture Notes in Physics **401**, Springer, p. 377

Hubeny, I., Lanz, T. 1995, ApJ, 439, 875

Kunasz, P. B., Hummer, D. G. 1974, MNRAS, 166, 19

Lites, B. W., Rutten, R. J., Kalkofen, W. 1993, ApJ, 414, 345

Magain, P. 1986, A&A, 163, 135

Mihalas, D. 1978, Stellar Atmospheres, W. H. Freeman and Co., San Francisco (second edition)

Mihalas, D., Auer, L. H. 1970, ApJ, 160, 1161

Narain, U., Ulmschneider, P. 1990, Space Sc. Rev., 54, 377

Narain, U., Ulmschneider, P. 1996, Space Sc. Rev., 75, 453

Nordlund, Å. 1982, A&A, 107, 1

Olson, G. L., Auer, L. H., Buchler, J. R. 1986, J. Quant. Spectrosc. Radiat. Transfer, 35, 431

Rybicki, G. B. 1972, in R. G. Athay, L. L. House, G. Newkirk (eds.), Line Formation in the Presence of Magnetic Fields, High Altitude Observatory, NCAR, Boulder, p. 145

Rybicki, G. B., Hummer, D. G. 1991, A&A, 245, 171

Scharmer, G. B. 1981, ApJ, 249, 720

Scharmer, G. B., Carlsson, M. 1985, J. Comput. Phys., 59, 56

Stein, R. F., Schwartz, R. A. 1972, ApJ, 177, 807

Werner, K. 1986, A&A, 161, 177

Werner, K. 1987, in W. Kalkofen (ed.), Numerical Radiative Transfer, Cambridge University Press, Cambridge U.K., 67

Werner, K. 1989, A&A, 226, 265

Basic Concepts
in Solar Magnetohydrodynamics

Robert W. Walsh

School of Mathematical and Computational Sciences,
University of St Andrews,
St Andrews, KY16 9SS, UK.
robert@dcs.st-and.ac.uk

1 Introduction

Most of the observed structure and interesting behaviour on the Sun is produced by the magnetic field. In particular, the solar atmosphere is not static and uniform but is a complex and highly inhomogeneous environment where there are many examples of the solar plasma interacting with this magnetic field.

In the following we shall use the theory of magnetohydrodynamics (MHD) to investigate several examples of solar phenomena. Firstly, the fundamental MHD equations are introduced along with some simplifying assumptions (Section 2). In Section 3, there is a short discussion on different wave modes produced in a magnetised plasma while Section 4 deals with magnetohydrostatic solutions to the MHD equations. Finally, Section 5 outlines several MHD theories of coronal heating mechanisms.

2 Basic Equations

Magnetohydrodynamics is the study of the interaction between a magnetic field and a plasma using a simplified set of Maxwell's equations along with Ohm's Law, the ideal gas law and equations for mass continuity, motion and energy.

2.1 Maxwell's Equations

These are

$$\nabla \times \mathbf{B} = \mu \mathbf{j} + \frac{1}{c^2}\frac{\partial \mathbf{E}}{\partial t} \tag{1}$$

$$\nabla . \mathbf{B} = 0 \tag{2}$$

$$\nabla \times \mathbf{E} = -\frac{\partial \mathbf{B}}{\partial t} \tag{3}$$

$$\nabla . \mathbf{E} = \frac{\rho_e}{\varepsilon} \tag{4}$$

where $\mathbf{E}$ is the electric field strength, $\mathbf{B}$ is the magnetic field strength, ρ_e is the charge density, $\mathbf{j}$ is the current density, μ is the magnetic permeability, ε is the permittivity of free space, t is the time and c is the speed of light in a vacuum ($\sim 3 \times 10^8 \text{ms}^{-1}$). These equations are simplified under certain assumptions (Priest, 1982):

- The plasma is treated as a continuous medium (valid as long as the lengthscales considered greatly exceed any internal plasma lengthscales, for example, the ion gyroradius) and is assumed to be in thermal equilibrium.
- μ and ε are assumed to be constant (and taken to have the vacuum values $\mu_0 = 4\pi \times 10^{-7} \text{Hm}^{-1}$ and $\varepsilon_0 = 8.854 \times 10^{-12} \text{Fm}^{-1}$ in the Solar context).
- Most of the other plasma properties are supposed to be isotropic except the thermal conduction, κ, which is much greater along the magnetic field direction than perpendicular to it (see Section 2.6).
- An inertial frame of reference is used. Rotational effects may become important when considering very large structures. Also, the plasma is treated as a single fluid system.
- A simplified Ohm's Law is used;

$$\mathbf{j} = \sigma \left(\mathbf{E} + \mathbf{v} \times \mathbf{B} \right) \tag{5}$$

where $\mathbf{v}$ is the plasma velocity and σ is the electrical conductivity which is assumed to be a constant.
- Flow, sound and Alfvénic velocities are much smaller than the speed of light and therefore relativistic effects can be ignored. Thus, the term $\partial \mathbf{E}/\partial t$ in (1) can be neglected to give

$$\nabla \times \mathbf{B} = \mu_0 \mathbf{j}. \tag{6}$$

2.2   The Induction Equation

Using equation (5) to eliminate $\mathbf{E}$ from equation (3) and with equation (6) and the triple vector product, we obtain

$$\frac{\partial \mathbf{B}}{\partial t} = \nabla \times (\mathbf{v} \times \mathbf{B}) + \eta_0 \nabla^2 \mathbf{B}, \tag{7}$$

where $\eta_0 = (\mu_0 \sigma)^{-1}$ is the constant magnetic diffusivity. Equation (7) is known as the **Induction Equation** and it links the evolution of the magnetic field to the plasma.

If v_0 and l_0 are typical velocity and lengthscale values for our system, then the ratio of the two terms on the right hand side of (7) give the **Magnetic Reynolds Number** $R_m = l_0 v_0 / \eta_0$. Thus, for example, in an active region where $\eta_0 = 1 \text{m}^{-2} \text{s}^{-1}$, $l_0 = 700 \text{km} \approx 1$ arcsec and $v_0 = 10^4 \text{ms}^{-1}$, we find $R_m = 7 \times 10^9 \gg 1$. Therefore, the magnetic field is frozen to the plasma and the electric field does not drive the current but is simply $\mathbf{E} = -\mathbf{v} \times \mathbf{B}$.

However, if the lengthscales of the system are reduced, the diffusion term will become important and the fieldlines are allowed to "slip" or diffuse through the plasma. This leads to the possibility of the magnetic fieldlines reconnecting (see Priest, 1996 for a review on MHD reconnection).

2.3   The Plasma Equations

The motion of the magnetic field is coupled to the behaviour of the plasma by the presence of the velocity term in equation (7) and in the following equations for mass continuity, motion and energy.

2.4   Mass Continuity

In an MHD system, mass must be conserved;

$$\frac{D\rho}{Dt} + \rho \nabla . \mathbf{v} = 0, \tag{8}$$

where

$$\frac{D}{Dt} = \frac{\partial}{\partial t} + \mathbf{v} . \nabla,$$

is the total derivative and ρ is the plasma density.

2.5   Motion

The equation of motion for the plasma can be written as

$$\rho \frac{D\mathbf{v}}{Dt} = -\nabla p + \mathbf{j} \times \mathbf{B} + \rho \mathbf{g} + \rho \nu \nabla^2 \mathbf{v}, \tag{9}$$

where p is the plasma pressure. The terms on the right hand side of the equation can be separated into :

- $-\nabla p$ — a plasma pressure gradient.
- $\mathbf{j} \times \mathbf{B}$ — a Lorentz force. From equation (6), and using the triple vector product, this becomes

$$(\mathbf{B} . \nabla) \frac{\mathbf{B}}{\mu_0} - \nabla \left(\frac{B^2}{2\mu_0} \right).$$

The first of these terms represents the change of $\mathbf{B}$ along a particular field line and therefore is a **magnetic tension force** whose strength is proportional to B^2. The second term is a **magnetic pressure force** with the magnetic pressure given by $B^2/2\mu_0$.
- $\rho \mathbf{g}$ — this is the effect of gravity where g is the local gravitational acceleration at the surface of the Sun (taken to be ~ 274 ms^{-2}).
- $\rho \nu \nabla^2 \mathbf{v}$ — this is the effect of viscosity on an incompressible flow. ν is the coefficient of kinematic viscosity which is assumed to be uniform throughout the plasma (Spitzer, 1962).

2.6   The Energy Equation

The fundamental energy equation is written as

$$\frac{\rho^\gamma}{\gamma - 1}\frac{D}{Dt}\left(\frac{p}{\rho^\gamma}\right) = \nabla.(K\nabla T) - L \tag{10}$$

where T is the plasma temperature and γ is the ratio of specific heats ($= 5/3$). K is the tensor of thermal conduction which can be split into components along and across the magnetic field,

$$\nabla_\parallel.\left(\kappa_\parallel \nabla_\parallel T\right) + \nabla_\perp.\left(\kappa_\perp \nabla_\perp T\right).$$

Along the magnetic fieldlines conduction is mainly by electrons and Braginski (1965) gives $\kappa_\parallel = \kappa_0 T^{5/2}$ W m^{-1} deg^{-1} with $\kappa_0 = 10^{-11}$ for the corona. Conduction across the fieldlines is mainly by ions and at coronal temperatures,

$$\frac{\kappa_\perp}{\kappa_\parallel} \approx 10^{-12}.$$

Thus, for the Sun with its strong magnetic field, a good approximation is that the vast majority of conducted heat occurs along the field and thus the conduction term can simply be written as $\kappa_0\nabla_\parallel.\left(T^{5/2}\nabla_\parallel T\right)$. L is the loss-gain function which has the form

$$L(\rho, T) = \rho^2 Q(T) - H(\mathbf{B}, \rho, T, s, t). \tag{11}$$

$Q(T)$ is the optically thin radiative loss function and H is the coronal heating function. $Q(T)$ has been calculated by several authors and is approximated

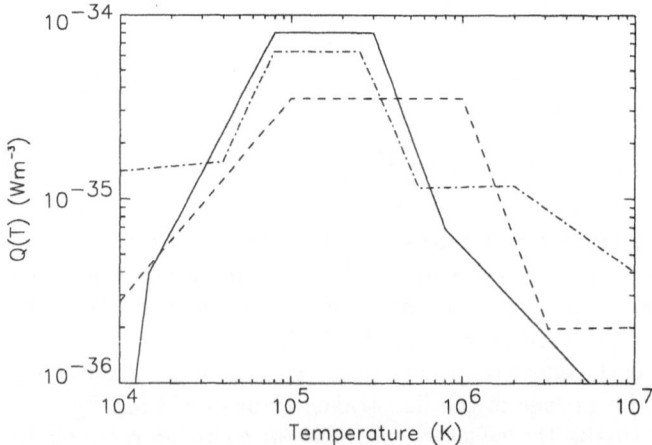

Fig. 1. Comparison of (—) Hildner (1974), (.-.-.) Rosner *et al* (1978) and (- -) Cook *et al* (1989) profiles for the radiative loss function

by a piecewise continuous function,

$$Q(T) = \chi T^{\alpha}, \tag{12}$$

where χ and α are constants within any particular range of temperature for the piecewise fit (see Figure 1).

The "unknown" coronal heating term may depend on a number of different physical parameters (magnetic field, plasma density etc..) as well as having a dynamic component associated with it. The different heating theories and forms of H will be discussed in Section 5.

2.7 Perfect Gas Law

For simplicity, the perfect gas law will be used;

$$p = \frac{R}{\tilde{\mu}}\rho T, \tag{13}$$

where R is the molar gas constant $(8.3 \times 10^3 \text{ m}^2 \text{ s}^{-2} \text{ deg}^{-1})$ and $\tilde{\mu}$ is the mean molecular weight with $\tilde{\mu} = 0.6$ in the ionised corona.

2.8 Summary of Equations

Thus, the fundamental equations considered are

$$\frac{\partial \mathbf{B}}{\partial t} = \nabla \times (\mathbf{v} \times \mathbf{B}) + \eta_0 \nabla^2 \mathbf{B}, \tag{14}$$

$$\frac{D\rho}{Dt} + \rho \nabla . \mathbf{v} = 0, \tag{15}$$

$$\rho \frac{D\mathbf{v}}{Dt} = -\nabla p + \mathbf{j} \times \mathbf{B} + \rho \mathbf{g} + \rho \nu \nabla^2 \mathbf{v}, \tag{16}$$

$$\frac{\rho^{\gamma}}{\gamma - 1} \frac{D}{Dt}\left(\frac{p}{\rho^{\gamma}}\right) = \kappa_0 \nabla_{\parallel} . \left(T^{5/2} \nabla_{\parallel} T\right) - \rho^2 \chi T^{\alpha} + H, \tag{17}$$

$$p = \frac{R}{\tilde{\mu}}\rho T. \tag{18}$$

Further details on the MHD Equations can be found in Priest (1982) and Priest (1994).

3 Magnetohydrodynamic Waves

The solar atmosphere is a dynamic environment containing features that are in continual motion. Photospheric granulation provides a constant forcing mechanism for the overlying atmospheric layers with the possibility of generating waves that propagate upwards where they may be dissipated in the chromosphere or possibly the corona. In a gas, we have sound waves which propagate in all directions. However, in a magnetised plasma, we have the possibility of a range of wave types as are described below.

3.1   Sound Waves

Consider a uniform plasma at rest with no magnetic field, a pressure p_0 and density ρ_0. Introducing a small disturbance produces a velocity $\mathbf{v_1}$ and a new pressure $p_0 + p_1$ and density $\rho_0 + \rho_1$. Thus for the equations we have,

$$\rho\frac{D\mathbf{v}}{\partial t} = -\nabla p, \tag{19}$$

$$\frac{\partial \rho}{\partial t} + \nabla.(\rho\mathbf{v}) = 0, \tag{20}$$

$$\frac{p}{\rho^\gamma} = K = \frac{p_0}{\rho_0{}^\gamma}, \tag{21}$$

where motions are so rapid that no heat exchange is occuring. If we linearise (19) to (21), neglecting squares and products of small quantities, we get

$$\rho_0\frac{\partial \mathbf{v_1}}{\partial t} = -\nabla p_1, \tag{22}$$

$$\frac{\partial \rho_1}{\partial t} + \rho_0\nabla.\mathbf{v_1} = 0, \tag{23}$$

$$p_1 = c_s{}^2\rho_1, \tag{24}$$

where $c_s{}^2 = \gamma p_0/\rho_0$ is the sound speed squared. Equations (22) - (24) can be combined to give

$$\frac{\partial^2 \rho_1}{\partial t^2} = c_s{}^2\nabla^2\rho_1, \tag{25}$$

- a wave equation. If we Fourier analyse an arbitary disturbance into three dimensional components of the form

$$\rho_1 = \text{const.}\exp\left[i\left(\mathbf{k.r} - \omega t\right)\right], \tag{26}$$

where ω is the wave frequency and $\mathbf{k.r} = k_x\hat{\mathbf{x}} + k_y\hat{\mathbf{y}} + k_z\hat{\mathbf{z}}$ is the wave vector, then, from Equation(25), we obtain a **dispersion relation**,

$$\omega^2 = k^2 c_s{}^2 \tag{27}$$

for a wave propagating in all directions with a phase speed $c_s = \omega/k$.

3.2   Alfvén Waves

For a wave travelling along a fieldline, we have (with an analogy with an elastic string), an Alfvén speed,

$$v_A = \sqrt{\frac{\text{tension}}{\rho}} = \frac{B}{\sqrt{\mu\rho}}. \tag{28}$$

Consider an ideal plasma at rest with a uniform field $\mathbf{B}_0 = B_0\hat{\mathbf{z}}$ and density ρ_0. A disturbance introduces a velocity $\mathbf{v}_1$ and affects the other variables such that $\mathbf{B}_0 \rightarrow \mathbf{B}_0 + \mathbf{B}_1$ and $\rho_0 \rightarrow \rho_0 + \rho_1$. As in Section 3.1, linearisation of the pressureless MHD equations gives

$$\frac{\partial \mathbf{B}_1}{\partial t} = \nabla \times (\mathbf{v}_1 \times \mathbf{B}_0), \tag{29}$$

$$\rho_0 \frac{\partial \mathbf{v}_1}{\partial t} = (\nabla \times \mathbf{B}_1) \times \frac{\mathbf{B}_0}{\mu}, \tag{30}$$

$$\frac{\partial \rho_1}{\partial t} + \rho_0 \nabla . \mathbf{v}_1 = 0, \tag{31}$$

$$p_1 = c_s^2 \rho_1. \tag{32}$$

Considering solutions of the form (26), the above are reduced to

$$-\omega \mathbf{B}_1 = (\mathbf{B}_0 . \mathbf{k}) \mathbf{v}_1 - \mathbf{B}_0 (\mathbf{k} . \mathbf{v}_1), \tag{33}$$
$$-\mu \rho_0 \omega \mathbf{v}_1 = \mathbf{B}_1 (\mathbf{B}_0 . \mathbf{k}) - \mathbf{k} (\mathbf{B}_1 . \mathbf{B}_0), \tag{34}$$

and from $\nabla . \mathbf{B} = 0$, $\mathbf{k} . \mathbf{B}_1 = 0$. Thus, for Alfvén waves propagating along $\mathbf{B}_0$, we have $\mathbf{k}$ parallel to $\mathbf{B}_0$ and we assume that $\mathbf{v}_1$ is perpendicular to $\mathbf{k}$ ($\mathbf{k} . \mathbf{v}_1 = 0$). Then, from (33), $\mathbf{v}_1$ is parallel to $\mathbf{B}_1$ and $\mathbf{B}_1 . \mathbf{B}_0 = 0$. This implies that $\omega^2 = k^2 v_A^2$.

Generally, the Alfvén waves may propagate at an angle θ to $\mathbf{B}_0$ which simply gives the dispersion relation

$$\omega^2 = k^2 v_A^2 \cos^2 \theta. \tag{35}$$

A phase speed diagram is shown in Figure 2 with the maximum phase speed being in the direction of $\mathbf{B}_0$; there is no propagation perpendicular to the initial field direction.

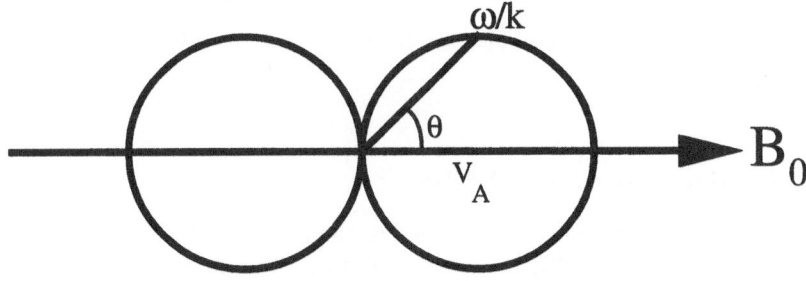

Fig. 2. Phase diagram for Alfvén Waves.

3.3   Compressional Alfvén Waves

If we consider $\mathbf{k}.\mathbf{v}_1 \neq 0$ then from (33) and (34) we get,

$$\mu\rho_0\omega^2\mathbf{v}_1 = \left[(\mathbf{B}_0.\mathbf{k})\,\mathbf{v}_1 - \mathbf{B}_0\,(\mathbf{k}.\mathbf{v}_1)\right](\mathbf{B}_0.\mathbf{k}) - \mathbf{k}.\left[(\mathbf{B}_0.\mathbf{k})\,(\mathbf{v}_1.\mathbf{B}_0) - B_0{}^2\,(\mathbf{k}.\mathbf{v}_1)\right]. \tag{36}$$

Then, $\mathbf{B}_0.(36)$ gives

$$\mu\rho_0\omega^2\,(\mathbf{v}_1.\mathbf{B}_0) = 0 \tag{37}$$

and $\mathbf{k}.(36)$ with (37) produces

$$\mu\rho_0\omega^2\,(\mathbf{k}.\mathbf{v}_1) = k^2 B_0{}^2\,(\mathbf{k}.\mathbf{v}_1) \tag{38}$$

or rather

$$\omega^2 = k^2 v_A{}^2, \tag{39}$$

- a dispersion relation for compressional Alfvén waves (see Figure 3). These waves propagate equally in all directions and since $\mathbf{k}.\mathbf{v}_1 \neq 0$, p_1 and ρ_1 are generally non-zero.

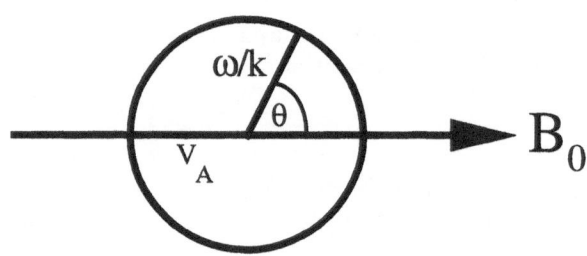

Fig. 3. Phase diagram for Compressional Alfvén Waves.

3.4   Magnetoacoustic Waves

If pressure fluctuations are included in the MHD equations by adding a $-\nabla p_1$ term to Equation (29), then the sound and compressional Alfvén wave are coupled together to give **two magnetoacoustic waves** with a dispersion relation of the form,

$$\omega^4 - \omega^2 k^2 \left(c_s{}^2 + v_A{}^2\right) + c_s{}^2 v_A{}^2 k^4 \cos^2\theta = 0. \tag{40}$$

The smallest root for w^2/k^2 gives the *slow* mode and the largest root, the *fast* mode. The phase speed diagram is sketched in Figure 4.

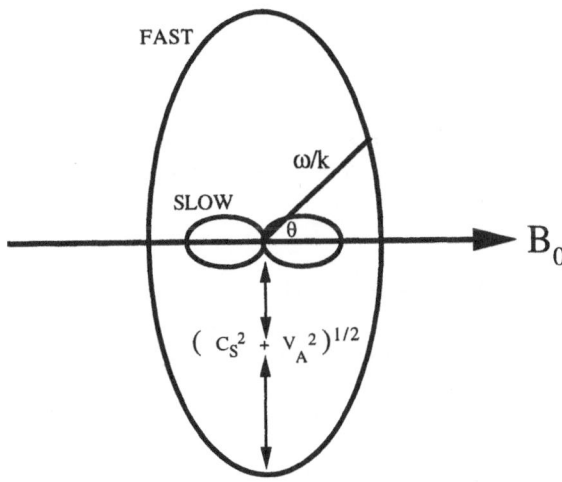

Fig. 4. Phase diagram for Magnetoacoustic Waves.

4   Magnetohydrostatics

An equilibrium structure with no flow is in *magnetohydrostatic force balance* such that

$$0 = -\nabla p + \mathbf{j} \times \mathbf{B} + \rho\mathbf{g}, \tag{41}$$

with $\nabla.\mathbf{B} = 0$, $p = R\rho T/\tilde{\mu}$ and $\mathbf{j} = \nabla \times \mathbf{B}/\mu$. Consider this balance under a number of different assumptions.

4.1   Vertical Magnetic Field

Suppose we have a uniform, vertical magnetic field $\mathbf{B} = B\hat{\mathbf{z}}$. Gravity acts vertically downwards $(-g\hat{\mathbf{z}})$ and therefore, in the $\hat{\mathbf{z}}$ direction,

$$0 = -\frac{dp}{dz} - \rho g, \tag{42}$$

and with the gas law,

$$\frac{dp}{dz} = \frac{-\tilde{\mu}pg}{RT}. \tag{43}$$

Integrating gives,

$$p = p_0 \exp\left(-\int_0^z \frac{\tilde{\mu}pg}{RT}dz\right), \tag{44}$$

where p_0 is the pressure at the base $(z = 0)$ of the fieldline. If the variation of temperature with height $T(z)$ is known, then (44) determines the pressure and density. In particular, if the temperature is assumed to be uniform ($T = T_0$) then

$$p = p_0 e^{-z/h}, \tag{45}$$

where $h = RT_0/\tilde{\mu}g$ is known as the pressure scale height. Down in the photosphere where $T_0 \approx 5000\text{K}$, $h \approx 150\text{km}$ whereas in the corona, $T_0 \approx 10^6\text{K}$, giving $h \approx 30\text{Mm}$.

4.2   Dominant Magnetic Field

If the pressure gradients and gravity are negligible, we have

$$0 = \mathbf{j} \times \mathbf{B}, \tag{46}$$

so that the current must be parallel to the magnetic field. In other words,

$$\nabla \times \mathbf{B} = \alpha \mathbf{B}, \tag{47}$$

where α is a function of position within a *force-free magnetic field*. Taking the divergence of (47) we find $\mathbf{B}.\nabla\alpha = 0$. That is, the rate of change of α in the direction of the magnetic field is zero. Thus, α is constant along a given fieldline (but can be different from one fieldline to the next). If we assume that α is uniform everywhere in the medium, then the curl of (47) gives

$$\left(\nabla^2 + \alpha^2\right) \mathbf{B} = 0. \tag{48}$$

These are called *constant* α or *linear force-free fields*.

4.3   Potential Fields

A particular case of interest is when the current vanishes such that

$$\nabla^2 \mathbf{B} = 0. \tag{49}$$

If $\mathbf{B} = \nabla A$ so that $\nabla \times \mathbf{B} = 0$ is satisfied identically, then $\nabla.\mathbf{B} = 0$ gives Laplaces Equation,

$$\nabla^2 \mathbf{A} = 0, \tag{50}$$

such that many of the general results associated with potential theory can be applied in this context. In particular,

- If the normal field component B_n is imposed on the boundary S of a volume V, then the potential solution in V is unique.
- If B_n is imposed on the boundary S, then the potential field is the one with the minimum magnetic energy.

This has certain implications. For example, it is known that during a solar flare, the normal field component at the photosphere remains virtually unchanged. As enormous amounts of energy are released during the eruption, the magnetic structure cannot be at potential. The excess magnetic energy could arise from a sheared force-free field.

An example of a potential field can be shown in two dimensions by considering the separable solutions to $A(x, z) = X(x)Z(z)$ so that $\nabla^2 A = 0$ implies,

$$\frac{1}{X}\frac{d^2 X}{dx^2} = \frac{-1}{Z}\frac{d^2 Z}{dz^2} = -n^2 \tag{51}$$

where n is a constant. A possible solution to (51) is

$$A = \left(\frac{B_0}{n}\right)\sin(nx)e^{-nz} \tag{52}$$

such that

$$B_x = \frac{\partial A}{\partial x} = B_0 \cos(nx)e^{-nz}, \tag{53}$$

$$B_z = \frac{\partial A}{\partial z} = -B_0 \sin(nx)e^{-nz}, \tag{54}$$

and is sketched in Figure 5 - a possible model for a potential arcade.

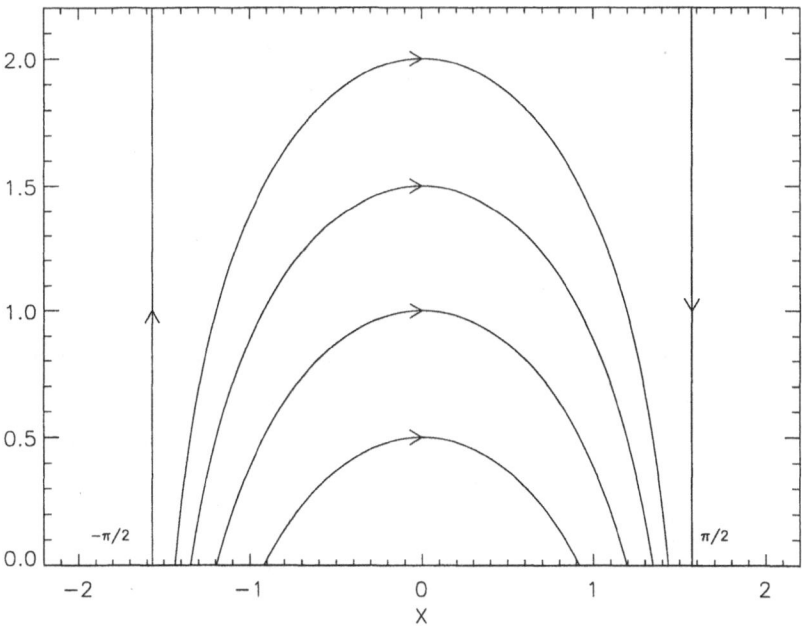

Fig. 5. The magnetic field for a potential arcade model.

4.4   Non-Constant α Fields

In two dimensions ($\mathbf{B}(x, z)$), the components of the magnetic field can be written in terms of the flux function,

$$B_x = \frac{\partial A}{\partial z}, B_y(x, z), B_z = -\frac{\partial A}{\partial x} \qquad (55)$$

so that $\nabla.\mathbf{B} = 0$ is immediately satisfied. We wish to model a situation where the footpoints of the field are anchored down on some surface (the photosphere at $z = 0$, say) (see Figure 6). If we project the resulting field onto the $x - z$ plane, then $dx/B_x = dz/B_z$ or rather,

$$\frac{\partial A}{\partial x}dx + \frac{\partial A}{\partial z}dz = 0, \qquad (56)$$

or $dA = 0$ such that $A =$ constant. Calculating the current density gives,

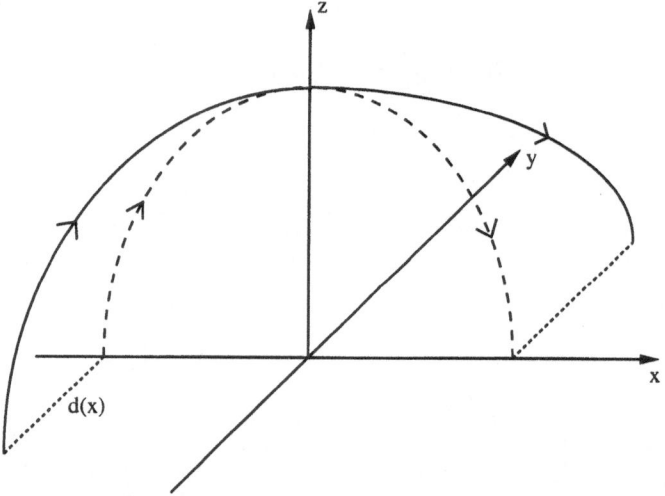

Fig. 6. A magnetic fieldline (solid) with footpoint displacement $d(x)$ with its projection onto the $x - z$ plane.

$$j_x = -\frac{1}{\mu}\frac{\partial B_y}{\partial z}, \qquad (57)$$

$$j_y = \frac{1}{\mu}\left(\frac{\partial B_x}{\partial z} - \frac{\partial B_z}{\partial x}\right), \qquad (58)$$

$$j_z = \frac{1}{\mu}\frac{\partial B_y}{\partial x}, \qquad (59)$$

such that the Lorentz Force reads as

$$\nabla^2 A \frac{\partial A}{\partial x} + B_y \frac{\partial B_y}{\partial x} = 0, \tag{60}$$

$$\frac{\partial B_y}{\partial x} \frac{\partial A}{\partial z} - \frac{\partial A}{\partial x} \frac{\partial B_y}{\partial z} = 0, \tag{61}$$

$$\nabla^2 A \frac{\partial A}{\partial z} + B_y \frac{\partial B_y}{\partial z} = 0. \tag{62}$$

Now, (61) is $\nabla B_y \times \nabla A = 0$; therefore, these vectors must parallel. However, they are perpendicular to the surfaces B_y =constant and A = constant, thus allowing us to say that $B_y = B_y(A)$. If we write $\partial B_y / \partial z = (dB_y/dA)(\partial A/\partial z)$, then (62) becomes,

$$\nabla^2 A + B_y \frac{dB_y}{dA} = 0 \tag{63}$$

which is known as the *Grad-Shafranov Equation*. This non-linear equation in A has some analytical solutions ($B_y(A) = $ constant, $cA^{1/2}, cA, e^{-2A}$ and A^{-1}); otherwise (63) must be solved numerically.

4.5   Application of Magnetohydrostatic Balance to the Internal Structure of a Prominence

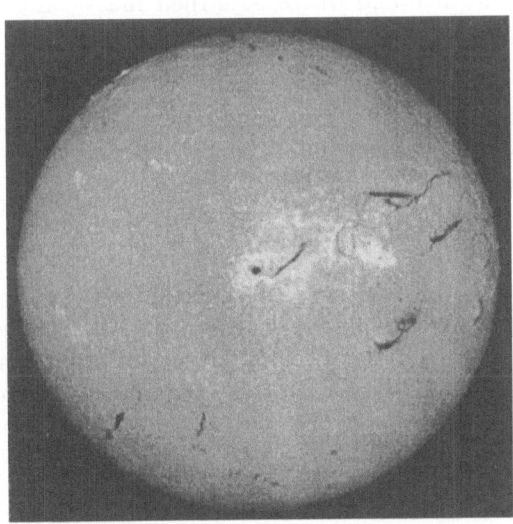

Fig. 7. Typical Hα image of the solar disc where filaments show up as dark ribbons of absorption (Space Environment Laboratory, Boulder, Colorado).

Quiescent prominences are large, cool dense structures which are observed as dark, "snake-like" Hα absorption features (filaments) on the solar disc (see

Table 1. Summary of Observed Quiescent Prominence parameters. Figures in brackets denote "typical" values.

Parameter	Value
Length	60 - 600 Mm (200)
Height	10 - 100 Mm (50)
Width	4 - 15 Mm (6)
Density	$10^{16} - 10^{17}$ m^{-3}
Temperature	5000 - 8000 K
Field	3 - 30 G (5)
Field Angle	$\approx 20°$ to axis
Lifetime	1 - 300 days

Figure 7). Table 1 gives a brief outline of the physical parameters associated with prominences and an excellent review on the many aspects of prominence research can be found in Tandberg-Hansen (1974). The main theoretical questions to be answered can be summarized as

- What is their magnetic structure (both inside and outside the prominence)?
- How are they formed and where does their mass come from?
- What causes the footpoint and fine structure observed?
- Why do they erupt outwards?

Let us consider a model for the internal structure of a prominence where the magnetic force will balance the effect of gravity on the mass of the prominence. Thus, in magnetohydrostatic equilibrium,

$$0 = -\nabla p + \mathbf{j} \times \mathbf{B} + \rho \mathbf{g}. \tag{64}$$

Note that outside the prominence is dominated by the magnetic field. Thus, consider the case when the prominence temperature is constant ($T = $ constant), $B_y = $ constant, $p_z = p_z(y)$, $B_z = B_z(y)$ with the boundary conditions that $P \to 0$ and $B_y \to \pm B_{z0}$ as $y \to \infty$. The $\hat{\mathbf{y}}$ component of the force-balance equation gives

$$-\frac{\partial}{\partial y}\left(p + \frac{B_z{}^2}{2\mu}\right) = 0, \tag{65}$$

which gives

$$p + \frac{B_z{}^2}{2\mu} = f(z), \tag{66}$$

or rather with the above boundary conditions,

$$p = \frac{B_{z0}{}^2 - B_z{}^2}{2\mu}. \tag{67}$$

The $\hat{z}$ component of (64) gives,

$$-\frac{\partial}{\partial z}\left(p + \frac{B_z{}^2}{2\mu}\right) - \rho g + \frac{B_y}{\mu}\frac{\partial B_z}{\partial y} = 0. \tag{68}$$

However, given (67), the total pressure term in (68) is zero. Thus, by using the gas law, (68) can be integrated as

$$B_y \int \frac{dB_z}{(B_{z0}{}^2 - B_z{}^2)} = \frac{g}{2RT} \int dy. \tag{69}$$

Putting $L = 2RTB_y/gB_{z0}$, we have

$$B_z = B_{z0} \tanh\left(\frac{y}{L}\right), \tag{70}$$

if $B_z = 0$ at $y = 0$. Thus, the pressure can also be specified from (67). Figure 8 shows the internal pressure profile as well as sketching the resulting magnetic field. Several generalisations of this solution have been constructed including spatial variations of the temperature (Milne *et al*, 1979) as well as a slow change in the value of B_y with height (Ballester and Priest, 1987).

5   Coronal Heating Theories

The nature of the coronal heating mechanism is still a fiercely debated topic. Although it has been accepted that the Sun's magnetic field plays a vital part, neither theory nor observations have yet been able to unambiguously settle this argument. Any coronal heating model should be able to explain

- the energy losses from both quiet and active regions;
- the variation of the corona's temperature with the solar cycle;
- the heating of coronal loop structures, formation of X-ray bright points and the presence of coronal holes.

However, definitive evidence for the dominance of a particular heating theory has been difficult to obtain due to a number of reasons (Zirker, 1993);

- there may be several different machanisms operating at once;
- the energy release sites may be too small to resolve with present instrumentation. MHD suggests that small scales must be generated for dissipation.
- if the heating is in discrete events (see Section 5.2), the frequency of these events may be too high to resolve.
- any unique signature may be destroyed during the thermal release of energy.

In the following we will investigate several different theoretical mechanisms suggested to deposit heat in the corona thereby balancing the energy losses due to thermal conduction, radiation and plasma mass loss.

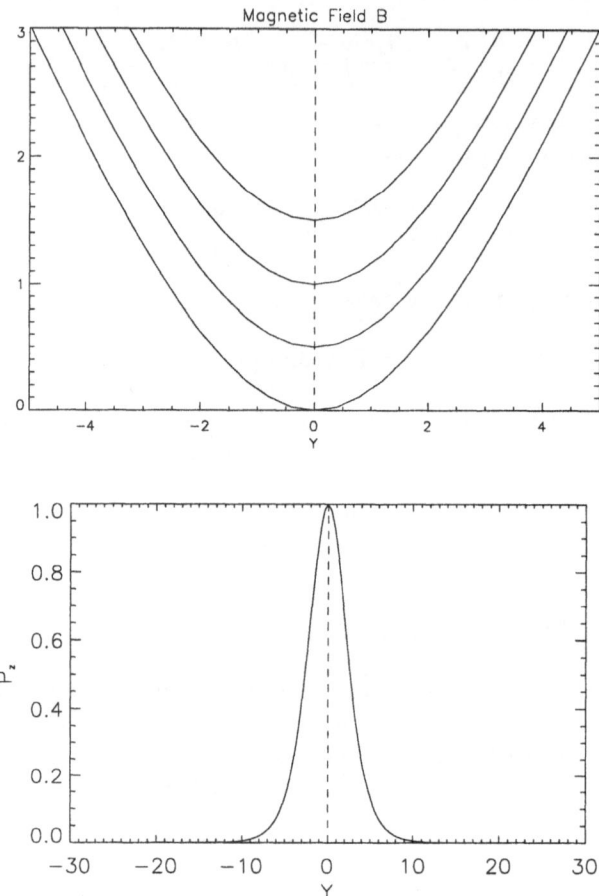

Fig. 8. Internal magnetic field and pressure for prominence support (with arbitary values for B_{z0}, T and p). Field direction is from left to right.

5.1   Heating Coronal Loop Structures

Coronal Loops It has been found that the solar corona consists of a large variety of loop structures that are thought to outline the magnetic field. Observations from the Soft X-ray Telescope (SXT) on Yohkoh (Strong, 1994) show a range of loop sizes and shapes with active regions appearing as bright tangles of magnetic field lines surrounded by a network of large scale loops in the Sun's quiet regions. These recent observations indicate that active region loops are in constant motion, forming then reforming. They appear to be slowly moving outwards from the Sun's surface, carrying plasma with them with the magnetic flux tubes confining the ionised plasma due to the high conductivity in the corona. Exotic loop geometries (twisted, sheared, kinked or cusped) appear possible with the magnetic structures showing changes on timescales that range from seconds to months. Table 2 gives a summary of some of the observed coronal loop parameters.

In order to investigate coronal loop dynamics, it is asssumed that the plasma $\beta = 2\mu p/B^2$ (given as the ratio between the plasma pressure and the magnetic pressure) is very much less than unity. Thus, the three dimensional set of MHD equations is reduced to a set of one dimensional equations *along the fieldlines* (see Walsh, Bell and Hood, 1995). An important aspect of the modelling is the form of the heating term. In the following we consider the importance of the spatial and temporal nature of the heat deposition and how it affects the thermal structure along the coronal loop as you travel from the chromosphere ($\approx 10^4$K), through the transition region ($\approx 10^5$K) and into the corona (2×10^6K).

Table 2. Summary of observed coronal loop parameters (Smith, 1997). Figures in brackets denote "typical" values.

Parameter	Value
Footpoint Separation	1.5 - 500 Mm (100)
Height	40 - 560 Mm (50 - 100)
Width	0.7 - 30 Mm (3 - 10)
Density	$0.2 - 20 \times 10^9$ cm^{-3} (2)
Density Enhancement	1.5 - 16 (3 - 10)
Apex Temperature	$2.4 - 6 \times 10^6$K (2)
Magnetic Field	5 - 300 G (100)
Aspect Ratio	0.01 - 0.25
Plasma β	0.003 - 0.01
Alfvén Speed	2000 km s^{-1}
Sound Speed	200 km s^{-1}

Dependence on Position The importance of where the heat is deposited along the loop structure can be demonstrated from Figure 9 where half of a 60Mm loop is shown and we assume that the loop is symmetric about the midpoint (at $s = 0$). The total heat deposition occurs at one of three separate locations ((a) near the top of the loop; (b) somewhere midway along the loop and (c) near the loop footpoints) with the corresponding thermal structure shown. As well as the obvious larger temperature at the apex for case (a) as compared to case (c) say, the deposition in this latter case also creates a different temperature profile *along the loop itself*. In particular, the deposition of energy within the "legs" of the magnetic loop structure leads to a very uniform temperature plateau along most of the loop length. That is not the scenario for cases (a) and (b).

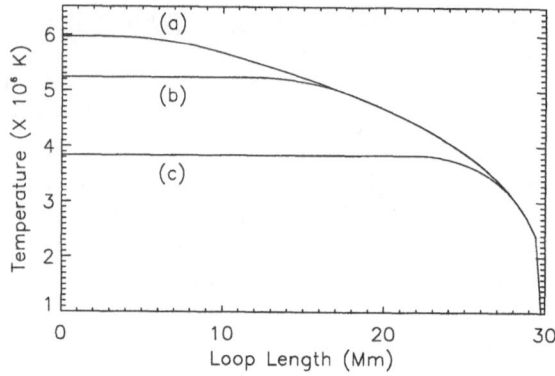

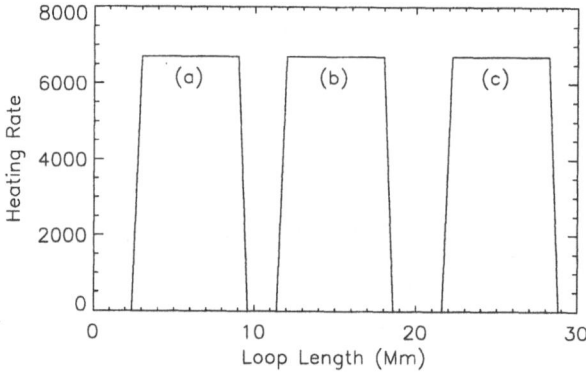

Fig. 9. Thermal equilibria along a symmetric, 60Mm coronal loop structure where the heat deposition occurs either at (a) near the loop summit; (b) midway along the loop length or (c) in the loop legs near the footpoints.

Priest *et al* (1997) have taken this idea a stage further by measuring the temperature along a large coronal loop ($\approx$ 700Mm) observed by Yohkoh. Assuming that radiation can be neglected, the authors set up a simple analytic model for the balance between conduction and the "unknown" heating term. Figure 10 shows the comparison between the observed loop temperature and the theoretical models. It appears that a uniform heating model best fits the observed temperature values.

Dependence on Time The latest observations of the solar atmosphere reveal it to be dynamic environment where there are brightenings and intensity changes right down to the temporal resolution of the instrument. Thus, it is very likely that the coronal heating mechanism will also be time dependent. As a simple introduction to the importance of the temporal nature of the

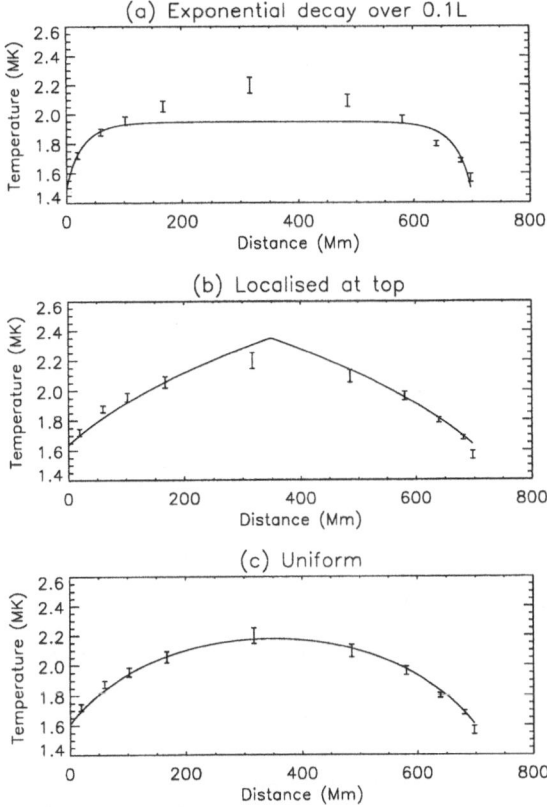

Fig. 10. Observed temperature in MK as a function of distance at ten points along the loop with the errors shown. Also included are the best fit model temperatures for heating that is (a) decaying from the feet over a tenth of the loop length, (b) concentrated at the summit and (c) uniform (Priest, *et al*, 1997).

heating term, consider the same coronal loop structure as outlined in the previous Section but in this case let the heat deposition be uniform along the loop *but vary in time* such that

$$H = H_0(1 + \sin(\omega t)), \tag{71}$$

where $H_0 = 50$ ($\approx 2 \times 10^{18}$ J s^{-1} or 2×10^{25} ergs s^{-1}) is the average heat deposited in the loop per heating cycle and ω is the heating frequency. Consider the variation of the temperature at the apex (T_0) of the loop as the heat deposition varies in time. In Figure 11 the change in T_0 is plotted against H/H_0 for a heating period of approximately 500s. The steady state T_0 values for constant H values are shown by the dashed line.

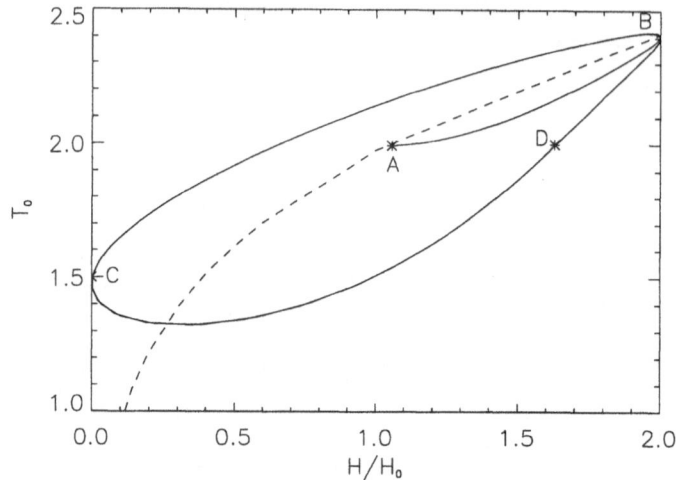

Fig. 11. Variation of the summit temperature T_0 (in units of 10^6K) with heating function H/H_0 for a heating period of $\approx$ 500s (Walsh, Bell and Hood, 1996).

At A, $t = 0$ and T_0 is at its equilibrium value. As $H(t)$ initially increases, T_0 rises and stays relatively close to the thermal equilibrium curve; at B, $H = 2H_0$, the maximum value of the heating. As the heating term decreases, T_0 falls and moves away from the equilibrium line — it reaches its minimum value just after $H = 0$ at C. By the time the temperature is at point D, T_0 has once again reached a value comparable to its steady state temperature but it has now been caught in a cyclic orbit. Thus, for this value of the frequency, the summit temperature "locks" into a quasi-steady orbit about its steady state value and is not evolving simply through a set of static equilibria but follows quite a distinct and separate path.

5.2   Small Scale Heating Events

Recent interest has centred on an idea suggested by Parker (1988) that the corona is heated by the cumulative effect of many small, localised bursts of energy corresponding to magnetic reconnection of the magnetic field. This arose from work that illustrated the spontaneous and unavoidable formation of tangential discontinuities in a force-free magnetic field whose photospheric footpoints are subjected to bounded, continuous displacements and shuffling. Thus, neighbouring flux tubes are wound and wrapped around each other in increasingly complicated patterns. Magnetic field reconnection occurs at the braiding boundaries creating heat and plasma flows (Parker 1987). Parker termed the event a "nanoflare" ($\approx 10^{24}$ erg per event) and cited the results of Porter *et al* (1984) and Porter *et al* (1987) on the discovery of localised, persistent and impulsive brightenings in some of the brightest UV sites in active regions.

The definitive detection of these energy release events has been very difficult to obtain. Various names have been given to what appear to be a wide range of possible transient phenomena in the solar atmosphere; X-ray, EUV and UV bright points, coronal jets, small-scale eruptive filaments, microflares, active region transient brightenings, explosive events and blinkers to name but a few. Some of the most recent observations are detailed in Table 3. Several authors have modelled the effect of these transient events on coronal

Table 3. Characteristics of the wide range of small-scale transient phenomena observed in the corona for approximate duration (D), lengthscale (L), energy released (E), calculated Doppler velocities (DV) and frequency of occurence (F) of the events over the solar disk.

Event	D (s)	L (Mm)	E (erg)	DV (km s^{-1})	F (s^{-1})
Blinkers					
Harrison (1997)	300	18	4.4×10^{25}	35-45	11
Explosive Events					
Dere (1994)	60	1.5	-	150	600
Innes *et al* (1997)	240	12-24	-	100	-
Micro/Nanoflares					
Parker (1988)	20	2	10^{24}	-	-
Moore *et al* (1994)	180	-	10^{27}	-	-
Schmieder *et al* (1994)	360	6	10^{28}	-	-
Porter *et al* (1995)	60	2-10	10^{27}	-	-
Active Region Transient Brightenings					
Schimizu (1995)	120 - 420	5 - 40	$10^{25} - 10^{29}$	-	-
Active Region Blinkers					
Ireland *et al* (1997)					
He I	600	14	3×10^{27}	-	-
O V	600	14	3×10^{25}	-	-
Mg IX	300	6	2×10^{24}	-	-
Fe XVI	300	5	3×10^{24}	-	-

plasma contained within a coronal loop-type model (see Kopp and Poletto, 1993; Cargill, 1994; Walsh, Bell and Hood, 1997). However, Hudson (1991) noted that if we extrapolate the observed rate of flares (10^{32} erg per event) down to the as yet undetected nanoflare energies (10^{24} erg), it is found that the contribution by these smaller events would be negligible. Thus, these smaller events must have a power scaling index that is much steeper (approximately -4) in order to contribute significantly to coronal heating.

Lu and Hamilton (1991) introduced the idea of Self-Organised Criticality (SOC) into the modelling of active region evolution. This situation arises from the competition between the external photospheric driver and the redistribution that occurs whenever local field gradients exceed some threshold value (via magnetic reconnection, releasing energy into the surrounding medium). This localised energy release can affect adjacent regions which may themselves go unstable and release energy - an avalanche occurs. Lu and Hamilton (1991) show that the total energy power index is approximately equal to -1.8, in line with flare observations. Vlahos *et al* (1995) investigate the role of anisotropy in the statistical flare model and this model is extended further in Georgoulis and Vlahos (1997). These authors find a separate power-law index (≈ -3.5) for many small "energy" events and that about 90% of the total energy released lies in these weaker flares.

Walsh and Georgoulis (1997) add some physics to this "statistical" flare model by investigating the response of solar plasma contained with a coronal loop to random, localised events generated by the SOC Model. Figure 12 shows the variation of the temperature along half the loop length with respect to time as the heating events occur. They find that, for a 20Mm loop, a typical "hot" thermal structure can be maintained.

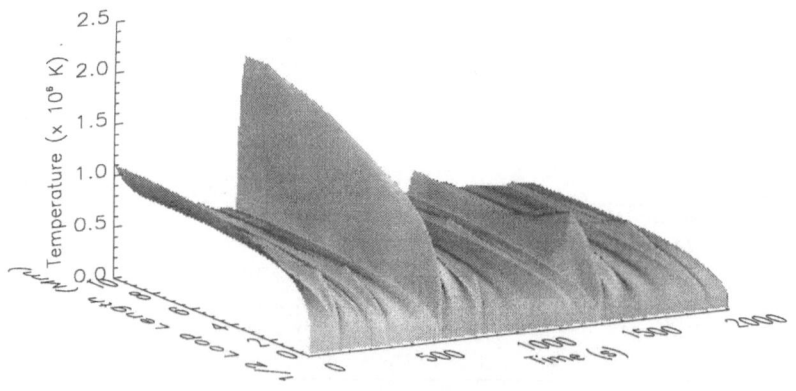

Fig. 12. Evolution of the temperature along the coronal loop for the first 2000 s of the numerical simulation (Walsh and Georgoulis, 1997).

5.3 Phase-Mixing in Dissipative Alfvén Waves

Phase Mixing (Heyvaerts and Priest, 1983; Ireland and Priest, 1997) occurs when the photospheric footpoints of the magnetic fieldlines are oscillating with a fixed frequency. Since each field line in an inhomogeneous atmosphere has its own Alfvén speed, the waves propagate at different speeds and soon move out of phase (Figure 13). Large spatial gradients build up until dissipation smooths them out and extracts the energy. In this situation a field line will only be heated where the dissipation is important. It should be noted that this heat source will be time dependent since there is no reason why the footpoints will continue to be moved with the same frequency. The frequency will vary and so will the position and amount of heat deposited.

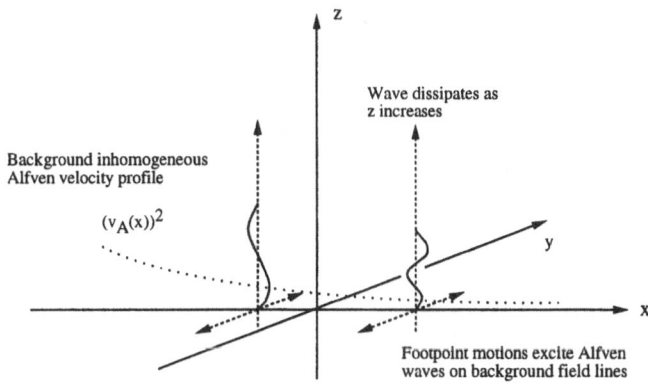

Fig. 13. Sketch for Phase-Mixing of Alfvén Waves (Ireland and Priest, 1997).

The inhomogeneity in the background Alfvén speed means that Alfvén waves on neighbouring field lines have different wavelengths. Therefore, as the wave progresses in z, waves on neighbouring field lines move out of phase relative to each other. Thus, consider the wave equation

$$\frac{\partial^2 v}{\partial t^2} = v_A^2(x)\frac{\partial^2 v}{\partial z^2},\tag{72}$$

which can be used to describe the phase-mixing process in an inhomogeneous, non-dissipative system. A possible solution is

$$v \approx \exp\left[-i(\omega t + k(x)z)\right],\tag{73}$$

where ω is the wave frequency and $k(x) = \omega/v_A(x)$. Thus,

$$\frac{\partial v}{\partial x} \sim v.z.\frac{dk}{dx},\tag{74}$$

such that, the inclusion of a non-uniform background Alfvén velocity creates gradients in the $x-$direction that increase with z. Also note that through

dk/dx, sharper gradients can appear at lower heights as dependent on the inhomogeneity of the plasma.

For a dissipative system, consider an initially vertical magnetic field (with $\mathbf{B}_0 = B_0(x)\hat{z}$ and $\rho_0 = \rho_0(x)$) whose photospheric footpoints are oscillating at frequency ω. This introduces perturbations $v(x, z, t)$ and $b(x, z, t)$ in the velocity and magnetic fields respectively. We shall consider phase-mixing only in the y-direction. From linearisation of the equation of motion and the induction equation, we obtain

$$\rho_0 \frac{\partial v}{\partial t} = \frac{B_0}{\mu} \frac{\partial b}{\partial z} + \rho_0 \nu_v \nabla^2 v, \tag{75}$$

$$\frac{\partial b}{\partial t} = B_0 \frac{\partial v}{\partial z} + \nu_m \nabla^2 b, \tag{76}$$

where ν_m is the magnetic diffusivity and ν_v is the kinematic viscosity. Equations (75) and (76) can be combined to give

$$\frac{\partial^2 v}{\partial t^2} = v_A^2(x) \frac{\partial^2 v}{\partial z^2} + (\nu_m + \nu_v) \left(\frac{\partial^2}{\partial x^2} + \frac{\partial^2}{\partial z^2} \right) \frac{\partial v}{\partial t}. \tag{77}$$

Thus, Equation (77) has a wave term $v_A^2(x) \partial^2 v / \partial z^2$ which is different on each fieldline and this generates the large horizontal gradients. These gradients are smoothed out by the damping term $\partial^3 v / \partial t \partial x^2$. The $\partial^3 v / \partial t \partial z^2$ term in (77) is not important for phase-mixing and will be neglected. If we assume a solution of the form,

$$v \sim \hat{v}(x, z) \exp\left[-i(\omega t + k(x)z)\right], \tag{78}$$

and also assume *weak damping*,

$$\frac{1}{k} \frac{\partial}{\partial z} \ll 1 \tag{79}$$

with *strong phase-mixing*,

$$\frac{z}{k} \frac{\partial k}{\partial x} \gg 1, \tag{80}$$

then the solution to (77) reads as,

$$\hat{v}(x, z) = \hat{v}(x, 0) \exp\left[-\frac{1}{6} \left(\frac{k(x)z}{R_{Tot}^{1/3}} \right)^3 \right], \tag{81}$$

where

$$R_{Tot} = \frac{\omega}{\nu_m + \nu_v} \left(\frac{d}{dx} \log k(x) \right)^2. \tag{82}$$

Therefore, the Alfvén waves decay as $\exp(-z^3)$ under typical coronal conditions. Further investigations into phase-mixing can be found in Browning and Priest (1984) where the authors looked at the effect of the Kelvin-Helmhlotz instability on phase-mixing and Hood, Ireland and Priest (1996) who show that Equation (77) has a similarity solution for a background Alfvén velocity $v_A^2 \sim \exp(x/a)$.

5.4   Resonant Absorption

Alfvén Waves Consider oscillating the footpoints of a set of fieldlines at a given frequency. The Alfvén speed on each fieldline is different in a structured medium ($\omega(x)$). When the frequency of oscillation matches the local frequency of some continuum mode (ω_{excite}), the fieldline resonates and a large amplitude develops (Figure 14). Non-ideal MHD limits the growth of the resonant mode and dissipates the incoming wave energy into heat along the entire length of the fieldline (see Davila, 1987).

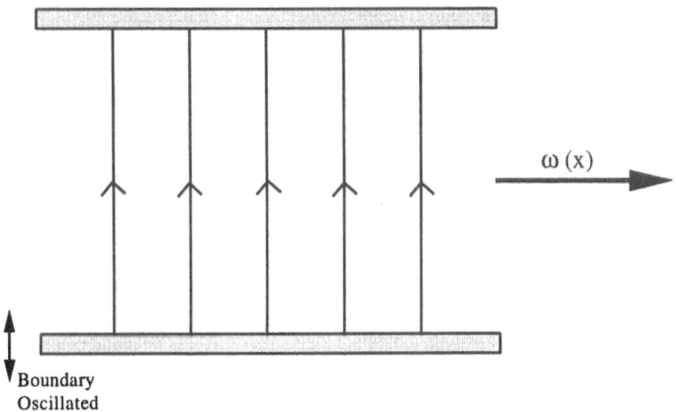

Fig. 14. Sketch for resonant absorption of Alfvén Waves.

Magnetoacoustic Waves Once again, consider a structured medium such that each fieldline has its own Alfvén speed but instead, investigate a wave incident on the side of this magnetic field (Figure 15a). Pressure must be continuous across the magnetic region. Thus, as the fieldline on the right-hand side oscillates in response to the incoming wave, this oscillation will be transmitted to the next fieldline to the left and the process continues. As this coupling occurs across the field, it is possible that the fieldline oscillation will once again match some resonate mode producing a resonate layer. Thus we observe a transmission, absorption and reflection of the incident oscillation (Figure 15b). For further ideas on resonant absorption see Ionson (1978); Hollweg (1984); Poedts, Belien and Goedbloed (1994); Wright and Rickard (1995) and Erdélyi and Goossens (1995).

The St.Andrews/RAL Loops Campaign This campaign was set up for the Coronal Diagnostic Spectrometer (CDS) (Harrison *et al*, 1995) on the

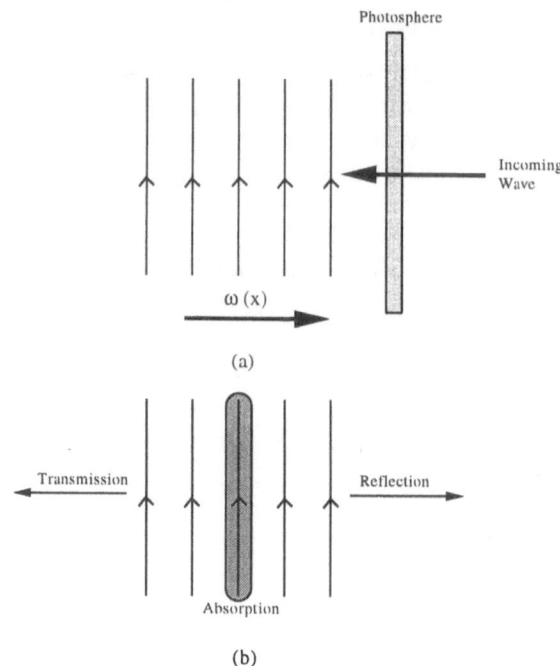

Fig. 15. Sketch for resonant absorption of Magnetoacoustic Waves.

Solar and Heliospheric Observatory (SOHO) with the aim of investigating the structure and activity of active region loops. One advantage of the CDS instrument is that it allows for the observation of the same portion of the Sun at *a wide range of temperatures at exactly the same time*.

Several long duration ($\approx$ 3000s), high cadence (15s) observations where made over physically different parts of an emerging active region and the results analysed for a wide range of possible wave periods. These results are displayed in Figure 16 and it appears that periods from 80s to 300s present in the chromospheric (He I) and transition region (O V) lines are not present in the coronal ones (Mg IX and Fe XVI). That begs the question: why do these low temperature oscillations not appear in the corona? A possible explanation can be found in the resonant absorption of slow magnetoacoustic waves. Using a simple one-dimensional model as outlined in the previous section and introducing the "structured medium" through a rapid density decrease from the chromosphere, through the transition region and into the corona, Figure 17 shows the percentage of the amount of the incident wave that is absorbed. Specific periods found in the time series analysis are added to show that approximately 80% of the incident wave is resonantly absorbed before reaching corona temperatures.

More details on this work can be found in Ireland *et al* (1997) and Erdélyi, Ireland and Walsh (1997).

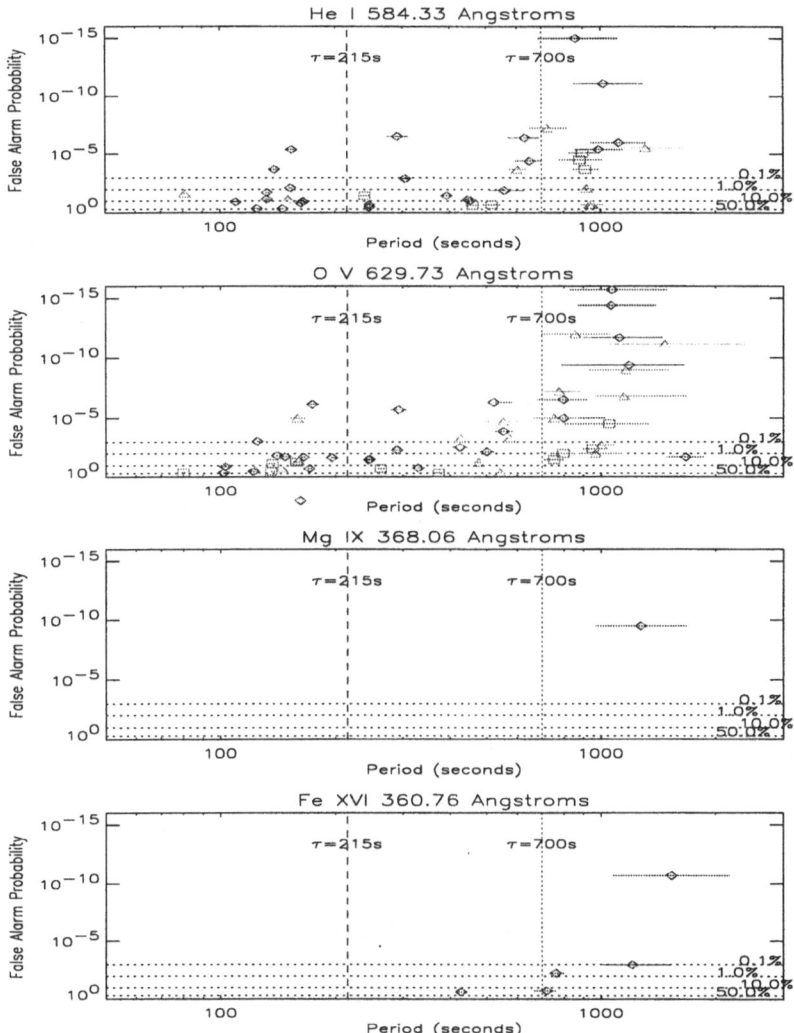

Fig. 16. Summary of time series analysis from Ireland *et al* (1997).

6   Conclusion

As we have seen, the Sun has a rich variety of phenomena that can be modelled using Magnetohydrodynamics. There have been many developments and advancements in MHD theory within the last decade particularly in the area of numerical simulations. With the latest images from Yohkoh, SOHO and the TRACE mission, we now have a timely opportunity for both theorists

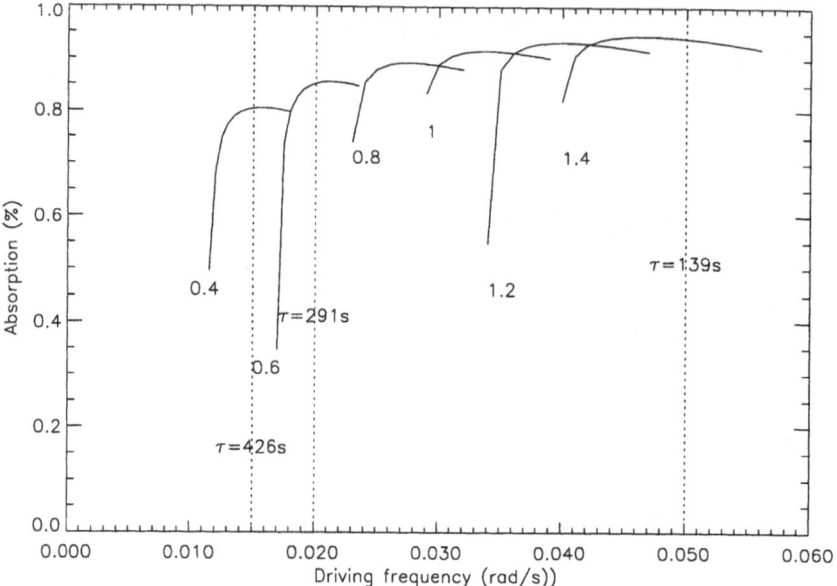

Fig. 17. Resonant Absorption of Slow Magnetoacoustic Waves from Erdélyi, Ireland and Walsh (1997).

and observers to work together to improve our knowledge and understanding of many aspects of our Sun.

7   Acknowledgements

RWW was funded for this work by a PPARC Research Fellowship. The author would wish to acknowledge the warm hospitality received during his visit to IAS for the Summer School and to thank E.R. Priest and J. Ireland in preparation of the manuscript.

References

Ballester, J.L. and Priest, E.R. (1987): *Solar Phys.*, **109**, 335
Braginsky, S.I. (1965): *Rev. Plasma Phys.*, **1**, 205.
Browning, P.K. and Priest, E.R. (1984): *A & A*, **131**, 283
Cargill, P.J. (1994): *Ap. J.*, **422**, 381
Cook, J.W., Cheng, C.-C., Jacobs, V.L. and Antiochos, S.K. (1989): *Ap. J.*, **338**, 1176
Davila, J.M. (1987): *Ap. J.*,**317**, 514
Dere, K.P. (1994): *Adv. Space Res.*, **14**, 13
Erdelyi, R. and Goossens, M. (1995): *A & A*, **294**, 575

Erdelyi, R., Ireland, J. and Walsh, R.W. (1997): *Ap. J.*, submitted
Georgoulis, M. and Vlahos, L. (1997): *Ap. J. Lett.*, in press
Harrison, R.A. *et al* (1995): *Solar Phys.*, **162**, 233
Harrison, R.A. (1997): *Solar Phys.*, in press
Heyvaerts, J. and Priest, E.R. (1983): *A & A*, **117**, 220
Hildner, E. (1974): *Solar Phys.*, **35**, 123
Hollweg, J.V. (1984): *Ap. J.*, **277**, 392
Hood, A.W., Ireland, J. and Priest, E.R. (1996):*A & A*, **318**, 957
Hudson, H.S. (1991): *Solar Phys.*, **133**, 357
Ionson, J.A. (1978): *Ap. J.*, **226**, 650
Innes, D.E., Inhester, B., Axford, W.I. and Wilhelm, K. (1997): *Nature*, in press.
Ireland, J. and Priest, E.R (1997): *Solar Phys.*, **173**, 31
Ireland, J., Walsh, R.W., Harrison, R.A. and Priest, E.R. (1997): *Ap.J.*, submitted
Kopp, R.A. and Poletto, G. (1993): *Ap. J.*, **418**, 496
Lu, E.T. and Hamilton, R.J. (1991): *Ap. J.*, **380**, L89
Milne, A.M., Priest, E.R., and Roberts, B. (1979): *Ap. J.*, **232**, 304
Moore, R.L., Porter, J.G., Roumeliotis, G, Tsuneta, S., Shimizu, T, Sturrock, P.A. and Acton, L.W. (1994): *Proc. of Kofu Symposium*, NRO Report No360, 89
Parker, E.N. (1987): *Ap. J.*, **318**, 876
Parker, E.N. (1988): *Ap. J.*, **330**, 474
Poedts, S., Belien, J.C. and Goedbloed, J.P. (1994): *Solar Phys.*, **151**, 271
Porter, J.G., Toomre, J. and Gebbie, K.B. (1984): *Ap. J.*, **283**, 879.
Porter, J.G., Fontenla, J.M. and Simnett, G.M. (1995): *Ap. J.*, **438**, 472
Porter, J.G., Moore, R.L. and Reichmann, E.J. (1987): *Ap. J.*, **323**, 380
Priest, E.R. (1982): *Solar Magnetohydrodynamics*, (D. Reidel Publ. Co., Dordrecht, Holland)
Priest, E.R. (1994): *Plasma Astrophysics*, **Saas-Fee 24**, eds. J.G. Kirk, D.B. Melrose and E.R. Priest, 1–109.
Priest, E.R. (1996): *Solar and Astrophysical MHD Flows*, ed. K.C. Tsinganos, 171–194
Priest, E.R., Arber, T.D., Heyvaerts, J., Foley, C.R. and Culhane, J.L. (1997): in preparation
Rosner, R., Tucker, W.H. and Vaiana, G.S. (1978): *Ap. J.*, **220**, 643
Schmieder, B., Fontenla, J., Tandberg-Hanssen, E. and Simnett, G.M. (1994): *Proc. of Kofu Symposium*, NRO Report No.360, 339
Shimizu, T. (1995): *PASJ*, **47**, 251
Smith, J. (1997): *PhD Thesis*, University of St.Andrews
Spitzer, L. (1962): *Physics of Fully Ionized Gases* (Interscience, New York)
Strong, K.T. (1994): *Proc. of Kofu Symposium*, NRO Report No.360, 53
Tandberg-Hansen, E. (1974): *Solar Prominences* (Reidel, Dordrecht, Holland)
Vlahos, L., Georgoulis, M., Kluiving, R. and Paschos, P. (1995): *A & A*, **299**, 897
Walsh, R.W., Bell, G.E. and Hood, A.W. (1995): *Solar Phys.*, **161**, 83
Walsh, R.W., Bell, G.E. and Hood, A.W. (1996): *Solar Phys.*, **169**, 33
Walsh, R.W., Bell, G.E. and Hood, A.W. (1997): *Solar Phys.*, **171**, 81
Walsh, R.W. and Georgoulis, M. (1997): in preparation
Wright, A.N. and Rickard, G.J. (1995): *Ap. J.*, **444**, 458
Zirker, J.B. (1993): *Solar Phys.*, **148**, 43

Heliospheric Plasma Physics: an Introduction

Marco Velli

Dipartimento di Astronomia e Scienza dello Spazio,
Università di Firenze, I-50125 Firenze, Italy

Abstract. A selection of topics in plasma physics and hydrodynamics relevant to the heliosphere is presented. The first three sections cover basic defining properties of a plasma, the essentials of particle orbit theory, including conservation of magnetic moment and the various drifts, and an introduction to kinetic theory, with an heuristic derivation of relaxation times. The fourth section is devoted to the hydrodynamic description of the solar wind, and is a pedagogical introduction to Parker's theory as well as to the methods of hydrodynamic and plasma stability. In the fifth and final section we return to the kinetic description of the solar wind plasmas and explore recent versions of the collisionless, or exospheric, models of wind acceleration, and compare their predictive merits and faults with those of the more widely studied fluid theories vis à vis in situ and remote observations of the corona and wind.

1 What is a plasma? Definitions and parameters

A plasma is a non-relativistic, globally neutral ensemble of charged particles in which the dominant interaction occurs through the collective self-consistent long range electromagnetic fields rather than the coulomb interaction between nearest neighbours, the kinetic energy of each particle being much greater than the potential energy due to neighbouring particles. The term "plasma" was first coined by Langmuir to indicate the portion of an arc discharge where electron and proton densities are high but the net space-charge density is negligible, a condition known as "quasi-neutrality". Plasmas are often dubbed the "fourth state of matter", although the process which leads to the ionization of neutral matter is not a phase transition. The Heliosphere is the plasma cavity created by the supersonic expansion of the solar corona into the interplanetary medium. The heliospheric plasma and the plasma in planetary magnetospheres embedded within, are the only natural plasmas for which accurate measurements of both macroscopic properties, such as velocity, temperature and densities as well as microscopic ones, such as the shape of the distribution functions, have been made in situ. Throughout these lectures I will refer to the solar corona and wind for examples and problems in plasma dynamics.

In a plasma, the electrostatic interaction between two particles is modified by the presence of other particles. A positive charge will attract negative charges and repel positive ones, leading to a screening of charges at great distances. In a plasma made up of an equal number of protons and electrons with densities $n_e = n_i = n$, temperatures $T_e = T_i = T$, the characteristic distance over which a charge is effectively neutralized is given by the Debye length λ_D,

$$\lambda_D = \left(\frac{kT}{4\pi ne^2}\right)^{1/2}. \tag{1}$$

The concept of charge screening is relevant only if the density is sufficiently large so that there are many particles in the volume defined by the Debye length; in other words, for our definition of a plasma to be satisfied the plasma parameter g must be

$$g = \frac{1}{n\lambda_D^3} \ll 1. \tag{2}$$

To be somewhat more precise, consider placing a test charge q in a point $\boldsymbol{x}_o$ in a plasma. Then, the electrostatic potential ϕ must satisfy Poisson's equation

$$\nabla^2\phi = -4\pi \sum_i e_i n_i - 4\pi q \delta^3(\boldsymbol{x} - \boldsymbol{x}_o) \tag{3}$$

while the densities of charged particles in equilibrium are determined by Boltzmann's equation

$$n_i = n_{i0}\exp\left(-\frac{e_i\phi}{kT}\right) \tag{4}$$

where the condition of charge neutrality implies that

$$\sum_i e_i n_{i0} = 0. \tag{5}$$

In the limit of small ratio of potential to thermal energies, $(e_i\phi/kT \ll 1)$ one finds

$$\nabla^2\phi = \phi/\lambda_D^2 - 4\pi q \delta^3(\boldsymbol{x} - \boldsymbol{x}_o). \tag{6}$$

The solution which satisfies the boundary condition of vanishing potential at infinity is

$$\phi = (q/r)\exp(-r/\lambda_D), \tag{7}$$

where $r = |\boldsymbol{x} - \boldsymbol{x}_o|$ and ,

$$\lambda_D^2 = \left(\frac{kT}{4\pi \sum_i n_{i0}e_i^2}\right). \tag{8}$$

Problem 1: verify that Equ.(7) satisfies equ.(6).

For a plasma to satisfy the quasi-neutrality condition almost everywhere one must have $\lambda_D/L \ll 1$ where L is a typical scale size of the system. Returning to one of the defining conditions for a plasma, that of small ratio of nearest neighbor potential energy to kinetic energy, one may write

$$\frac{\langle e^2/r \rangle}{\langle \frac{1}{2}mv^2 \rangle} \ll 1. \tag{9}$$

Recalling that the average kinetic energy of a particle is just the definition of temperature, and that the average interparticle distance is related to the density, i.e.,

$$\left\langle \frac{1}{2}mv^2 \right\rangle \sim kT, \quad \left\langle 1/r \right\rangle \sim n^{1/3}, \tag{10}$$

we find

$$\frac{e^2 n^{1/3}}{kT} \ll 1 \tag{11}$$

which again is nothing but the condition of small plasma parameter $g \ll 1$ previously derived for effective plasma shielding. Another immediate consequence of the above is that interparticle collisions will be dominated by small angle scattering, since the distance r_c at which two particles of equal mass will stop, i.e. the average impact parameter, divided by the average interparticle distance $r \sim n^{-1/3}$, scales as $g^{2/3}$. The probability of a large angle collision for two particles is therefore $p_2 \sim nr_c^3 \sim g^2 \ll 1$. For a three-particle collision, this becomes $p_3 \sim p_2^2 \sim g^4$ and so on, providing the basis for a small parameter expansion in the formal development of the kinetic theory of plasmas, as will be further discussed below. Above, we have neglected Lorentz force, (i.e., that $< F_{Coul.}/F_{Lor.} > \sim (v/c)^2$).

Problem 2: show that if the plasma is non-relativistic binary electrostatic interactions are much larger than the Lorentz force, (i.e., that $< F_{Coul.}/F_{Lor.} > \sim (v/c)^2$).

Consider now typical values for the plasma in the solar corona and solar wind at the earth's orbit: in the former case $n \simeq 10^8$ cm^{-3}, $T = 10^6$ K one obtains $\lambda_D \simeq 0.5$ cm, $g \simeq 10^{-7}$, while in the latter $n \simeq 10$ cm^{-3}, $T \simeq 10^5$ K and one finds $\lambda_D \simeq 5 \cdot 10^2$ cm, $g \simeq 10^{-9}$.

Up to now, we have described only spatial properties of the plasma shielding effect. The process has an associated time-scale which provides one of the fundamental frequencies in plasma dynamics. Consider a plasma composed of ions and electrons and neglect for the moment ion motion while electrons are displaced by a small distance x in one direction, then an electric field is generated with $E = 4\pi e n x$ causing electrons to feel a restoring force with equation of motion

$$m_e \frac{\partial^2 x}{\partial t^2} = -eE = -4\pi e^2 n x \tag{12}$$

and one immediately identifies harmonic oscillations with frequency

$$\omega_{pe} = \left(\frac{4\pi n e^2}{m_e} \right)^{1/2}. \tag{13}$$

These are known as electron plasma oscillations or Langmuir oscillations and do not propagate in a cold plasma.

Problem 3: show that if we consider the ions to have a finite mass, then oscillations occur with

$$\omega^2 = \omega_{pe}^2 + \omega_{pi}^2. \tag{14}$$

Considering once more typical values of densities for the solar corona and solar wind at 1 AU we find

$$\nu_{pe} \simeq 10^4\, n_e^{\frac{1}{2}} \simeq 10^8 \text{ Hz (Corona)}, \simeq 3\,10^4 \text{Hz (Wind at 1 AU)} \tag{15}$$

$$\nu_{pi} \simeq \left(\frac{m_e}{m_i}\right)^{1/2} \nu_{pe} \simeq 2\,10^6 \text{ Hz (Corona)}, \simeq 6\,10^2 \text{ Hz (Wind at 1 AU)} \tag{16}$$

2   Particle orbit theory: a synopsis

2.1   Cyclotron motion and the E × B drift

Though a plasma owes much of its behavior to collective effects, a basic understanding of the dynamics of individual particles in time and space varying electromagnetic fields is an essential prerequisite. The equation of motion of a non-relativistic particle of mass m and charge e may be written

$$m\frac{d\boldsymbol{v}}{dt} = e\left(\boldsymbol{E} + \frac{\boldsymbol{v}}{c} \times \boldsymbol{B}\right) \tag{17}$$

where $\boldsymbol{E}, \boldsymbol{B}$ are the electric and magnetic fields respectively. Taking components parallel and perpendicular to the magnetic field $\boldsymbol{B}$, this reads

$$\frac{dv_\parallel}{dt} = \frac{e}{m}E_\parallel, \tag{18}$$

$$\frac{d\boldsymbol{v}_\perp}{dt} = \frac{e}{m}\left(\boldsymbol{E}_\perp + \frac{\boldsymbol{v}_\perp}{c} \times \boldsymbol{B}\right). \tag{19}$$

Motion along the magnetic field is therefore a straightforward accelerated motion via the parallel electric field. For static fields, the perpendicular motion is conveniently separated into a time varying and time independent part

$$\boldsymbol{v}_\perp = \boldsymbol{v}_D + \boldsymbol{v}_1(t) \tag{20}$$

which upon substitution yields

$$\frac{d\boldsymbol{v}_1}{dt} = \frac{e}{m}\left(\boldsymbol{E}_\perp + \frac{\boldsymbol{v}_D}{c} \times \boldsymbol{B} + \frac{\boldsymbol{v}_1(t)}{c} \times \boldsymbol{B}\right). \tag{21}$$

By choosing $\boldsymbol{v}_D$ appropriately one may make the first two terms on the right hand side of Equ.(21) cancel, i.e.

$$\boldsymbol{E}_\perp + \frac{\boldsymbol{v}_D}{c} \times \boldsymbol{B} = 0$$

and taking the vector product with the magnetic field then yields

$$v_D = c\frac{E_\perp \times B}{B^2}.$$ (22)

The lowest order, time-averaged perpendicular motion is thus given by what is known as the "E cross B" drift. This motion, which does not depend on particle mass or charge and does not therefore cause a current flow in the plasma, is the fundamental "slow" fluid motion of a plasma in the so called MHD ordering of plasma theory. The magnetohydrodynamic model of a plasma, or MHD, essentially limits the motion across the magnetic field to the drift just defined, with all its implications (such as the fundamental Alfvén or frozen-in theorem, according to which field lines are carried by the fluid in a topology preserving fashion). The time dependent motion given by

$$\frac{dv_1}{dt} = \frac{e}{m}\left(\frac{v_1(t)}{c} \times B\right)$$ (23)

is best described by introducing orthogonal vectors e_2, e_3 perpendicular to the magnetic field. When the latter is constant, the motion is given by

$$v_1 = v_1\left(e_2 cos\Omega_c t - e_3 sin\Omega_c t\right)$$ (24)

where $\Omega_c = eB/mc$ is the cyclotron frequency. Electron and proton plasma frequencies may be expressed as

$$\Omega_{ce} = 1.76 \; 10^7 \; B \; \text{rad/sec} \qquad \Omega_{ci} = 9.58 \; 10^3 B \; \text{rad/sec},$$ (25)

where B is expressed in Gauss. Typical coronal values ($B = 10$ Gauss) are $\Omega_{ce} = 1.76 \; 10^8$ rad/sec, $\Omega_{ci} = 10^5$ rad/sec while in the solar wind at 1 AU one has $B = 10^{-4}$ Gauss and $\Omega_{ce} = 1.76 \; 10^3$ rad/sec, $\Omega_{ci} = 1$ rad/sec.

Problem 4: show that in the presence of a gravitational field of acceleration g one has in addition to the electric drift a gravitational drift given by

$$v_{Dg} = \frac{mc}{e}\frac{g \times B}{B^2}.$$ (26)

An interesting property of the gravitational drift is that velocity is proportional to mass and depends on charge so that a net current which is mostly carried by protons is generated

$$J = n_i ev_i - n_e ev_e \simeq n_i m_i c\frac{g \times B}{B^2}.$$ (27)

Problem 5: show that this is exactly the current required to maintain force balance across the magnetic field for a current carrying fluid of mass density $\rho = n_i m_i$.

2.2   Adiabatic invariants

Because electric and magnetic fields in space are never uniform or constant, there are in general no exact integrals of motion to which one can resort to simplify the understanding of charged particle dynamics. These are replaced in many situations by approximate integrals or adiabatic invariants, quantities that are conserved, provided the time/space variations in the fields are smaller than the characteristic time-scales of particle motion. Consider first the magnetic moment of a charged particle in a uniform but slowly time-varying magnetic field. The kinetic energy of perpendicular motion may be written

$$U_\perp = \frac{1}{2}mv_\perp^2 = \frac{1}{2}m\omega^2 r_\perp^2,$$

where $\omega, r_\perp$ are the instantaneous gyro-frequency and radius respectively. The induction electric field does work on the particle, since it is parallel to the instantaneous gyro-motion, the rate of energy gain/loss being given by

$$\frac{dU_\perp}{dt} = ev_\perp \cdot E \tag{28}$$

If the particle is gyrating rapidly, so that the relative change in magnetic field during one gyro-orbit is small, one may calculate the change in perpendicular energy by neglecting the change in the orbit caused by the electric field, so that over one gyro-period T

$$\Delta U_\perp = \int_0^T dt\, ev_\perp \cdot E \simeq \oint eE \cdot dx$$

and the latter integral may be calculated using the induction equation and Stokes' theorem

$$\Delta U_\perp \simeq e/c \int ds \frac{\partial B}{\partial t} \simeq \frac{e}{c}\pi r_\perp^2 \frac{dB}{dt}. \tag{29}$$

To obtain the gyro-averaged rate of change of the energy one must then divide by the gyro-period so that

$$\frac{dU_\perp}{dt} = \frac{\Delta U_\perp}{T} = \frac{U_\perp}{B}\frac{dB}{dt}. \tag{30}$$

It follows that the magnetic moment $\mu = U_\perp/B$, ratio of the gyro-energy to magnetic field, is conserved. One may show that μ is conserved also in a purely spatially varying magnetic field in which gradients have scales larger than the cyclotron radius of a particle. In the latter case, the total particle energy is also conserved, and this has an important consequence on the dynamics, as it allows the possibility of particle trapping. A typical example of particle trapping is given by magnetic mirror configurations, where an axially symmetric magnetic field is generated by two current rings placed parallel and concentrically to each other at some distance on an axis. The magnetic field increases from a minimum midway between the two rings to a maximum at the center of each ring. One may write for each particle

$$\frac{1}{2}mv^2 = \frac{1}{2}mv_\parallel^2 + \frac{1}{2}mv_\perp^2 = \frac{1}{2}mv_\parallel^2 + \mu B. \tag{31}$$

From Equ.(31) that if a particle starts off with a given total energy and magnetic moment in a region of low enough magnetic field, and moves in a direction of increasing field, it will be reflected when it reaches a position where

$$B = \frac{mv_0^2}{2\mu}. \tag{32}$$

Suppose now that the minimum and maximum values of the magnetic field in the configuration are given by B_m, B_M, then a particle will be trapped if its velocity in the position of minimum field forms an angle to the magnetic field $\theta > \theta_L$, where the loss-cone angle θ_L is defined as

$$sin(\theta_L) = \left(\frac{B_m}{B_M}\right). \tag{33}$$

Particles within the loss-cone are instead lost from the magnetic mirror configuration.

A coronal loop with both foot-points anchored in the photosphere is a typical configuration which presents a magnetic mirror because of the tapering of magnetic field lines in the photosphere and chromosphere. For trapped particles in such a configuration there exists a second adiabatic invariant related to the periodic motion in the direction parallel to the magnetic field,

$$I_2 = \oint mv_\parallel ds. \tag{34}$$

For example, type IV microwave bursts are believed to arise via gyro-synchrotron radiation from flare-produced mildly relativistic electrons which are trapped at coronal heights in an active region.

There is also a third adiabatic invariant arising from the periodic motion around flux tubes, which corresponds to the flux enclosed by the particle drifts around a trapping flux tube (see, e.g., Sturrock, (1994) §3 for a thorough discussion of adiabatic invariants and particle orbits).

2.3 Particle drifts: guiding center motion

We have already introduced the E cross B and gravitational drifts of charged particles. A general approach to the solution of the equations of motion, when gradients are small with respect to the cyclotron radius and the fastest time-scale is given by the gyro-motion around the magnetic field is, given by multiple scale analysis. One introduces the small parameter

$$\delta \simeq r_\perp \frac{|\nabla B|}{B} \simeq \frac{1}{\Omega_c} \frac{1}{B} \frac{dB}{dt} \ll 1, \tag{35}$$

and then carries out a formal expansion of the equations of motion, where one must take care of the ordering of the different terms: the particle position reads

$$r(t) = R(t) + \delta r_1(\gamma(t), t) + \delta^2 r_2(\gamma(t), t) + \ldots \tag{36}$$

where now $R(t)$ represents the position of the particle averaged over the rapid gyro-motion, i.e. the position of the guiding center of the particle orbit. Here $\gamma(t)$ represents the phase of gyration, and it will follow that $\gamma(t) = \Omega_c t$ to lowest order. Since the frequency is a large parameter, time-derivatives are $O(\delta)$ with respect to γ derivatives: another way of expressing this is to consider the inertial term in the equation of motion to be $O(\delta)$ compared with the electromagnetic term, i.e. to replace m/e with $\delta m/e$ in the formal expansion, and consider the gyro-phase time derivative to have an expansion

$$\dot{\gamma} = \frac{1}{\delta} \Omega_0(t) + \Omega_1(t) + \ldots. \tag{37}$$

The guiding center velocity therefore has an expansion of the form

$$\dot{R}(t) = \delta v_0(t) + \delta v_1(t) \ldots, \tag{38}$$

and clearly does not depend on the phase. To define the guiding center position unambiguously one imposes that the phase average of all successive terms in the position vector should vanish: $< r_i(\gamma, t) >_\gamma = 0$.

The electric and magnetic fields must also be expanded; consistency requires the parallel electric field to be small, i.e., its expansion must begin at $O(\delta)$, while the magnetic field reads

$$B(r, t) = B(R, t) + \delta r_1 \cdot \nabla B(R, t) + \ldots. \tag{39}$$

The particle velocity has a more complicated appearance because all previous expansions, in γ, r_i, v_i contribute

$$\frac{dr}{dt} = v_0(t) + \delta v_1(t) \ldots + \frac{\partial r_1}{\partial \gamma}(\Omega_0 + \delta \Omega_1 + \ldots) + \delta \frac{\partial r_1}{\partial t} + \ldots.$$

Upon substituting into Equ.(17), one finds the following equation at lowest order:

$$\Omega_0^2 \frac{\partial^2 r_1}{\partial \gamma^2} - \frac{e \Omega_0}{mc} \frac{\partial r_1}{\partial \gamma} \times B(R) = \frac{e}{m}\left(E(R) + \frac{v_0}{c} \times B(R)\right). \tag{40}$$

Clearly the left hand side of Equ.(40) depends on the phase, and its gyro-average vanishes identically, while the right hand does not. The general solution (with the appropriate choice of integration constants) therefore reduces to that obtained in Equ.(24) with the same choice of coordinate system,

$$r_1 = r_c\left(e_2 \sin\gamma + e_3 \cos\gamma\right),$$ (41)

where r_c is the gyration (Larmor) radius and

$$\Omega_0 = \Omega_c, \qquad v_{0\perp} = c\frac{E_\perp \times B}{B^2}.$$ (42)

At subsequent orders, one finds the same operator acting on r_i, that is, equations of type

$$\Omega_0^2 \frac{\partial^2 r_i}{\partial\gamma^2} - \frac{e\Omega_0}{mc}\frac{\partial r_i}{\partial\gamma} \times B = \frac{e}{mc}v_{i-1}\times B + F(v_0 \ldots v_{i-2}, r_1 \ldots r_{i-1}, \Omega_0 \ldots \Omega_{i-1}).$$ (43)

This equation determines the dependence of r_i on γ but also, via appropriate averages, the drift velocity v_{i-1}, the frequency Ω_{i-1} and the slow time dependence of r_{i-1}. This is because the RHS of Equ.(43) must vanish when multiplied by all the independent solutions to the homogeneous equation and averaged (i.e. it must be orthogonal to the kernel of the operator on the LHS). As previously, taking the average of Equ.(43) leads to

$$\frac{e}{mc}v_{i-1} \times B = - < F_\perp >_\gamma, \quad < F_\parallel >_\gamma = 0,$$ (44)

which determines the perpendicular component of v_{i-1} and the parallel component of v_{i-2}. Taking the scalar product and averaging respectively with r_1 and $\partial r_1/\partial\gamma$ then determines Ω_{i-1} and r_{i-1}. Let us determine the first order guiding center velocity or drift:

$$\frac{e}{mc}v_1 \times B = \frac{dv_0}{dt} - \frac{e}{m}E_\parallel b + \frac{e\Omega_0}{mc}\left\langle r_1 \cdot \nabla B \times \frac{\partial r_1}{\partial\gamma}\right\rangle.$$ (45)

Problem 6: derive Equ.(45); show also, using the explicit form of r_1, that the last term in bracket equals

$$\left\langle r_1 \cdot \nabla B \times \frac{\partial r_1}{\partial\gamma}\right\rangle = \frac{r_c^2}{2}\nabla|B(R)|.$$ (46)

Substitution of the above in Equ.(45) finally leads to

$$v_{1\perp} = \frac{b}{\Omega_0} \times \frac{dv_0}{dt} + \frac{\Omega_0 r_c^2}{2B}b \times \nabla B,$$ (47)

where $b = B/B$ is the unit vector in the field direction. The first term in Equ.(47) is called the "inertial drift" and the second the "gradient-B drift". Taking the scalar product of Equ.(45) with b on the other hand yields (since $1 = b \cdot b$ variations in b are perpendicular to b),

$$m\frac{dv_{0\parallel}}{dt} = eE_\parallel - \mu b \cdot \nabla B(R)$$ (48)

where we have once again introduced the magnetic moment $\mu = \frac{1}{2}r_c^2\Omega_c^2$; the magnetic mirror force therefore appears explicitly on the RHS.

Problem 7: consider a 10^6K proton at the surface of the Sun in a 1 G magnetic field. Assuming the temperature to be isotropic compare the mirror force, calculated for a spherically symmetric purely radial magnetic field, with the gravitational force. What magnetic field magnitude is required for the two to be equal?

3 Basics of kinetic theory

A complete definition of the state of a plasma composed of N particles is given by the probability density P in a $6N$ dimensional phase space defined by position and momentum of each particle, with volume element

$$d^3r^1 d^3p^1 \ldots d^3r^N d^3p^N = d1d2\ldots dN,$$

so that the probability of finding the first particle in the element $d1$, the second in the element $d2$....the Nth in the element dN is given by

$$dW = Pd1\ldots dN$$

and the probability density satisfies the normalization constraint

$$\int P(1,\ldots,N,t)d1\ldots dN = 1.$$

This complete description is however essentially useless, since one is really interested in a statistical description describing the probability of finding 1,2,... particles in a given volume of phase space. This information is given by the single particle distribution function

$$F_1(1) = N \int P(1,\ldots,N,t)d2\ldots dN,$$

the two-particle distribution function

$$F_2(1,2) = N^2 \int P(1,\ldots,N,t)d3\ldots dN$$

and so on. These distributions are normalized so that

$$\int F_1(1)d1 = N, \quad \int F_2(1,2)d1d2 = N^2, \quad \ldots.$$

One also defines the correlation functions $C_1(1)$, $C_2(1,2)$, $C_3(1,2,3)$ to describe the non factorizable parts of the multi-particle distribution function:

$$F_1(1) = C_1(1), \quad F_2(1,2) = C_1(1)C_1(2) + C_2(1,2), \tag{49}$$
$$F_3(1,2,3) = C_1C_1C_1 + C_1C_2 + C_3,\ldots. \tag{50}$$

where in the last expression a sum over permutations of arguments is implicitly assumed in the C_1C_2 contribution. Clearly, the s-particle correlation vanishes as long as there are at least two statistically independent particles.

The basic equation of kinetic theory is the continuity of probability density in phase space

$$\frac{\partial P}{\partial t} + \sum_{i=1}^{N} \frac{\partial}{\partial r_i}(\dot{r}_i P) + \frac{\partial}{\partial p_i}(\dot{p}_i P) = 0, \tag{51}$$

which thanks to Hamilton's equations

$$\dot{r}_i = -\frac{\partial H}{\partial p_i}, \qquad \dot{p}_i = \frac{\partial H}{\partial r_i} \tag{52}$$

becomes Liouville's theorem

$$\frac{dP}{dt} = \frac{\partial P}{\partial t} + \dot{r}_i \cdot \frac{\partial P}{\partial r_i} + \dot{p}_i \cdot \frac{\partial P}{\partial p_i} = 0 \tag{53}$$

To find how the 1-particle distribution function evolves one must integrate over $d2 \ldots dN$; the resulting equation involves an integral of the two-particle distribution function of the type (if we limit ourselves to consider only electrostatic interactions between particles)

$$\frac{dF_1}{dt_1} = e^2 \int \left(\frac{\partial}{\partial r_1} \frac{1}{|r_1 - r_2|} \cdot \frac{\partial}{\partial p_1} \right) C_2(1,2) d2, \tag{54}$$

where the subscript in the total time derivative indicates that derivative operators on F_1 are limited to particle 1 coordinates. This is the first of a chain of equations in what is known as the BBGKY hierarchy (Lifshitz and Pitaevskii, (1981), Akhiezer et al. (1975)). When the plasma is not far from a state of statistical equilibrium and the plasma parameter is small, one may show that the pair correlation function $C_2(1,2)$ is of order g with respect to the single particle distribution function, while the three-particle correlation function is of order g^2. If one neglects the 2-particle correlation function entirely Equ.(54) reduces to the Vlasov equation:

$$\frac{\partial F_1}{\partial t} + v_1 \cdot \frac{\partial F_1}{\partial r_1} + \frac{e}{m} \left(E + \frac{v_1}{c} \times B \right) \cdot \frac{\partial F_1}{\partial v_1} = 0 \tag{55}$$

This equation, together with Maxwell's equations, forms a closed system which is the fundamental model describing the dynamics of collisionless plasmas, i.e. for plasmas in which the mean free path (as defined below) is of the same order as the typical scale of the plasma configuration. Since Equ.(55) expresses constancy of the distribution function along particle orbits, it follows that any function of the constants of the motion forms a solution. A typical example is given by the collisionless modeling of the solar wind expansion, as we shall see in the final sections of this lecture, where the constants of the motion considered will be the total particle energy and the magnetic moment.

Volume in phase-space is preserved by the Vlasov equation, and the entropy, defined as $S = -\int F\log(F)\, d^3p\, d^3v$, remains constant.

If 2-particle correlations, and hence collisions, are taken into account, then the term on the RHS of Equ.(54) couples via $C_2(1,2)$ into the next order equation in the BBGKY hierarchy, see, e.g., Akhiezer et al. (1975). First consider a plasma in thermal equilibrium at temperature T in the absence of external electro-magnetic fields: the probability density then follows the Gibbs distribution; taking only into account the dominant electrostatic interaction, which depends exclusively on the relative position $r = |r_2 - r_1|$ of two particles, the particle distribution functions may then be shown to be separable in velocity and coordinate space, and in particular the one-particle distribution assumes the Maxwellian form

$$F_1 = F_m = n\left(\frac{m}{2\pi kT}\right)^{3/2}\exp\left(-\frac{mv^2}{2kT}\right). \tag{56}$$

Neglecting three-particle correlations (terms which are of order g^2 and $g^2\log(g)$), the solution for $C_2(1,2)$ may be written as

$$C_2(1,2) = F_m(v_1)F_m(v_2)\Psi(r), \tag{57}$$

$$\Psi(r) = -\frac{e^2}{rkT}\exp\left(-\frac{r}{\lambda_D}\right), \quad r \gg n^{-1/3} \tag{58}$$

$$\Psi(r) = -1 + \exp\left(-\frac{e^2}{rkT}\right). \quad r \ll n^{-1/3} \tag{59}$$

One therefore sees explicitly how at distances greater than the Debye length the probability of finding two particles is exponentially reduced. If one now considers a plasma which is not in thermodynamic equilibrium, but for which the departure from thermal equilibrium is weak, one can still neglect third order correlations, but the solution for the pair-correlation function will depend explicitly on the form of the single-particle distribution. The derivation is somewhat tedious and we refer to Akhiezer et al. (1975), §1.3, and Lifshitz and Pitaevskii, (1981), §1.3, §4.1, for the details. The result is the Boltzmann equation,

$$\frac{\partial F_1}{\partial t} + v_1 \cdot \frac{\partial F_1}{\partial r_1} + \frac{e}{m}\left(E + \frac{v_1}{c}\times B\right)\cdot\frac{\partial F_1}{\partial v_1} = \left(\frac{\partial F_1}{\partial t}\right)_c \tag{60}$$

where the derivative on the RHS, known as the collision integral, in the Landau formulation (i.e. in the hypothesis of weak momentum exchange in the Coulomb interaction between particles) is expressed as

$$\left(\frac{\partial F_1}{\partial t}\right)_c = -\frac{\partial s_i}{\partial p_i},$$

i.e. as a divergence of a particle flux in momentum space (p_i is the ith component of the particle 1 momentum). The vector s has components

$$s_i = 2\pi e^4 L \int \left(F_1(1)\frac{\partial F_1(2)}{\partial p_{2j}} - F_1(2)\frac{\partial F_1(1)}{\partial p_{1j}} \right)\left(\frac{\delta_{ij}}{v} - \frac{v_i v_j}{v^3} \right) d^3 p_2, \qquad (61)$$

where $v_i = (v_1 - v_2)_i$ is the relative velocity, and L is the Coulomb logarithm $L = \log r_M/r_m$, the logarithmic measure of the maximum to minimum impact parameter for Coulomb scattering.

The distribution function F_1 obeying Boltzmann's equation still contains much more information than what is often required. Consider for example Maxwell's equations for the electric and magnetic fields; to solve for the self-consistent fields, the sources, i.e., charge density and current density must be specified. Such quantities are obtained as the zeroth and first order velocity moments of the distribution function F_1. Fluid descriptions of the plasma are constructed by taking such moments of Equ.(60): a closure problem presents itself again, since the advection term couples the evolution of nth order moments of F_1 to that of the $n+1^{th}$. The best known fluid plasma closures are the cold plasma approximation, in which all second order and higher moments are assumed to vanish, see e.g, Boyd and Sanderson (1969) §3.6 for a discussion of the limitations of such a model, and Magnetohydrodynamics. Here the basic assumption is that the E cross B drift is of the same order as the thermal velocity of the plasma; the distribution functions are assumed to be Maxwellian functions, and the pressure is assumed to be isotropic. Higher order moments, including the heat flux and viscous stresses, vanish in this case, and the electric field obeys

$$E + \frac{v}{c} \times B = 0,$$

i.e., ohmic dissipation is neglected to lowest order. For a detailed discussion of this and other closure approximations, such as those leading to the drift and gyrokinetic equations, we refer to Hazeltine and Meiss (1991), §4.1-4.6 and Boyd and Sanderson (1969), §3.

Coulomb collisions among particles allow the different ion species to relax to a common temperature. Though one may use Equ.(60,61) to derive equations for the relaxation times (Braginskii (1966)) and the transport co-efficients for a Maxwellian plasma, we briefly summarize the results here giving only a heuristic justification. Consider first the ion-ion and electron-electron collision frequencies. These may be written as $\nu = n\sigma v$ where σ is the Coulomb cross-section, which may be estimated via the closest approach for particles colliding exactly at the thermal speed: $\sigma = \pi r^2$ where $r = 2e^2/mv^2$. The collision frequencies then read

$$\nu_{ii} = \left[\frac{4\sqrt{\pi}L}{3}\right]\frac{ne^4}{m_i^{1/2}(kT_i)^{3/2}} \qquad (62)$$

$$\nu_{ee} = \left[\frac{4\sqrt{\pi}L}{3}\right]\frac{ne^4}{m_e^{1/2}(kT_e)^{3/2}} \qquad (63)$$

where the coefficient in square brackets comes from a precise evaluation with Maxwellian distributions. These give the estimated relaxation times ($\tau = 1/\nu$) for each individual distribution function. The electron-ion relaxation time is however much smaller, since the energy of an electron remains essentially unchanged in a collision with an ion, while the change in ion energy will be of the order of m_e/m_i times the electron energy in one collision. This means that the electron-ion temperature equilibration time is

$$\tau_{ei} = \frac{m_i}{m_e}\tau_{ee} = \frac{m_i}{m_e\nu_{ee}}.$$

In other words we have the relations $\tau_{ee} : \tau_{ii} : \tau_{ei} = 1 : (m_e/m_i)^{\frac{1}{2}} : (m_e/m_i)$. This allows an intermediate fluid closure in which both ions and electrons are treated as separate fluids coupled by collisions (the two-fluid model). The estimates of relaxation times given above may be used to introduce non-ideal effects, or transport coefficients, in the MHD equations, as listed below:

$$\frac{\partial \rho}{\partial t} + \nabla \cdot \rho \boldsymbol{v} = 0, \tag{64}$$

$$\rho\left(\frac{\partial \boldsymbol{v}}{\partial t} + \boldsymbol{v} \cdot \nabla \boldsymbol{v}\right) = -\nabla p + \frac{1}{c}\boldsymbol{j} \times \boldsymbol{B} - \nabla \cdot \boldsymbol{P}, \tag{65}$$

$$\frac{\partial p}{\partial t} + \boldsymbol{v} \cdot \nabla p + \gamma p \nabla \cdot \boldsymbol{v} = (\gamma - 1)\left(\nabla : \boldsymbol{\kappa}\nabla T + \eta \boldsymbol{j}^2\right), \tag{66}$$

$$-\frac{1}{c}\frac{\partial \boldsymbol{B}}{\partial t} = \nabla \times \boldsymbol{E} \tag{67}$$

$$\nabla \times \boldsymbol{B} = 4\pi \boldsymbol{j}/c, \tag{68}$$

$$\boldsymbol{E} + \boldsymbol{v} \times \boldsymbol{B}/c = \eta \boldsymbol{j} \tag{69}$$

$$\nabla \cdot \boldsymbol{B} = 0. \tag{70}$$

Here ρ, p are the density and pressure respectively, $\boldsymbol{v}, \boldsymbol{E}, \boldsymbol{B}$ velocity, electric and magnetic fields, $\boldsymbol{j}$ is the current, γ the ratio of specific heats, while $\boldsymbol{P}$ is the viscous stress tensor, and η and κ are the resistivity and thermal conduction, which for a plasma are also generally tensors with differing components along and across the magnetic field. Ohm's equation (69) essentially arises from the electron fluid equation of motion

$$m_e n \frac{d\boldsymbol{v}_e}{dt} = -\nabla p_e - en\boldsymbol{E} - \frac{en}{c}\boldsymbol{v}_e \times \boldsymbol{B} - m_e n \nu_{ee}(\boldsymbol{v}_e - \boldsymbol{v}_i), \tag{71}$$

where indices i, e refer to the electron and ion (proton) speeds respectively. If one neglects electron pressure and inertia, one is left with Equ.(69) where the resistivity $\eta = 4\pi\nu_{ee}/\omega_{pe}^2$. Thermal conduction along the magnetic field may be estimated in a similar way as $\kappa = n\lambda v$, where v is the thermal speed of the plasma electrons, (in the perpendicular direction gyro-frequency effects strongly inhibit thermal conduction, see e.g. Golant et al. (1980), §9). The mean free path is defined as usual as $\lambda = 1/n\sigma = v/\nu_{ee}$ so $\kappa \sim \kappa_0 T^{5/2}$ for a plasma with Coulomb collisions. For a high temperature plasma such as the solar corona, thermal conduction becomes very large, and this is one of the key elements that lead to the prediction of a very extended corona and subsequently a wind, as discussed in the next section. The MHD model is abundantly used to discuss the physics of the outer solar atmosphere and solar wind (see e.g. the chapter on MHD in this book as well as Priest, (1984)).

Before going on to discuss the hydrodynamics of the solar wind expansion we close this section with a remark on the question of relaxation times and dissipation as it should occur in order to account for coronal heating. Coulomb collisions for a test particle in a plasma in thermodynamic equilibrium lead to the following form for the drag

$$\frac{dv}{dt} = -\frac{v}{\tau} = -\phi(v) \tag{72}$$

where for high speeds one has the cross section derived above and $\phi(v) \sim v^{-2}$, while for low velocities it may be shown that (e.g. Sturrock, (1994) §10.4, §10.5) $\phi(v) \sim v$. In the presence of an electric field the equation of motion becomes

$$\frac{dv}{dt} = \frac{e}{m}E - \phi(v). \tag{73}$$

For an electric field which is not too strong, there will be two equilibrium points where the velocity remains constant, one at low speeds and one at high speeds. It is easy to conclude that the high speed equilibrium point must be unstable, since for a small increase of velocity the drag will decrease and vice-versa. In fact the whole region in velocity space where $\phi(v)$ is a decreasing function is potentially unstable to what is known as the runaway effect. To estimate the strength of the electric field required for the runaway to become an important process it is sufficient to estimate the intensity required for an electron at the thermal speed to acquire a speed comparable to the thermal speed between successive collisions. In one mean free path, a particle starting at the thermal speed v will end up with a velocity $V \simeq eE\lambda/mv$. If we now impose $V = v$ we find

$$E_D = \frac{2\pi n e^3 L}{kT}. \tag{74}$$

The value of the Dreicer electric field for typical values of coronal plasma parameters $n = 10^9$ cm^{-3}, $T = 10^6$K (the Coulomb logarithm $L \simeq 20$) is $E_D \simeq 10^{-7}$ in cgs units. Consider now the electric fields which must be present in the solar corona to account for the average heat flux, lost, estimated at $\phi = 10^6$ erg/cm^2/sec, if the dominant dissipation is indeed ohmic, collisional dissipation: at scales on which the currents dissipate, the electric field may be estimated as $E = \eta j$. For a coronal volume of height H and surface S the volumetric heating rate is then $\epsilon = \phi/H$, implying that $\eta j^2 f = \epsilon$, where f is a filling factor giving the fraction of the volume over which currents effectively dissipate. Little is known about this filling factor, but the observations of very fine scale structure in the corona would seem to indicate that it must be fairly small, say $f \simeq 0.1$. Then $E = \eta j = (\eta \epsilon/f)^{1/2}$. Taking a volume of height 10^5 km and typical coronal values for the resistivity, one obtains $E \simeq 1$ statvolt/cm (cgs).

Problem 8: check the arguments presented above.

This field is about seven orders of magnitude larger than the Dreicer field, and even revising estimates on coronal heating, filling factors, in order to reduce this enormous ratio, one is still forced to conclude that the dissipation scales in the coronal plasma are essentially collisionless. On the other hand accelerated particles are produced copiously in the solar corona and are observed via the radio noise they produce (e.g., type III bursts, waves at the electron plasma frequency driven by accelerated electron beams) as well as in situ in the solar wind.

4   Hydrodynamics of the solar wind expansion: why the solar wind is supersonic

Although knowledge of a solar influence at the Earth's orbit dates back from Lord Carrington's observations in the second half of the last century, that aurorae often occurred several hours after white light solar flares, the first direct indication of a continuous outflow of fast particles from the Sun came from Biermann's investigation in the fifties of the shape of the cometary ion tails, from which he deduced an average speed of around 475 km/sec for this flow. In 1957 Chapman showed how a static conductive corona starting at 10^6 K at the Sun should maintain a high density out to far distances (in fact, after an initial decrease, the density should increase outward again!), and in 1958 Parker argued, on the basis of the unreasonably high pressures that static solutions yielded at large distances, that "probably it is not possible for the solar corona, or, indeed, perhaps the atmosphere of any star, to be in complete hydrostatic equilibrium out to large distances". He then proceeded to show that a viable solution yielding negligible pressures at infinity consisted in a flow accelerating continuously and becoming supersonic at large distances. Consider hydrostatic balance for a spherically symmetric atmosphere with gravity

$$\frac{\partial p}{\partial r} = -m_p n \frac{g}{r^2}, \tag{75}$$

where g/R_0 is the gravitational acceleration at the solar surface (and R_0 the solar radius) and we have normalized distances with solar radius. Also m_p is the proton mass so that the mass density $\rho = m_p n$. Recalling that $p = nkT$ we may integrate to find that

$$nkT = n_0 kT_0 \exp\left(-\int_1^r dr \frac{m_p}{kT} \frac{g}{r^2}\right) \tag{76}$$

so that a static spherically symmetric extended atmosphere with a temperature profile decreasing with distance less rapidly than $1/r$ requires a finite pressure at infinity to be confined, the same being true if the atmosphere 'evaporates' a subsonic flow, or breeze. The subsequent Parker - Chamberlain debate on supersonic/subsonic evaporation was cut short by the in situ measurement of the steady, supersonic wind by the Luna 2 (1959) and Explorer (1961) spacecrafts (Hundhausen, 1972). However, Mestel (quoted in Roberts and Soward, 1972), first remarked that it would not take a large fall in coronal temperature for the pressure of the local interstellar medium (ISM) to be sufficient to suppress the solar wind entirely. Indeed, the pressure of the ISM, $p_{ISM} \simeq 1.24 \ 10^{-12}$ dyne/ cm^2 would suffice to confine a $4 \ 10^5$ K static corona with base density 10^9 cm^{-3}. So although correct, the argument for a supersonic wind does not appear to be as strong on the basis of pressure arguments only. In reality, the dependence of spherically symmetric, non-rotating atmospheres with flows on changes of the external conditions is somewhat more subtle, as shown by Velli (1994), and it is worthwhile to discuss the problem in detail. For the sake of analytical simplicity, we will consider only isothermal flows, i.e. flows for which the temperature may be considered constant out to great radial distances.

4.1   Stationary isothermal flows: breezes, winds, accretion

The equations of motion for one-dimensional, spherically symmetric, stationary isothermal flow neglecting self-gravity may be written in the form

$$\frac{\partial}{\partial r}\left(\rho v r^2\right) = 0, \quad p = c^2 \rho \tag{77}$$

$$v\frac{\partial v}{\partial r} = -\frac{1}{\rho}\frac{\partial p}{\partial r} - \frac{g}{r^2} \tag{78}$$

where v is the velocity, c the constant sound speed. For a static atmosphere, the pressure profile is given by $p = p_0 \exp(-g/c^2 + g/rc^2)$ which, as discussed above, implies a non-vanishing asymptotic value for the pressure at large distances $p_\infty^s = p_0 e^{-g/c^2}$. In terms of the mach number $M = v/c$ the flow equations may be written (a prime denoting radial derivatives throughout this section)

$$\left(M - \frac{1}{M}\right)M' = \frac{2}{r} - \frac{g}{r^2 c^2} \tag{79}$$

which may be integrated and expressed in two equivalent ways

$$\frac{1}{2}\left(M^2 - M_0^2\right) - \log\left(\frac{M}{M_0}\right) = 2\log r + \frac{g}{rc^2} - \frac{g}{c^2}. \tag{80}$$

$$M^2/2 + \log p - \frac{g}{rc^2} = M_0^2/2 + \log p_0 - \frac{g}{c^2}, \tag{81}$$

where M_0 is the base Mach number. The second form is essentially the conservation of energy flux, where for an isothermal atmosphere the enthalpy is expressed as $\log p$ instead of $\gamma p/(\gamma - 1)$. Eq.(79) has a singular point at the sonic point, $r = g/2c^2$, $M = 1$. Solutions to the above equations may be represented in the (M, r) phase plane illustrated in Fig. 1, which, following the symmetry of Equ.(79) is symmetric in the sign of M. The diagram is divided into 4 (8, considering positive and negative M) areas (labeled I-IV) by the two critical (transonic) solutions which cross at the sonic point $r = g/2c^2$, $M = 1$. Single valued continuous flow profiles $M(r)$ which are subsonic for all r, the breezes, lie below both transonic curves (region I). Among flows which are subsonic at the atmospheric base the accelerating transonic has the special property that density and pressure tend to zero at large distances: because of the small but finite values of the pressure of the ambient 'external' medium, a terminal shock transition, connecting to the lower branch of the double valued solutions filling region II will in general be present (see e.g. Holzer and Axford, 1970). The jump conditions across such a shock are found from conservation of mass and momentum across the shock, which read (superscripts -,+ denote the solution immediately upstream and downstream from the shock respectively)

$$\rho^- M^- = \rho^+ M^+, \tag{82}$$

$$\rho^+ (M^+)^2 + \rho^+ = \rho^- (M^-)^2 + \rho^-, \tag{83}$$

from which one finds immediately $M^+ M^- = 1$. This gives a way to graphically construct shock transitions; it is sufficient to plot the curves corresponding to $1/M$ for the transonics (dashed lines in Fig. 1), and connect the transonic with the double valued curve in region II where the dashed line intersects them: such solution is given by curve W. The downward transonic in itself is not a possible solution for outflows, since a continuous transition from supersonic to subsonic flows is unstable. However, it plays the same role the Parker wind solution plays for inward directed accretion flows: for negative M solutions, the same construction leads to accretion shocks in the flow (McCrea, 1956), this time in the region labeled IV, and one such transition is shown in Fig. 1 (curve A). For given base values of the pressure, the position of the shock is uniquely determined by the pressure of the interstellar medium, and the distance from the critical point to the shock decreases as the pressure increases: conservation of mass across the shock immediately gives the asymptotic pressure in terms of the upstream Mach number M^- as

$$p_\infty = p_0 M^- \exp\left(M_*^2 - 2\frac{g}{c^2} - M^{-2} + \frac{1}{M^{-2}}\right)/2, \tag{84}$$

where M_* is the base Mach number of the upward transonic; p_∞ is a monotonically decreasing function of M^-, which is itself, obviously, a monotonically increasing function of the shock position r_s, so that increasing p_∞ decreases r_s. When p_∞ reaches a value $p_\infty^c = p_0 \exp(M_*^2/2 - g/c^2)$, the shock distance $r_s = r_c = g/2c^2$ and the discontinuity in the flow velocity reduces to a discontinuity in the derivative of the profile. This is the fastest possible, or critical, breeze, made up of the section of upward transonic below r_c and the section of downward transonic beyond r_c. For the breeze solutions, with

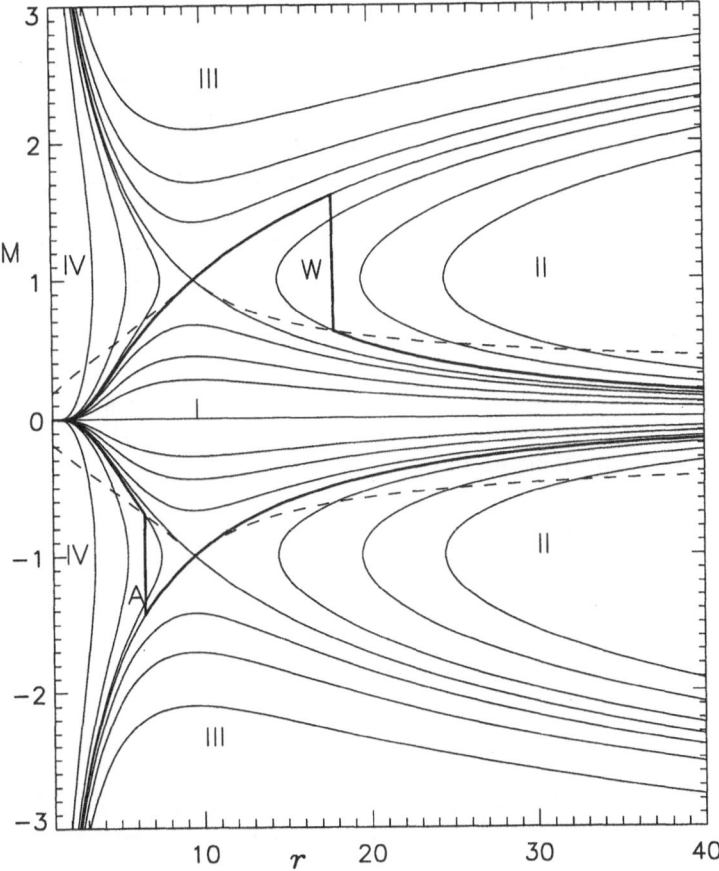

Fig. 1. The (M,r) phase plane. The continuous curves are appropriate both for positive and negative M. The dashed line intersection with double valued curves defines the shock position for winds (region II, curve W) or accretion flows (region IV, curve A).

a base Mach number M_0 such that $M_* > M_0 \geq 0$, the asymptotic pressure is easily calculated to be

$$p_\infty = p_0 \exp\left(M_0^2/2 - g/c^2\right) \geq p_\infty^s. \tag{85}$$

It follows that the pressure required to confine a breeze *increases* with increasing base Mach number, and is greater, if only slightly, than that of a static atmosphere. The limiting value of p_∞ is again p_∞^c. For a given base pressure and asymptotic pressures $p_\infty^c > p_\infty \geq p_\infty^s$ it therefore appears that *two* possible stationary outflow solutions exist, a supersonic shocked wind and a subsonic breeze.

When such conditions occur, it is frequently the case that one of the solutions may be unstable, i.e. that any small perturbation may lead the flow to evolve away from stationarity. In this case, it is the breeze solutions that are unstable: to prove this we introduce small perturbations (sound waves) and linearize the equations of motion around the stationary state. We will apply boundary conditions which allow the configuration to evolve from one to the other of the stationary solutions we have found: i.e., the perturbing sound waves will leave the pressure (and density) unperturbed at the atmospheric base and infinity, i.e. they will be standing waves. It is convenient to introduce characteristic variables $y^\pm = \hat{m} \pm \hat{p}$, where $\hat{p} = \tilde{p}/p$ is the adimensional normalized pressure perturbation and $\hat{m}$ is the Mach number (velocity) fluctuation. In these variables an outward (inward) propagating sound wave has $y^- = 0$ ($y^+ = 0$). Assuming a time dependence $y^\pm = y^\pm(r)exp(-i\omega + \gamma)t$ the linearized equations become

$$(M \pm 1)y^{\pm\prime} - i(\omega + i\gamma)y^\pm + \frac{1}{2}(y^\pm + y^\mp)\frac{M'}{M}(M \mp 1) = 0. \qquad (86)$$

Problem 9: prove Equ.(86), and show that in a uniform flow one has stable propagating waves with dispersion relation $\omega^\pm = (M \pm 1)k$ where k is a radial wave-number, i.e. a spatial dependence $\sim exp(ikr)$ is assumed.

In the presence of a non-uniform but stable flow, Equ.(86) describes wave propagation and reflection, and a conserved flux, the wave action flux, exists (in a static medium, the wave energy flux is conserved); when there are mass motions, this is replaced by the wave action flux: see, e.g. Kadomtsev, (1983) §4 for a discussion of wave energy and Velli, 1993 and references therein for a discussion of wave-action conservation). When $\gamma \neq 0$ the wave action evolution equation becomes

$$\left[\frac{(M+1)^2}{M}|y^+|^2 - \frac{(M-1)^2}{M}|y^-|^2\right]' + 2\frac{\gamma}{M}\left[(M+1)|y^+|^2 - (M-1)|y^-|^2\right] = 0, \qquad (87)$$

the first square bracket being proportional to the wave action.

Problem 10: prove Equ.(87). Hint: one must multiply the equations for inward and outward waves by the appropriate factor and then sum to remove *the y^+, y^- coupling terms.*

Notice that for $|M| < 1$ the term in the second square bracket is positive definite. Integrating this equation between 1 and r and imposing the boundary condition that the pressure perturbation vanish at the extremes, we find the following estimate for γ:

$$\gamma = \frac{2\left(|y^+|_0^2 - |y^+|_r^2\right)}{\int_1^r dr \, M^{-1}\left[(M+1)|y^+|^2 - (M-1)|y^-|^2\right]}, \tag{88}$$

where $|y^+|_0^2$, $|y^+|_r^2$ are the fluctuation amplitudes at the atmospheric base and r respectively. It follows then that if the perturbation amplitude is non vanishing at the base but tends to 0 at great distances, the flow is unstable. Now for $\omega = 0$ and large r Equ.(86) have, for breeze velocity profiles, leading order asymptotic solutions

$$y^+ \sim \frac{e^{\mp\gamma r}}{r} \qquad y^- \sim \pm \frac{e^{\mp\gamma r}}{2\gamma r^2} \tag{89}$$

so the boundary conditions are satisfied either by the first solution, if γ is positive, or the second, if γ is negative. In both cases the amplitudes tend to zero at great distances, the numerator of Equ.(88) is always positive, γ is also positive, and, provided eigenmodes exist, breeze solutions are unstable. Numerical solutions show that this is indeed the case: in Fig. 2 we plot the growth rate of the instability as a function of base Mach number. The growth rate is largest for high values of the base Mach numbers but both the static atmosphere ($M_0 = 0$) and the critical breeze ($M_0 = M_*$) are marginally stable. In the latter case the perturbation equations become singular at the sonic point, because the phase speed of the inward propagating wave vanishes there: an additional regularity condition must be imposed in the stationary equations, effectively isolating the region below the sonic point from the region beyond it. This is the mathematical reason behind the stability of flows with a continuous subsonic/supersonic transition.

The breeze instability is driven by the unfavorable stratification (Equ.(85)). Imagine a static atmosphere, and let the pressure at infinity increase: clearly an *inflow*, not an outflow, is expected to result. That accretion breezes are stable follows immediately from the analysis presented above; the denominator in Equ.(88) changes sign, so the only consistent way to satisfy the boundary conditions is to choose the second solution in Equ.(89), implying a negative value for γ. In fact, the stationary equations are symmetrical in M, while the perturbation equations are invariant under a change in sign of both *M and γ.*

Given that breezes are unstable, we see that even in the pressure range $p_\infty^c > p_\infty \geq p_\infty^s$ the only stable outflow is a supersonic wind with a terminal

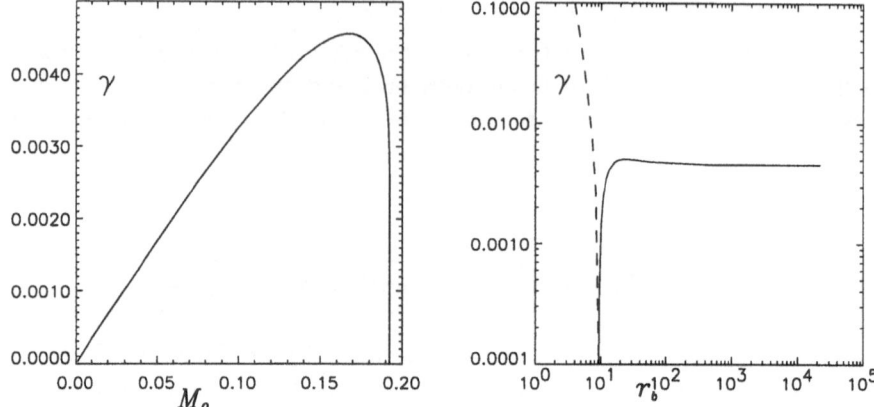

Fig. 2. left: Growth rate γ of breeze instability as a function of base Mach number ($g/c^2 = 5.0$). Marginal stability is obtained for $M_0 = 0$, $M_0 = M_*$. Right: Maximal growth rate for breezes (continuous line) and subsonic accretion (dotted line) as a function of the position of the outer boundary r_b. The two lines join in $r_b = r_s$, where $\gamma = 0$.

shock. What happens if the pressure difference between the coronal base and the distant medium varies? As p_∞ increases, the shock moves inward, decreasing in amplitude as the critical point is approached. When the critical breeze is reached, there is no neighbouring outflow solution capable of sustaining a higher pressure at infinity. The only possibility for the flow is to collapse into its symmetrical ($M \to -M$) critical breeze accretion profile, which is also marginally stable. As the pressure is increased further, an accretion shock is formed *below* the sonic point, connecting the symmetrical of the downward transonic to one of the double valued curves in region IV, (as shown by the curve labelled A in Fig. 1). For $p_\infty > p_\infty^c$ there is a unique shocked accretion flow (McCrea, 1956), and the shock position moves inward from the critical point as the pressure is increased beyond p_∞^c (if the pressure is too high, the shock may occur below the normalization radius chosen). Consider now what happens if, starting from a shocked accretion flow, the pressure at the surface increases, or alternatively the pressure of the ISM decreases. The shock moves outwards, but this time, as p_∞ decreases below p_∞^c, the flow can evolve with continuity into subsonic accretion. As p_∞ decreases further, the accretion-breeze velocities decrease, but when p_∞ decreases below p_∞^s, the flow must accelerate again into a supersonic shocked wind.

The stratification produced by breezes, though globally unstable, is not locally unstable everywhere: for example, below the critical point the pressure in breezes decreases with height more rapidly than in the static case. Inspec-

tion of Equ.(81) actually shows that this is true out to the radius r_s where
the Mach number of the flow has decreased to the same level as the base
Mach number, which may be calculated by imposing $M = M_0$ in Equ.(81),
i.e. $2\log r_s + g/r_s - g = 0$. This equation is independent of base Mach number,
which also means that at this height the pressure is the same for all breezes,
while below this radius, the pressure at a given height is a monotonically
decreasing function of base Mach number. As the boundary conditions are
imposed at closer and closer distances r_b the growth rate of the instability is
reduced, and marginal stability is obtained when $r_b = r_s$. Imposing boundary
conditions below this radius stabilizes the breezes, but consequently desta-
bilizes subsonic accretion, as is shown in Fig. 2, where the maximal growth
rate for breezes (continuous line) and accretion (dashed line) as a function
of r_b is plotted. For large values of g this marginal stability radius depends
exponentially on g as $r_s \simeq \exp(g/2) - g/2$.

When the boundary is at $r_b < r_s$ the equilibrium flows still present an
hysteresis-type cycle in terms of the enthalpy jump between r_b and the coronal base, but in a reversed order with respect to that previously described:
supersonic accretion is blown into supersonic winds as the base pressure is
increased beyond a critical value, while an outflow breeze phase exists before
the collapse to accretion as the pressure at the outer boundary is increased
beyond the value appropriate to a static atmosphere.

In generalizing to polytropic or other more realistic equations for the
energy, some attention is necessary since the density and pressure may fall
to 0 at a finite distance, and transonic flows do not exist for all polytropic
indices ($\gamma \leq 3/2$ below the sonic point is a necessary condition Parker, 1963).
With these caveats, the discussion of the isothermal case is easily generalized.
The energy equation now becomes (as the sound speed varies, c_0 is its base
value)

$$v^2/2 + c^2/(\gamma - 1) - g/r = v_0^2/2 + c_0^2/(\gamma - 1) - g.$$

For breezes the asymptotic behavior $v \sim 1/r^2$ still holds, so that in fact we
may write

$$c_\infty^2/(\gamma - 1) = v_0^2/2 + c_0^2/(\gamma - 1) - g$$

which shows that the temperature at great distances from the central object
increases with the base Mach number, up to the value which, for a given
base density and pressure, gives a transonic flow (Holzer and Axford, 1970).
Conservation of energy across the shock then implies that independently of
the asymptotic pressure, c_∞ is always the same. It is still true that given the
base density and pressure, for a range of pressures at great distances between
that of the static atmosphere and that of the critical breeze there are two
solutions, an unstable breeze (or stable accretion) and a shocked wind, but
now the thermodynamic state of the distant medium is different, the breeze
having a higher density and lower temperature.

This concludes our discussion of the hydrodynamics of radial flows under
gravity; for a review of recent fluid modeling of the solar wind as well as

in-situ measurements and observations see the solar wind chapter, in this book.

5   Kinetic models of the solar wind expansion

In fluid models of the solar wind, a fundamental role is played by transport coefficients such as the heat conduction, whose properties are well defined only if there are small gradients in the system and the distribution function does not depart significantly from a Maxwellian. For example, the conditions for the applicability of the $\kappa \sim \kappa_0 T^{5/2}$ thermal conduction law, which relies on collisions, is that the mean free path be much smaller than the characteristic scale of the temperature gradient, a condition which is verified below about 15 solar radii. Also, we have seen that the conditions for the electric field to become greater than the Dreicer electric field are easily overcome in the solar corona. Finally, distribution functions in the solar wind, show significant anisotropies in temperature as well as accelerated populations (see e.g. Marsch 1991).

To begin the study of kinetic models of coronal evaporation consider first the separate behavior of electrons and protons with identical temperatures in the solar gravitational fields. From hydrostatic equilibrium one finds that

$$\frac{\partial p_e}{\partial r} = -m_e n_e \frac{g}{r^2}, \tag{90}$$

$$\frac{\partial p_i}{\partial r} = -m_i n_i \frac{g}{r^2}, \tag{91}$$

from which it follows that the electron scale height is much larger than the proton scale height

$$n_e = n_{e0}\exp\left(-\frac{gm_e}{kT}\left(1 - \frac{1}{r}\right)\right), \tag{92}$$

$$n_i = n_{i0}\exp\left(-\frac{gm_i}{kT}\left(1 - \frac{1}{r}\right)\right), \tag{93}$$

and therefore quasi-neutrality is violated. What happens is that the charge separation creates a polarization electric field, known as the Pannekoek-Rosseland (PR) field, which lifts the protons and pulls the electrons back in order to ensure quasi-neutrality.

Problem 11: show that the PR field is given by

$$E_{PR} = \frac{m_p - m_e}{2eR_0}\frac{g}{r^2}. \tag{94}$$

Because the mean-free path of particles increases with height due to the decrease in density, one may estimate a radius above which collisions will no longer be important. This leads to the separation of the solar corona into two regions: a collisional barosphere and a collisionless exosphere, whose base is taken to be at that distance r_0 where the mean free path λ is equal to the scale height H. For typical coronal values, this exobase is located between 2 and 10 solar radii. One may now use Vlasov's equation and a Maxwellian distribution function at the exobase to determine how the distribution evolves with height.

Problem 12: show that the solution is given trivially by the fact that F is only a function of the constants of the motion, so that $F(r,v) = F_0(r_0,v_0)$.

Using the PR field and a 10^6 K temperature Chamberlain obtained a mean flow speed at the earth's orbit of around 10 km/sec., and this was the main reason he defended the "breeze" hypothesis against the Parker "wind" model. Subsequent modifications which allowed for an energy-dependent exobase, yielded an increase of the speed to about 140 km/sec but still much less than the speed observed. A major breakthrough was however made by Lemaire and Scherer, (1971) and Jockers (1970), who realized that the polarization field in a flowing atmosphere must be much larger than the PR field. The reason is that otherwise the Sun would continuously charge itself: compare the proton and electron escape velocities

$$V_p = \left(2\frac{g}{r_0} - 2\frac{e\phi_E(r_0)}{m_p}\right)^{1/2},\tag{95}$$

$$V_e = \left(2\frac{g}{r_0} - 2\frac{e\phi_E(r_0)}{m_e}\right)^{1/2},\tag{96}$$

where the potential from the PR electric field (which vanishes at infinity) is given by

$$\phi_E(r) = \frac{m_p - m_e}{2e}\frac{g}{r}.\tag{97}$$

It follows that $m_e V_e^2 = m_p V_p^2$ in $r = r_0$; one may then integrate over the Maxwellian distribution function to find the flux of particles which have $v > V$ and therefore contribute to the mass flux in the wind. The result is that the ratio of electron to proton flux is $F_e/F_p = \sqrt{m_p/m_e} \simeq 42$ (yes! you have the answer to the famous question first posed by Deep Thought in Adams, 1979), and the Sun would become positively charged.

Problem 13: prove the result above.

In reality one must adjust the electric field so as to satisfy both quasi-neutrality and conservation of current, instead of taking as given the PR field (which was derived assuming a static stratification). The Vlasov equation is again a function only of the constants of motion, which, taking the radial magnetic field into account, are the total energy (sum of kinetic, polarization potential, and gravitational) and the magnetic moment. Charge conservation and the Vlasov equation are solved iteratively until convergence is achieved. Lemaire and Scherer showed that for an exobase at $3.5R_0$ and the usual 10^6 K corona the actual potential was about twice as large as the PR potential (600 Volts vs 270 Volts), and that with such a potential in fact the effective escape speed for protons falls to zero, i.e., all outgoing protons escape, while electrons are divided into three categories: escaping, ballistic, and trapped. They obtained an outflow velocity of 240 km/sec at Earth's orbit. The major pitfall of such models as compared to observations is the dramatic decrease in perpendicular to parallel proton and electron temperatures, which comes from the conservation of magnetic moment in a radially expanding magnetic field ($B \sim 1/r^2$). In reality, the same problem would arise also in the absence of magnetic fields because of conservation of angular momentum, which would also lead for individual particles to $v_\perp \sim 1/r$. Nonetheless, Hundhausen (1972) was able to conclude his discussion by saying that the predictions of the exospheric models "are in better agreement with observations than the predictions of basic fluid models". Unfortunately, while in the subsequent twenty five years much attention has been given to fluid modeling, little has been done to overcome the temperature anisotropy problem in exospheric models, a problem which must find its solution in the interaction of the plasma turbulence observed in the wind with the distribution functions. For an introduction to the theory of plasma waves and instabilities we refer to the following books aimed at various levels: Hasegawa, (1975), Kadomtsev, (1965), Kadomtsev, (1983), Sagdeev and Galeev (1969), Stix (1962), Nicholson, (1983), Lifshitz and Pitaevskii, (1981), Akhiezer et al. (1975).

Recently, the Lemaire and Scherer (1971) model has been updated by Maksimovic et al. (1997a), in that a distribution function for protons and electrons more closely resembling those observed *in situ* has been taken as initial condition for the Vlasov equation. This class of distribution functions, the κ distribution, has, with respect to a Maxwellian, an extended high energy tail. Interest in this kind of distribution has been sparked by the possibility of explaining the question of the heating of the corona and expansion of the wind via a very straightforward process known as velocity-filtration (Scudder, 1992 a,b). For a Maxwellian distribution function, which seen as a function of kinetic and potential energies is separable as well as self-similar, the temperature (second order moment of the distribution) does not change with height (gravitational or electric potential). If on the other hand there is an excess of high energy particles, these will climb through the gravitational potential and preferentially survive at greater heights, leading to a wider (hotter) distribution. Though the question of whether such distributions may be created at chromospheric levels remains an open one, it is true that κ function distributions fit the solar wind electron distributions remarkably well, as shown in Maksimovic et al. (1997b), and the correlation distribution-width wind speed has the proper sign, i.e. wider distributions "go faster", as predicted by the theory, which can easily produce speeds at 1 AU in excess of 500 km/sec, with all the drawbacks and simplifications discussed above. It therefore seems that kinetic solar wind models are reaching a competitive stage, though the ultimate aim should be to construct a theory which converges to fluid closures and is both in agreement with present observations, and is predictive as concerns possible future missions, such as the solar probe.

Acknowledgements I would like to thank J.C. Vial and K. Bocchialini for giving me the opportunity to present these lectures as well as providing an enjoyable atmosphere at Medoc during the summer school. Special thanks to Karine also for her careful reviewing and editing of the manuscript.

References

Adams, D. (1979): *The Hitchhiker's Guide to the Galaxy* (Pan Books LTD, London)

Akhiezer, A., Akhiezer, I., Polovin, R., Sitenko, A. and Stepanov K., (1975):*Plasma Electrodynamics* (Pergamon Press, Oxford)

Boyd, T.J.M. and Sanderson, J.J., (1969): *Plasma Dynamics* (Nelson, London)

Braginskii, S. I., (1966): Transport Processes in a Plasma. Reviews of Plasma Phys., **1**, 205.

Golant, V.E., Zilinskij, A.P. and Sacharov, S.E., (1980): *Fundamentals of Plasma Physics* (Wiley & Sons, NewYork)

A. Hasegawa, (1983): *Plasma Instabilities and Nonlinear Effects* (Springer Verlag, Berlin)

Hazeltine, R.D. and Meiss, J.D., (1991): *Plasma Confinement*, (Addison-Wesley, Reading)

Holzer, T. E. and Axford, W. I. (1970): Theory of Stellar Winds and Related flows, Ann. Rev. Astr. Ap. **8**, 30

Hundhausen, A.J. (1972): *Coronal Expansion and Solar Wind* (Springer Verlag, Berlin)

Jockers, K. (1970): Solar Wind Models Based on Exospheric Theory, A. & A. **6**, 219

Kadomtsev, B. (1965): *Plasma Turbulence* (Academic Press, London)

Kadomtsev,B. (1983): *Phenomènes Collectifs dans les Plasma* (MIR, Moscow)

Krall, N. and Trivelpiece, A., (1973): *Principles of Plasma Physics* (McGraw-Hill, New York)

Lemaire, J. and M. Scherer, (1971): Kinetic Models of the Solar Wind, J. Geophys. Res. **76**, 7479

Lifshitz, E., and Pitaevskii, L., (1981):*Physical Kinetics* (Pergamon Press, Oxford)

Maksimovic, M. Pierrard, V. and Lemaire, J. (1997a): A Kinetic Model of the Solar Wind with Kappa Distribution Functions in the Corona A. & A. **324**, 725

Maksimovic, M. Pierrard, V. and Riley, P. (1997b): Ulysses Electron Distributions Fitted with Kappa Functions Geophy. Res. Letts. **24**, 9

Marsch, E. (1991): Kinetic Physics of the Solar Wind Plasma, *Physics of the Inner Heliosphere II* (Springer Verlag, Heidelberg) 45

McCrea, W. (1956): Shock Waves in Steady Radial Motion Under Gravity, Ap. J. **124**, 461

Nicholson, D., (1983): *Introduction to Plasma Theory* (Wiley & Sons, NewYork)

Parker, E. 1963: *Interplanetary Dynamical Processes* (Interscience, New York)

Priest, E.R., (1984): *Solar Magnetohydrodynamics*, (Kluwer, Dordrecht)

Roberts, P. H. and Soward, A. M. (1972): Stellar Winds and Breezes, Proc. Roy. Soc. London **328** A, 185

Sagdeev R.Z. and Galeev, A.A. (1969): *Nonlinear Plasma Theory* (W.A. Benjamin, New York)

Scudder, J. (1992a): On the Causes of Temperature Change in Inhomogeneous Low-Density Astrophysical Plasmas, Astrophys. J. **398**, 299

Scudder, J. (1992b): Why All Stars Should Possess Circumstellar Temperature Inversions, Astrophys. J. **398**, 319

Stix, T.H. (1962): *The Theory of Plasma Waves* (McGraw-Hill, New York)

Sturrock, P., (1994): *Plasma Physics* (Cambridge University Press, Cambridge)

Velli, M. (1993): On the Propagation of Ideal, Linear Alfvén Waves in Radially Stratified Stellar Atmospheres and Winds, A.& A. **270**, 304

Velli, M. (1994): From supersonic winds to accretion: comments on the stability of stellar winds and related flows, Ap. J., **432**, L55

Instrumentation: Spectroscopy

Philippe Lemaire

Institut d'Astrophysique Spatiale, Unité mixte Université-CNRS,
bâtiment 121, Université de Paris-Sud,
F-91405 Orsay Cedex, France

Abstract. Our knowledge of the the solar atmosphere comes from the light analysis. The ultraviolet wavelength is the privileged spectral range to study the solar plasma in several stages of ionization emitted from the chromosphere to the corona. Starting with basic notions of spectroscopy, an overview of important parameters is given. Then, after a short description of classical spectrometers, a detailed analysis of the properties of the mountings used on SOHO is performed; some examples are shown. I discuss how to separate the real solar signal from the instrumental contribution.
From what has been learnt on the Sun during the last two decades, I make a prospective for future solar spectrometers developments.

1   Basic equations

What information do we have in light? The light is the result of a distribution of photons that can be collected on a detector through an optical system which makes the integration over a frequency range (spectral range and resolving power) and over a time range (time resolution). The light spectrum is composed of lines and continuum representing the local condition of the source that emits it. Line and continuum are the result of transitions between 2 levels of energy, respectively bound-bound transition and bound-free (and/or free-free) transition.

The spectral line intensity distribution can be represented by a Voigt function, a combination of gaussian and Lorentz profiles. At the ν frequency:

$$\Phi(\nu) = \frac{a}{\pi^{3/2}\Delta\nu_D} \int_{-\infty}^{\infty} \frac{e^{-x^2}}{(u-x)^2 + a^2} dx \tag{1}$$

where $u = \frac{\nu - \nu_0}{\Delta\nu_D}$ and $a = \frac{\Gamma}{4\pi\Delta\nu_D}$

The Doppler width:

$$\Delta\nu_D = \frac{\nu_0}{c}\sqrt{\frac{2kT}{m} + V_t^2} \tag{2}$$

takes into account the temperature T, the atomic mass m, the velocity of light c (k, Boltzmann's constant), and the turbulent velocity V_t.
The Lorentzian width Γ is the sum of the collisional and natural widths.

Note. For details, we refer to Emerson (1996) or other textbooks (Thorne (1988)).

2   Grating spectrometers

In this section only a few grating spectrometers using entrance slits are studied.

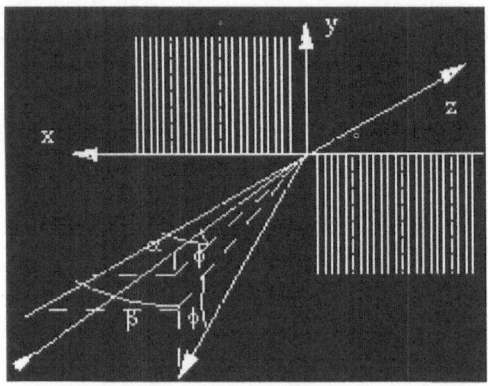

Fig. 1. Grating incidence and diffraction scheme.

2.1   Basic formulae

The ruled grating works as a multi-slit combination of diffraction patterns. The condition that the contributions from all rulings interfere constructively for a light of wavelength λ (see Figure 1) is

$$m\lambda = \sigma cos\phi(sin\alpha + sin\beta) \tag{3}$$

where
m is the order number, σ is the distance between two adjacent grooves, ϕ is the incidence angle along the grooves, α is the incidence angle in the plane perpendicular to the grooves and β is the diffracted angle in the same plane. Most of the spectrometers are used "on-plane", with $\phi = 0$. To improve the efficiency of the gratings in a spectral range the shape of the grooves has a triangular profile, and δ, the so-called blazed angle is the angle between the facet and the plane of the grating. In the Littrow configuration, when the incidence and diffraction angles are normal to the facets ($\alpha = \beta = \delta$) the

efficiency is maximum and the blaze wavelength is defined as

$$\lambda_b = \frac{2\sigma sin(\delta)}{m}$$

The grating grooves can be produced by a mechanical ruling or by hologra-
phy. The mechanical ruling is performed with a very fine diamond tool on a
soft coating, the positionning of the tool is controlled by an interferometer.
The holographic grating is the result of a laser interferogram registered in a
photon-sensitive resin deposited on the grating blank; then the resin is chem-
ically developped to remove parts of the interferogram. A ionic beam can be
used to transfer the interferogram in the layer below the resin (etching and
blaze generation processes).

Following the objectives to be reached either processes have their own advantages:

- ruled gratings can have better blaze angles (and efficiency); low number of
grooves by mm can be produced with very high blaze angles (e.g. 63°) to
work in high orders with very high spectral resolution (echelle),

- holographic gratings have very low scattering and a pseudo-blaze can be
produced. Gratings with a very large number of lines by mm and large size
are available (e.g. 6000 l/mm, 300 mm diameter).

Spectral resolution The limiting theoretical resolving power is given by

$$R = \frac{\lambda}{\delta\lambda} = mN \tag{4}$$

where $\delta\lambda$ is the the spectral resolution and N is the total number of grooves
illuminated ($N = \frac{W}{\sigma}$, W width of illuminated part of the grating).

The angular dispersion comes from equation

$$\frac{d\beta}{d\lambda} = \frac{m}{\sigma cos(\beta)} \tag{5}$$

Grating efficiency The relative distribution of the intensity from a pure
monochromatic line of wavelength λ across the different orders is given by

$$I = \frac{sin^2(N\gamma')}{sin^2(\gamma')} \times \frac{sin^2\gamma}{\gamma^2} \tag{6}$$

with $\gamma' = \frac{\pi\sigma}{\lambda}(sin\beta + sin\alpha)$, $\gamma = \frac{\pi b}{\lambda}(sin\beta + sin\alpha)$, and $b = \sigma cos\delta$.

The relative distribution of intensity across grating orders at wavelength
1216 Å for the SUMER spectrometer (Wilhelm et al. (1995)) is displayed
in Figure 2 (parameters: 3600 grooves/mm, blaze wavelength 1170 Å, and 20
mm illuminated part of the grating). The central part of the SUMER 1216 Å
diffraction pattern in the grating focal plane is shown in Figure 3.

Sometimes it is important to know the relative contribution of the wings of
the diffraction pattern to the wings of strong lines. By example, taking into

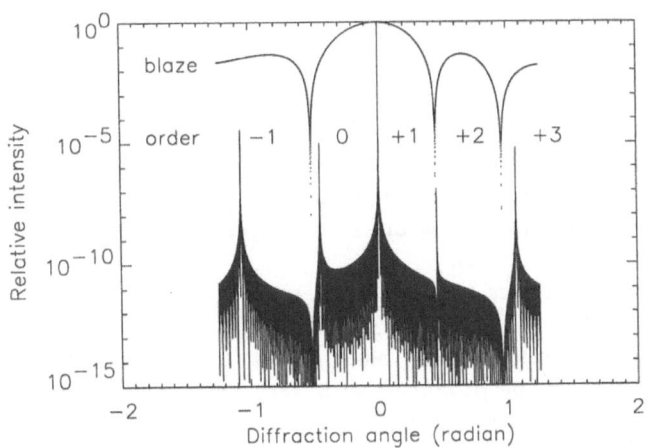

Fig. 2. Intensity pattern of the H I Lyα (1216 Å) wavelength for the SUMER/SOHO blazed grating (blaze contribution - upper curve; relative intensity distribution across grating orders - lower curve).

account of the grating parameters of UVCS/SOHO Ly-α channel (2400 l/mm, 12.85°, blazed for Ly-α, Kohl et al. (1995)) and knowing the grating size illuminated we can show on Figure 4 that there is no theoretical limitation to observe the Thompson scattering of the H I Lyα line at 2.5 $R_\odot$. The 10^{-4} Thompson scattering level expected is well above the diffracted wing further than 5 Å.

2.2   Properties of some spectrometer mounts

Rowland spectrometer The most simple mount involves only one spherical concave grating which collects the light coming from a slit and focusses the monochromatic beams on the Rowland circle of diameter the radius of the mirror blank of the grating.

This mounting is very efficient in the ultraviolet and extreme ultraviolet (UV/EUV) as only one surface is used. Although the spectral resolution depends of the number of grooves/mm and focal length, the main disadvantage is the astigmatism of spectral lines. A source point on the entrance slit gives elongated monochromatic lines perpendicular to the dispersion plane, and consequently the angular resolution along the slit is very poor.

The astigmatism is reduced by engraving the grating on a toroidal surface blank (Figure 5). The Normal Incidence Spectrometer (NIS) of CDS/SOHO (Harrison et al. (1995)) and the UVCS/SOHO spectrometer (Kohl et al.

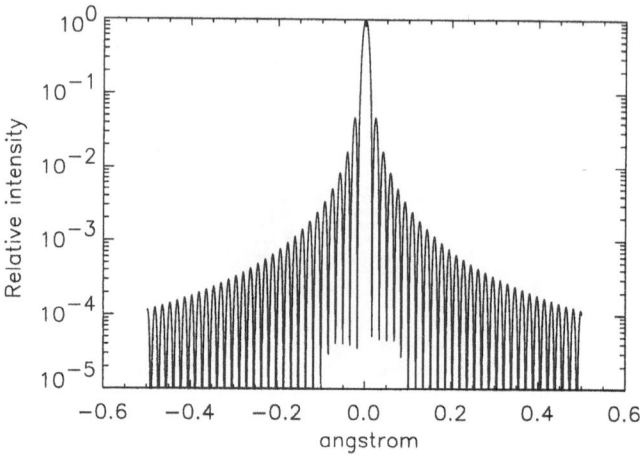

Fig. 3. Intensity pattern of the H I Lyα (1216 Å) wavelength for the SUMER/SOHO blazed grating.

(1995) use toroidal gratings.

In these cases, the grating has two radii of curvature: R_h in the dispersion plane and R_v perpendicular to this plane. The astigmatism is corrected in two positions in the diffraction plane symmetrical to the grating normal $\pm\beta_0$ through the equation

$$\frac{R_v}{R_h} = cos\alpha \times cos\beta_0 \tag{7}$$

Wadsworth mount Another use of a spherical grating illuminated by a collimated beam is the Wadsworth mount (Figure 6). The diffracted converging beams give mochromatic stigmatic images in the plane of the grating focus located near the normal to the illuminated section of the sphere. The main advantage is the slow variation of the astigmatism with the distance to the nominal focal point (see Table 1). Disadvantages come from the need to add another optical surface to collimate the incident beam on the grating and the variation of the focal length with the wavelength.

The SUMER/SOHO spectrometer which will be described later uses the Wadsworth mounting and the adjustment of grating focal length as a function of wavelength is done by moving the grating along its normal. The HRTS (High Resolution Telescope Spectrometer) instrument which has flown several times on rockets and was on SPACELAB2 (1985) Shuttle flight is a tandem Wadsworth mount. The first spherical grating provides monochromatic collimated beams which are collected and diffracted with additive dispersion by the second spherical grating (Bartoe and Brueckner (1975))(Figure 6).

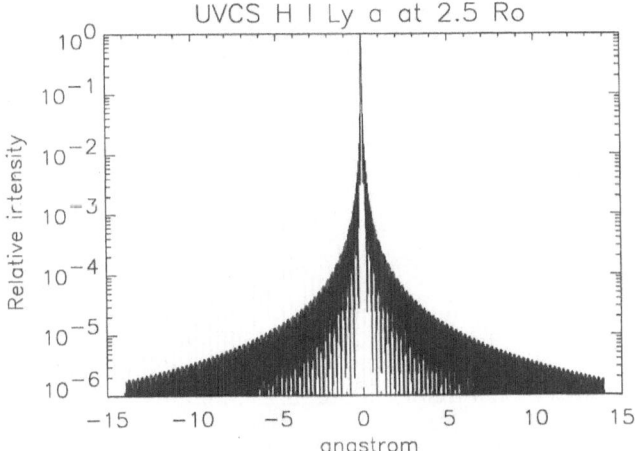

Fig. 4. Intensity pattern of the H I Lyα (1216 Å) wavelength for the UVCS/SOHO blazed grating established for 2.5 $R_\odot$ observation.

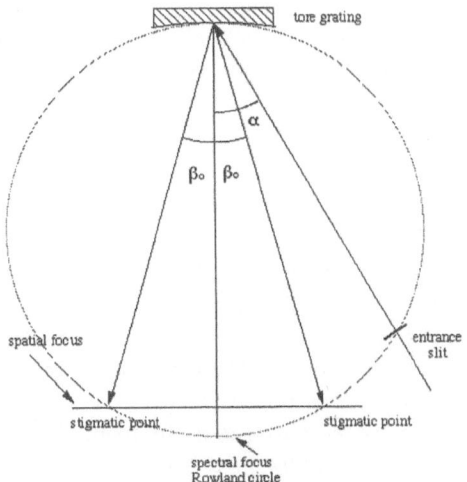

Fig. 5. Grating on a toric surface with 2 stigmatic points, a flat spatial focus and a curve spectral focus along Rowland circle.

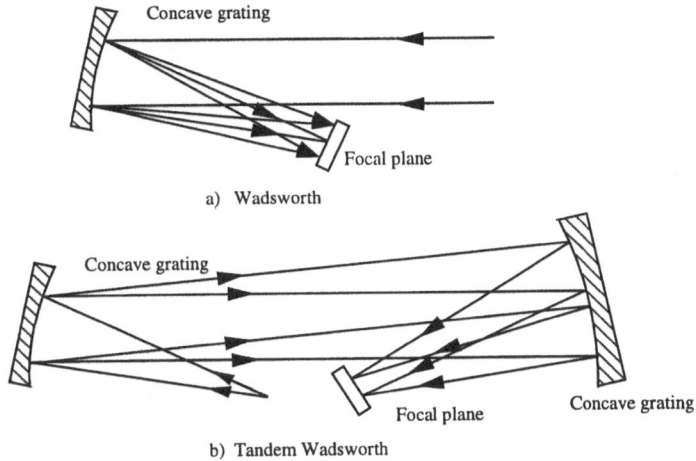

a) Wadsworth

b) Tandem Wadsworth

Fig. 6. Wadsworth (a) and tandem Wadsworth (b) mountings.

A comparison between Rowland and Wadsworth characteristics is shown in Table 1.

2.3 Detectors

The quality of a spectrometer is often limited by the detector characteristics. On SOHO, in the FUV/EUV wavelength range open detectors (without entrance window) are used such as XDL (Cross Delay lines, Siegmund et al. (1994)) on SUMER and UVCS, 1-D SPAN (1 dimensional Spiral Anode) on GIS/CDS, and VDS (Viewfinder Detector Subsystem) on NIS/CDS (Harrison et al. (1995)). The first element of the detector is the microchannel plate (MCP) assembly (see Figure 7) which converts the photons in electrons and amplifies the number of electrons. The MCP gain is related to the voltage applied; VDS is used with a very low gain (600 V to 950 V) while other detectors are used with gains reaching 2×10^7 (voltage above 4000 V). To increase the efficiency of the detectors a photocathode can be deposited in front of the MCP (such as KBr for SUMER and UVCS). In the XDL detector, the electrons cloud generated by an effective photon through the MCP is collected on two anodes along X and Y directions acting as delay lines. The localisation of the cloud centroid along X and Y lines is given by the time difference of the signal measured on the anodes at the extremities of the line. With very fast reading electronics the XDL is a photon counting system with very low noise (in the order of 10^{-5} $count\ sec^{-1}\ pixel^{-1}$). The limitations come from the maximum number of $counts\ sec^{-1}\ pixel^{-1}$ and from the total number of $counts\ sec^{-1}$ on the detector.

Table 1. *Rowland and Wadsworth characteristics.*

Parameter	Rowland	Wadsworth
source distance	$R\cos\alpha$	∞
image distance	$R\cos\beta$	$\frac{R\cos^2\beta}{\cos\beta+\cos\alpha}$
astigmatism	$\sin^2\beta + \sin^2\alpha\left(\frac{\cos\beta}{\cos\alpha}\right)$	$\sin^2\beta$
coma	$\frac{\sin\beta\,\tan^2\beta+\sin\alpha\,\tan^2\alpha}{2R^2}$	$\frac{\sin\beta}{2R^2}\left(\frac{\cos\alpha\,(\cos\beta+\cos\alpha)}{\cos^2\beta}\right)$

MCP/CCD detectors have a low gain MCP to produce a cloud of electrons on a phosphor which fluoresces in the visible. The visible photons are transferred to the CCD (Charge Coupled Device) through an optical lens (VDS/CDS) or a bundle of very thin optical fibers (SERTS - Solar EUV Rocket Telescope and Spectrograph (Neupert et al. (1992), 1997 flight). The detector is able to record a high number of photons in a pseudo photon counting mode. The limitations are given by the spreading of the image through the optical transfer (for instance the VDS/CDS lens used to image the phosphor onto the CCD introduces scattered light) and the noise level of the CCD electronics readout can reach the equivalent of 0.1 *count sec^{-1} pixel^{-1}*.

3 SUMER/SOHO spectrometer

3.1 Optical scheme

The SUMER optical scheme is displayed in Figure 8. The SUMER spectrometer is described in Wilhelm et al. (1995). The parallel beams issued from any point of the solar surface are collected and focussed by the one surface (off-axis parabola) telescope onto the spectrometer entrance slit plane. Selection of the solar area is performed by a pseudo-rotation of the parabola around its focus. The entrance pupil baffle limits a fixed effective surface of 9 x 13 cm^2 on the mirror. The divergent beam issued from the slit is collimated by an off-axis parabola, deflected by a plane mirror and sent at adjustable incidence (scan mirror) onto the spherical concave grating. The monochromatic diffracted beams are focussed onto the plane of the detectors. Detector A is on the normal of the grating and fulfills the optimum conditions of the

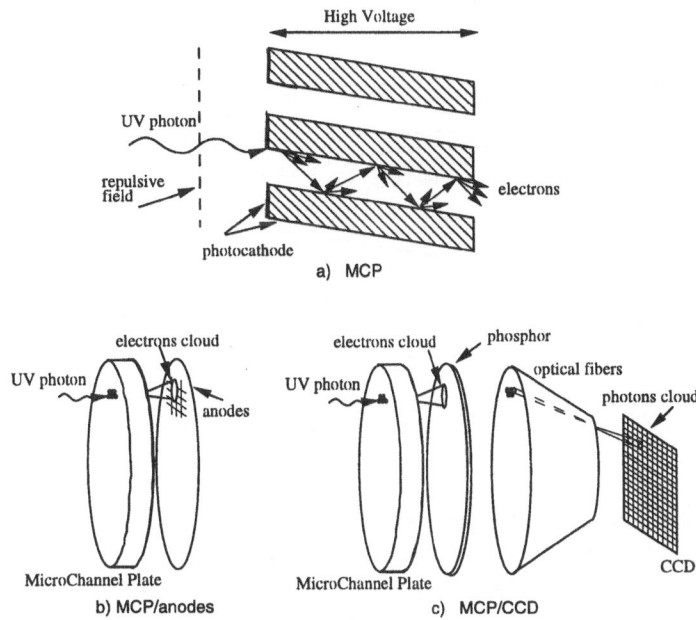

Fig. 7. Detector schemes; a) MCP; b) MCP/anodes; c) MCP/CCD

Wadsworth mount; detector B is slightly off the normal and can be postioned at two best focal planes, one for optimum spectral resolution and the other one for optimum angular resolution (along the slit image). Each detector has 1024 x 360 pixels (about 26.5 x 26.5 μm size); half of the MCP of each detector (central part) has increased sensitivity with a KBr photocathode and at each extremity few pixels (about 40) are protected with an attenuator.

A set of slits (0.3 x 120, 1 x 120, 1 x 300 and 4 x 300 $arcsec^2$) can be selected. The slit plane can also be adjusted to select the optimal telescope focus. The wavelength selection is performed by the scan mirror and the grating focal length is simultaneously adjusted by moving the platform supporting the grating and the scan mirror.

All the optical surfaces are made from a deposit of CVD-SiC on a SiC blank (UTOS^{TM}, USA), and the 3600 lines/mm holographic grating was transfered by etching in the CVD-SiC (Jobin & Yvon^{TM}, France).

3.2   Angular and spectral resolutions

In the laboratory the telescope and the collimator have been tested to determine its Point Spread Function (PSF, Saha et al. 1996). The image of the entrance pupil given by the collimator mirror is diaphragmed by the aper-

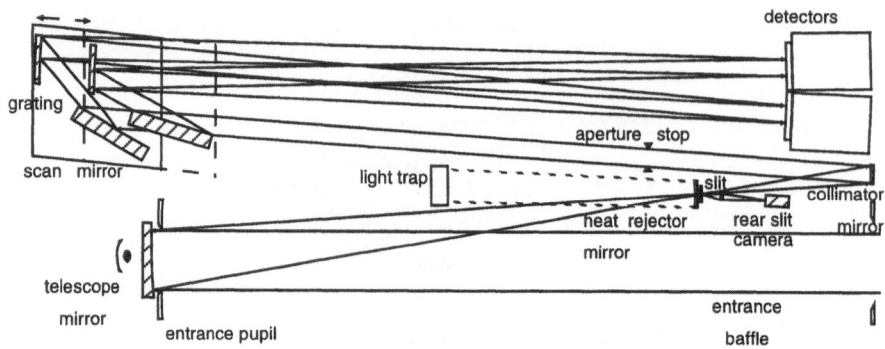

Fig. 8. SUMER optical scheme

ture stop (or Lyot stop) to eliminate the pupil diffraction pattern. The image qualities of the mirrors are limited by the shape of the surface and the residual local slopes and microroughness (by reference to a perfect surface).

The subpixel responses of the detector along spatial and spectral directions to the direct illumination of a $10\mu m$ hole have been measured by Siegmund et al. (1994). In the computation of the effective resolution the 26.5 μm pixel size sampling had to be taken into account.

Spectral resolution The spectral resolution is computed with:
- one arcsecond slit width,
- the collimator PSF,
- the grating diffraction figure,
- the optical aberrations inside the spectrometer (from ray tracing),
- the detector subpixel response and pixel sampling.

The results are displayed in Figure 9a and b. The PSF for detector A and B are shown in Figure 9a, while in b the detector A PSF is superposed by least square computation with a gaussian fit plus a constant. The gaussian fit optimised to the central core of the line is a very bad approximation in the PSF wings. At 1084 Å, the Full Width at Half Maximum (FWHM) of detector A determined from the gaussian fit is 1.87 pix (or 0.0827 Å), and for detector B the FWHM is 2.57 pix (or 0.1134 Å). The focus of detector B is supposed to be optimized for the best spatial resolution.

Spatial resolution The spatial resolution along the slit is computed with:
- the telescope PSF,
- the collimator PSF,
- the optical aberrations inside the spectrometer (from ray tracing),

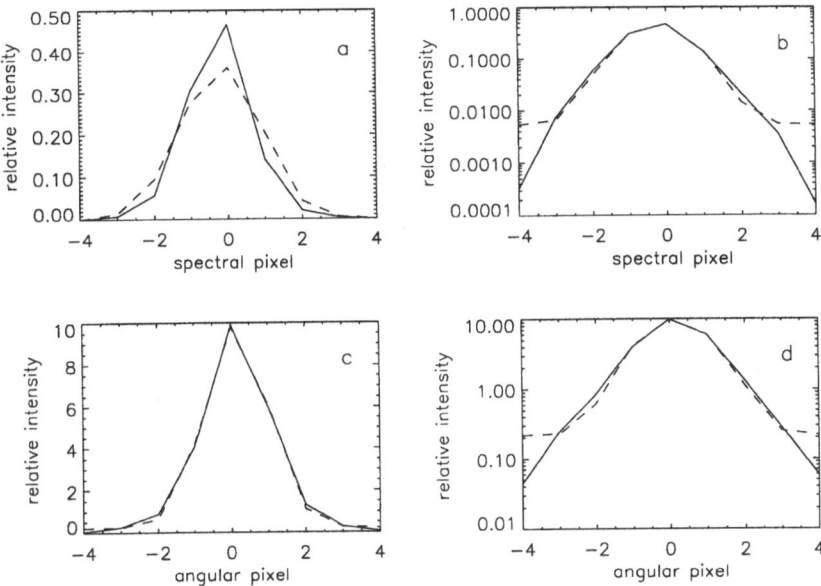

Fig. 9. SUMER computed angular and spectral resolutions at 1084 Å. Spectral resolution: a) detector A (solid) and B (dashed); b) detector A logarithmic scale computed (solid) and fitted by a gaussian (dashed). Angular resolution: computed (solid) and gaussian fit (dashed) in linear (c) and logarithmic scales.

- the detector subpixel response and pixel sampling.

The results are shown in Figure 9c and d, where the dashed lines correspond to the gaussian fitting (gaussian plus a constant). Here again the gaussian fit does not reproduce the total PSF wings. Only the detector A PSF is plotted, since the detector B optimized for the best spatial resolution has negligeable aberrations. The FWHM derived from the gaussian fit is 2.0 pixels (or 2.0 arcsec) at 1084 Å.

3.3 Scattering

Off-limb Knowing the PSF wings of the telescope up-to 3200 arcsec (Saha et al. 1996) and representing the solar disk emission as a flat disk of unity (per arcsec^2), the relative scattered light at 1236 Å(wavelength used for the laboratory measurement) above the limb can be computed as a function of distance from Sun center. The data are displayed in Figure 10 and are smaller than the observed values by about a factor 1.5 at 1200 arcsec and 1.1 at 1600 arcsec. The off-limb scattered light from the telescope is wavelength depen-

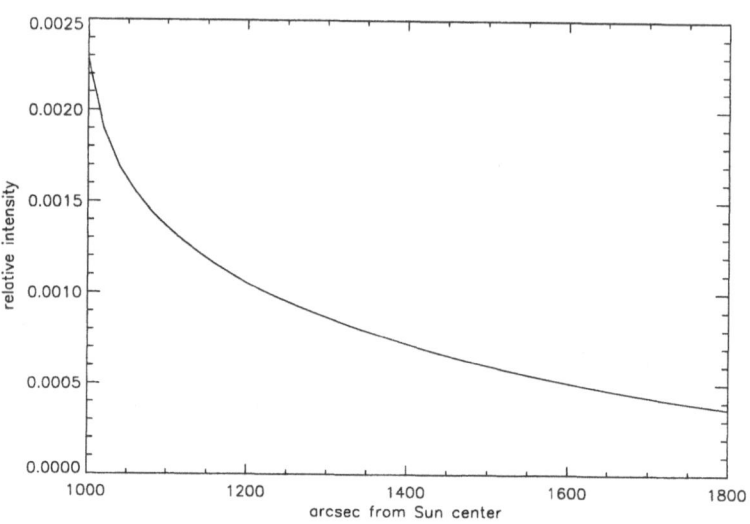

Fig. 10. Computed radial telescope scattering off-limb at 1236 Å.

dent. At 1700 arcsec the measured relative values are respectively 0.00063, 0.00038 and 0.00031 at 860, 1140 and 1460 Å.

On-disk Using the same scattering figure measured with the telescope it is possible to compute the contrast of a hole seen at Sun center by the telescope. With the hypotheses:
- the hole located at Sun center has a radius of 1, 3 and 6 arcseconds,
- the relative intensity in the hole is 10% of the flat solar disk,
in the focal plane of the telescope we obtain the results shown on Figure 11. For a hole radius greater than 2 arcseconds the scattered light leak from the disk into the hole contributes to about 1%.

3.4   Detector corrections

Counts number The detection of individual counts is limited by the speed of the readout electronics. When the total number of electrons clouds generated by effective photons in the entrance of the MCP exceeds 50000 s^{-1}, some are confused with neighbours and missed in the counting. The correction from this effect can be taken into account in using the programme

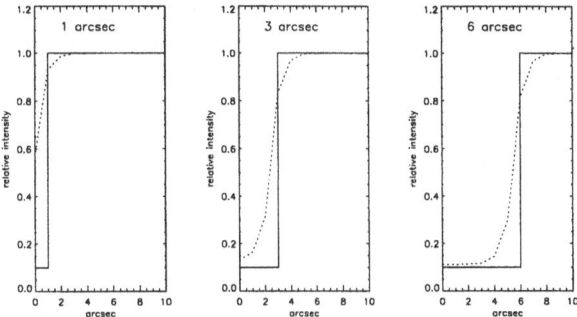

Fig. 11. Computed contribution of the SUMER telescope PSF to the smearing of structures in its focal plane at 1236 Å. Tests are made with holes of radius of 1, 3 and 6 arcseconds with 10% relative intensity: initial radial cut (solid lines) and after convolution with telescope PSF (dotted lines).

DEADTIME_CORR.PRO accessible through the webb in the SOHO software of SUMER (Contrib:[Wilhelm]).
When the local number of counts pixel^{-1} s^{-1} is greater than 5, there is a local gain depression which has to be corrected for in using the programme LOCAL_GAIN_CORR.PRO (at the same webb address).

Flat field The relative pixel to pixel response of the detector is not unity. To correct for the local variation from time to time an on-board flat field was created (seen as SUMFF_* in the data base) which gives the local correction to apply to the data. The original data used are generated by defocussing the spectrometer near the head of the H I Lyman continuum to obtain a pseudo-uniform illumination of the detector. Then a local running average over a small area of the detector is the local unity reference. The flat field correction can be applied in using SUM_FFIELD (Contrib:[Kucera]).
The flat field correction has to be carefully managed; local gain depressions have been seen after exposure to counts pixels^{-1} s^{-1} greater than 10 during several hours. The nearest (in time) flat field must be used.

Distortion The image of a rectangular grid onto the detector produces a distorded grid. The correction is managed by DESTRETCHN.PRO (Contrib:[Moran]) or similar programmes derived from this version. Here again, local gain depressions produced by high count rate can modify locally the distortion (during several days).
An accurate correction of the lines distortion is a necessary step to establish a precise positioning of lines in the wavelength scale to measure the Doppler

shifts. The accuracy of the reconstruction is generally better than 0.25 pixel. For long sequences of observation a drift of the whole detector by up to ±0.25 pixel has been noticed. This drift shows a correlation with the thermal cycling of the spectrometer (about 2 hours for data taken before march 1997).

3.5   Absolute radiometry

Providing calibrated intensities in line profiles is an important goal of the SUMER spectrometer. This will permit to compute emission measure, line ratio, radiative losses, etc.. and will constrain theoretical modelling.
The SUMER radiometric calibration has been established in few steps:
- laboratory relative calibration of individual components,
- absolute efficiency of the full spectrometer (Hollandt et al. (1996)) with a secondary standard source,
- in-flight calibration in using standard stars measured by IUE (International Ultraviolet Explorer) (Wilhelm et al. (1997)).
An example of the results obtained from the calibration is displayed on

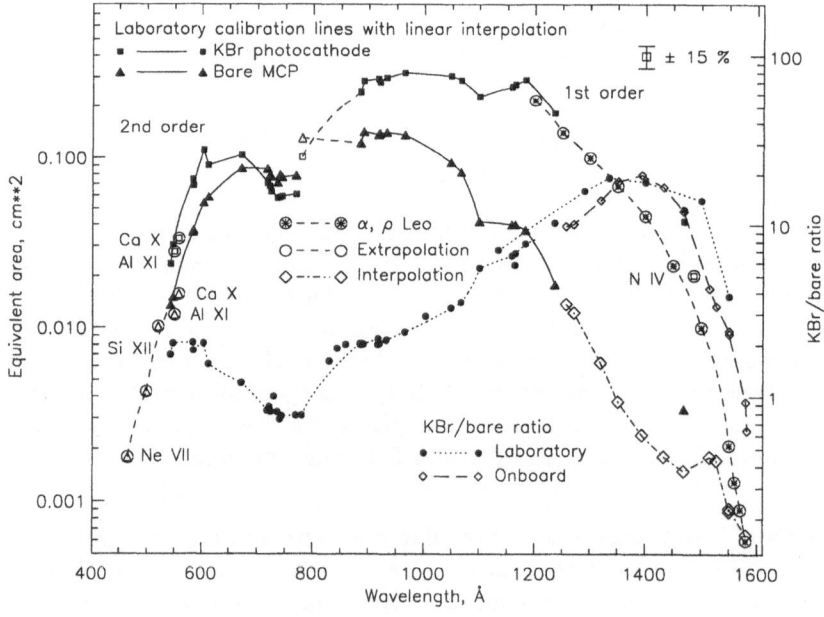

Fig. 12. Summary plot of the SUMER absolute efficiency with detector A (from Wilhelm et al. 1997).

Figure 12. The comparison between in-flight and laboratory calibration shows the photometric stability of the spectrometer.

4   Prospective

From the data acquired by the spectrometers on SOHO we can derive a few constraints on future instrumentations. Let us discuss at few domains which are of importance and, then, look at the instrumental requirements.

Angular resolution A lot of solar structures are not resolved with the present spectrometers. The SUMER spectrometer with 1.4 x 2.0 arcsec^2 best resolution is averaging over small structures. An improvement of a factor 10 (0.1 - 0.2 arcsec resolution) will give us a better estimate of the filling factor (and density). At the same time as we improve the resolution of structures the contrast between bright and dark structures will increase. To look at the same time at bright structures and neighbouring dark ones the PSF wings must be very very low. As an example, at 1000 Å the 0.1 arcsecond diffraction limit corresponds to a telescope of 25 cm diameter. The PSF of a 25 cm diameter telescope without obscuration (no secondary mirror) and with obscuration (a 7.5 cm diameter secondary mirror mount without spider) are displayed in Figure 13. For comparison the PSF of a square mirror of same area is shown along an axis parallel to the side and along a diagonal.

Spectral resolution The actual SUMER spectral resolution permits to have 3-5 pixels in the Full Width at Half-Maximum (FWHM) of most of the lines and an accuracy better than ± 2 km s^{-1}. An increase of a factor 2 in the sampling seems necessary to obtain the full capacity of the spectrometer, i.e. $\lambda/\delta\lambda = 4 \times 10^4$.

Temporal resolution The required temporal resolution is constrained by the temporal evolution of phenomena (increase of intensity of more than a factor 2 in few seconds), by their speed of movement (vertical and horizontal) and by the lifetime.
For instance an explosive event with an horizontal velocity of 70 km s^{-1} will cross a 0.1 arcsec resolution pixel in 1 second. So, an objective should be to make a small image with spectral line profiles in each point within 1 second; the size of this image must be about 1 arcsec^2 or more.
Trying to fulfill simultaneously the 3 objectives (spatial, spectral and temporal) requires to increase the sensitivity of the instrument (or equivalent area). This can be done by increasing the size of the telescope, reducing the number of optical surfaces and increasing the efficiency of optical surfaces and detector. Optimisation of each parameter is a technical challenge for an extreme ultraviolet instrument put in space.

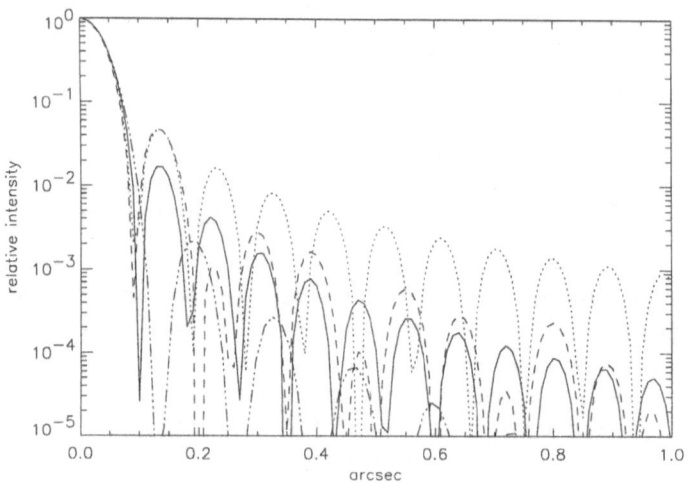

Fig. 13. PSF of a telescope at 1000 Å. Diameter 25cm without obstruction (solid) and with obstruction(10%, dashed). Square mirror with the same area along one axis parallel to the side (dotted) and along a diagonal (dash dot dot dot).

Although very difficult to manage in ultraviolet, other spectrometer types, such as Fourier Transform Spectrometers cannot be excluded. Developments in this domain are in progress (Roesler and Harlander 1991).

References

Bartoe, J.-D. F. , and Brueckner, G. E. (1975): J. Opt. Soc Am. **65**, 13

Emerson, D. (1996): *Interpreting Astronomical Spectra* (John Wiley and Sons)

Harrison, R.A., Sawyer, E.C., Carter, M.K., Cruise, A.M., Cutler, R.M., Fludra, A., Hayes, R.W., Kent, B.J., Lang, J., Parker, D.J., Payne, J., Pike, C.D., Peskett, S.C., Richards, A.G., Culhane, J.L., Norman, K., Breeveld, A.A., Breeveld, E.R., Al Janabi, K.F., McCalden, A.J., Parkinson, , J.H., Self, D.G., Thomas, P.D., Poland, A.J., Thomas, R.J., Thompson, W.T., Kjeldseth-Moe, O., Brekke, P., Karud, J., Maltby, P., Aschenbach, A., Bräuninger, H., Kühne, M., Hollandt, J., Siegmund, O.H.W., Huber, M.C.E., Gabriel, A.H., Mason, H.E., Bromage, B.J.I. (1995): The Coronal Diagnostic Spectrometer for the Solar and Heliospheric Observatory. Solar Phys. **162**, 233–290

Hollandt, J., Schühle, U., Paustian, W., Curdt, W., Kühne, M., Wende, B., and Wilhelm, K. (1996): Radiometric Calibration of the Telescope and Ultraviolet Spectrometer SUMER on SOHO. Applied Opt. **35**, 5125–5133

Kohl, J.L., Esser, R., Gardner, L.D., Habbal, S., Daigneau, P.S., Dennis, E.F., Nystrom, G.U., Panasyuk, A., Raymond, J.C., Smith, P.L., Strachan, L., Van Ballegooijen, A.A., Noci, G., Fineschi, S., Romoli, M., Ciaravella, A., Modigliani, A., Huber, M.C.E, Antonucci, E., Brnna, C., Giordano, S., Tondello, G., Nicolosi, P., Naletto, G., Pernechele, C., Spadaro, D., Poletto, G., Livi, S., Von der Lühe, O., Geiss, J., Timothy, J., Gloecker, G., Allegra, A., Basile, G., Brusa, R., Wood, B., Siegmund, O.H.W., Fowler, W. (1995): The Ultraviolet Coronagraph Spectrometer for the Solar and Heliospheric Observatory. Solar Phys. **162**, 313–356

Neupert, W.M., Epstein, G.L., Thomas, R.J., and Thomson, W.T. (1992): An EUV Imaging Spectrograph for High Resolution Observations of the Solar Corona. Solar Phys. **137**, 87

Roesler, F.L., and Harlander, J. (1991): Spatial Heterodyne Spectroscopy: Interferometric Performance at any Wavelength Without Scanning. Proc. SPIE **1318**, 234–243

Saha, T.T., Leviton, D.B., and Glenn P. (1996): Performance of ion-figured silicon carbide SUMER telescope mirror in the vacuum ultraviolet. Appl. Opt. **36**, 6421

Siegmund, O.H.W., Stock, J.M., Marsh, D.R., Gummin, M.A., Raffanti, R., Hull, J., Gaines, G.A., Welsh, B., Donakowski, B., Jelinsky, P., Sasseen, T., and Tom, J.L. (1994): Delay line detectors for the UVCS and SUMER instruments on the SOHO satellite. Proc. SPIE **2280**, 89-100

Schroeder, D.J. (1987): *Astronomical Optics* (Academic Press, Inc)

Wilhelm, K., Curdt, W., Marsch, E., Schühle, U., Lemaire, P., Gabriel, A.H., Vial, J.-C., Grewing, M., Huber, M.C.E., Jordan, S.D., Poland, A.I., Thomas, R.J., Kühne, M., Timothy, J.G., Hassler, D.M., and Siegmund, O.H.W. (1995): SUMER - Solar Ultraviolet Measurements of Emitted Radiation. Solar Phys. **162**, 189–231

Wilhelm, K., Lemaire, P., Feldman, U., Hollandt, J., Schühle, U., and Curdt, W. (1997): Radiometric Calibration of SUMER: Refinement of the Laboratory Results Under Operational Conditions on SOHO. Applied Opt. **36**, 6416–6422

Thorne, A.P. (1988): *Spectrophysics* (Chapman and Hall, London)

A SOHO User Manual

Petrus C. Martens

SOHO Science Operations Coordinator,
SOHO Experiment Operations Facility,
Code 682.3, Goddard Space Flight Center,
Greenbelt, MD 20771
USA

Abstract. This paper is intended to serve as the first version of the "SOHO User Manual", a "how to" guide for those interested in analyzing existing SOHO data, or proposing new SOHO observations. Questions addressed are, how to use the SOHO catalogs, where to find the appropriate data analysis software, how to request permission to use proprietary data, how to propose and prepare SOHO observing programs.

In an outlook to the future I will emphasize the potential for joint observations during the rising phase of the cycle in the extended SOHO mission, collaborations with Yohkoh, Ulysses, and TRACE, as well as with the ISTP spacecraft.

1 SOHO Data Libraries

1.1 Description

Four SOHO data libraries (archives) are under construction, one in the US, at the SOHO Experiment Analysis Facility (EAF) at Goddard Space Flight Center, and three in Europe, at MEDOC in Orsay, Rutherford Appleton Laboratory in the UK, and in Torino. The libraries at Goddard and Medoc are operational, and the one at Goddard is accessible via the Web. The SOHO data libraries contain the data from all the SOHO experiments, except for the MDI helioseismology data, that are stored in a separate facility at Stanford, and also retrievable via a Web interface.

The purpose of these data libraries is to facilitate multi-experiment SOHO data analysis, and to make the data easily accessible (via the Internet) for the widest possible group of scientific users. To achieve this goal the SOHO data library features are:

- a uniform data format (FITS)
- a uniform access to all experiment archives
- a campaign catalog that cross-links experiment data of coordinated observations
- a complete set of supporting synoptic data
- a complete set of analysis software, regularly updated with user contributions.

The data library is completed by a set of catalogs, with open access again via the Web. These catalogs are:

- SOHO Main Catalog
- Event Catalog
- Campaign Catalog
- Picture Catalogs:
    - SOHO Summary Data (A set of daily representative images)
    - Synoptic Data (Daily images from other observatories and spacecrafts)

For access see the SOHO Homepage at *http://sohowww.nascom.nasa.gov*, and press the button *Data Archive*. A more detailed description of SOHO science operations and data products can be found in St. Cyr et al. (1995).

1.2   Data Library Versus Data Archive

Data from space missions are often archived after the mission has been completed. The tremendous growth in storage and processing speed of workstations, as well as the explosive growth in the use of the Internet, has made it feasible and desirable to have a "live" archive, also called data library. The desirability is most easily understood by considering figure 1, which is a hypothetical curve of the data-use for an average space mission. Data use is a hard to measure quantity, but one may assume that it is roughly proportional to publication rate, and preceeds that by a year or two. Clearly it will take off rapidly as data become available after launch. It will even grow more as data enter the public domain, which is typically one year after the observation. One may assume it will continue growing at a smaller speed until around the end of the mission when a gradual decline sets in. Data use curves will of course vary from mission to mission, but it is to be expected that most of it has occurred two or three years after the completion of the mission, at which time a "traditional" archive would become operational.

Prior to launch the ESA and NASA SOHO Project Scientist's teams have set themselves the goal of having a SOHO data library operational at the start of SOHO's science operations. The SOHO archive was declared operational in early January 1997, well within the early phase of the mission. Thus the goal of preparing a "live" archive has been largely achieved.

1.3   Data Analysis Software

SOHO data analysis software is embedded in "SolarSoft". According to its Web page at *http://www.space.lockheed.com/solarsoft/ssw_whatitis.html*, "the SolarSoft system is a set of integrated software libraries, data bases, and system utilities which provide a "common" programming and data analysis environment for Solar Physics. The SolarSoftWare (SSW) system is built from

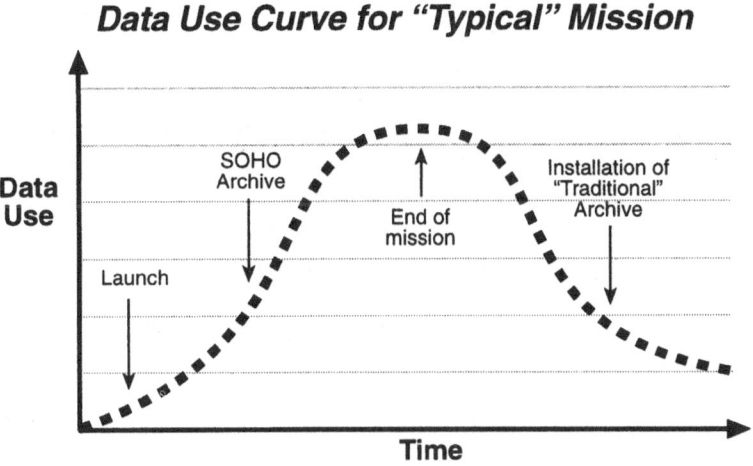

Fig. 1. A hypothetical data use curve for a space mission

Yohkoh, SOHO, SDAC and Astronomy libraries and draws upon contributions from many members of those projects. It is primarily an IDL based system, although some instrument teams can and do integrate executables written in other languages. The SSW environment should provide a consistent look and feel at various Co-I institutions (SOHO, Yohkoh, etc) which will facilitate sharing and exchange and minimize "coming-up-to-speed" time when doing research away from your home institute."

One of the primary goals of the SSW is to "promote the use of certain standards which facilitate coordinated data analysis".

Some of the capabilities of SSW are: time series analysis, spectral fitting, image and image cube (movies) display, IDL data manipulation (structure, string, array, mathematics...), FITS, solar limb fitting, grid overlay, and coordinate transformations.

Solarsoft accepts user contributed software, after testing, and if sufficiently documented in the required format.

1.4   Data Access

Access to SOHO data in the Public Domain is open to everyone via the Web interface given above, while access to proprietary data is limited in accordance with the rules set by the SOHO Science Working Team. For the current availability see: *http://sohowww.nascom.nasa.gov/whatsnew/archivenews.html*.
MDI data are public immediately and in the extended mission, after May 1998, the data from all US experiments (MDI, UVCS, LASCO) will be in the public domain immediately.

Other ways to gain access to SOHO data, or to propose new observations are:

- The official ESA/NASA Guest Investigator program with funding through national agencies in Europe. For more information see: *http://sohowww.nascom.nasa.gov/operations/guest_investigators/*
- Collaborations between SOHO and Ground Based Observatories or other spacecrafts, usually with exchange of data
- SOHO workshops and symposia, with open participation
- Collaborations through SOHO experiment teams on data analysis and new observing programs
- Visits to Goddard and MEDOC to participate in daily operations, develop observing programs, and for data analysis.

1.5   Searching for SOHO Observations

The starting point in finding SOHO observations are the SOHO catalogs, either the SOHO main catalog, campaign catalog, or the individual experiment catalogs. These are accessed via the SOHO Archive Web pages, at the URL given above, via Web forms.

If one is interested in multi-experiment coordinated observations, often in collaboration with other spacecrafts or Ground Based Observatories (GBO's), the campaign catalog is the best starting point. It can be searched on "keyword", for example the date and time of the observations, or a significant word in the name of the campaign, e.g. "filament", or "coronal hole". One can also search on the names of participating experiments or observatories, campaign type (Joint Observing Program, Intercalibration, collaboration, or campaign), the name of the campaign coordinator, or a word match in the campaign objective or comments fields.

This search will yield the campaign descriptions of the matching entries, and the Coordinated Observation ID #'s. These numbers can then be used as entries in the main catalog to retrieve the desired experiment data, according to the procedure described below.

For most specific searches the main catalog will be used directly. Here too searches are by "keyword", such as the time and date of observations, SOHO instrument, object of the observations (e.g. Sunspot), observing mode, wavelength, detector, coordinates of the target, campaign ID #, etc. Submitting the search will yield a page called the "Database Query Result", with links to the relevant experiment observing programs. For example, a recent search on the object keyword "Comet" produced 194 matching program numbers. Following these links one arrives at the file names of the data in the experiment catalogs, which then can be retrieved over the Web.

If the observing programs listed contain data that are not yet in the Public Domain, and one is not authorized by the instrument Principal Investigator to access these data, this will be indicated in the list for each relevant observing program, and further access to the data files will be blocked.

The Web interfaces for SOHO catalog searches and data retrieval are being enhanced to reflect lessons learned from a comparison with other on-line data libraries, and in response to comments and requests from the user community. The goal is to reach a stable interface in early 1998.

2   Planning of SOHO Observations

2.1   Preparing New Observations

Often, for a very specific research project, the exact data needed do not exist, and new observations have to be proposed. In this section I will present guidelines on how to propose new observations with the SOHO experiments.

For single experiment investigations the procedure is rather simple. Contact the PI of the experiment in question with a brief proposal and scientific justification, or apply to the SOHO Guest Investigator Program. Several experiments have set up a mechanism for submitting observing proposals on their Web pages (see *http://sohowww.nascom.nasa.gov/instruments.html*). If your observing proposal is accepted, the PI will link you with a technically competent team member to develop observing sequences. The observing sequences will need to be worked out in detail and tested, and the observations will be scheduled when appropriate in the experiment schedule.

For multiple experiment studies – anything from a collaboration between two SOHO experiments to full fledged multi-site, multi-spacecraft campaigns more coordination will be required

For SOHO observations one should contact the PI's of the relevant experiments and the Science Operations Coordinator (SOC) with a brief proposal and scientific justification, or apply to the Guest Investigator Program. Simultaneously one should contact the relevant people for other spacecrafts (Ulysses, Yohkoh, TRACE, etc.), and any Ground Based Observatories. If your proposal is accepted you will be linked up with technically competent SOHO team members to develop observing sequences, and your campaign will be reviewed and scheduled at the monthly planning meeting.

If the campaign is sufficiently complex, or if it is intended to be repeated several times, you will be asked to write it up as a SOHO Joint Observing Program (JOP), which will be posted on the SOHO Web pages. You will be assigned as JOP or campaign leader, and must negotiate the observing dates and times with all parties involved, and keep the SOC informed. Your final observing times will be listed on the Monthly Calendar after approval at the monthly meeting. In the weekly meeting prior to your observations, a detailed timeline for the SOHO observing program will be created, including your JOP. During the observations you must be either present at the EOF, or closely in touch, to coordinate target selection and deal with last minute issues. When the observations have been completed you are required to complete the campaign information in the campaign catalog.

In this section several resources for SOHO Planning have been introduced, such as the monthly calendar, the weekly timeline, the daily target list, the JOP list, and the campaign catalog. All of these can be found from the SOHO Operations Page at: *http://sohowww.nascom.nasa.gov/operations/*.

2.2   Ingredients of a SOHO JOP

A SOHO JOP proposal must contain the following information:

- Title, Author(s), Update History
- Participating SOHO Experiments, and Other Observatories
- Scientific Objective
- Scientific Justification
- Detailed Observing Sequences per Experiment
- Operational Considerations (e.g. direct S/C contact required, method for target selection, fall-back strategies)

For specific examples see again the SOHO Web pages. At present there are more than 70 "official" SOHO science JOPs, and 13 intercalibration JOPs.

2.3   Scientific Analysis

Once data have been obtained the JOP or campaign leader will normally be given proprietary data rights for a year or so by the PI's of the experiments involved. Feedback and guidance with the scientific analysis can be obtained from team members of the experiments involved or through presentations and discussions at the numerous SOHO related workshops and symposia. Informal workshops intended for multi-experiment (SOHO and non-SOHO) data analysis are quite frequent. A current list of meetings relevant for the SOHO community can be found by following the link from the SOHO Homepage.

One is requested to enter papers resulting from data analysis or theoretical work based on SOHO results in the SOHO Bibliography & Publications Database at *http://sohowww.nascom.nasa.gov/bibliography/*.

Little has been said in the above about the observations of the SOHO helioseismology experiments (GOLF, VIRGO, and MDI). The reason is that these experiments are mainly devoted to obtaining long time series of the same observable. Little or no coordination with other experiments is required for that. A very active program of scientific analysis of SOHO helioseismology data is being carried out, and those interested should either visit the experiment Web sites or visit one of the many helioseismology workshops and conferences to find out how to get involved.

3   Future Observations

SOHO will be operated for probably at least four more years beyond the end of it nominal mission in May 1998. This puts it in an excellent position to study the rising part of cycle 23. Some of the studies proposed for the rising part of the cycle are summarized below.

First there is the study of the large scale structure of the solar corona the goal of which is to understand the large-scale, stable, coronal structure of equatorial helmet streamers and polar coronal holes that can persist for several solar rotations at solar minimum. This study has been run for a whole month continuous in August 1996 (Whole Sun Month) and will be repeated several times during the remainder of the SOHO mission, to provide a unique data set for the study of the evolution of the large scale corona as a function of the phase of the Solar cycle.

Second there is the study, and the effort of early detection, of Earth-directed CME's, a key part of the overall International Solar Terrestrial Physics Program (ISTP) to study the Sun-Earth system. As the magnetic activity in cycle 23 rises, CME's will become more frequent and stronger, and the SOHO experiments, through their coordinated "CME-watch" observing programs, will be able to build up a statistically significant database of predictions versus occurrence for Earth-directed CME's. At the same time the collaborating ISTP missions will collect data to study the observed strength and impact on the magnetosphere of CME's as a function of the physical parameters at their Solar origin.

Thirdly, and somewhat related to the second goal, there is the question of the relation between the eruption of filaments, CME's, and X-ray flares. Even in the present, quiet part of the cycle, SOHO experiments have regularly observed filament eruptions and subsequent CME's, but large X-ray flares, detected by Yohkoh and GOES, have been relatively rare. As the cycle progresses X-ray flare activity will increase, and a key issue is whether CME's and filament eruptions will increase by the same percentage, or whether the nature of the filament eruption/CME/flare events changes in that relatively more energy is expended in high energy X-ray and gamma-ray emission.

Figure 2 shows the timelines for SOHO, Ulysses, Yohkoh, and TRACE. Yohkoh will keep on observing for as long as possible, and many of its observations are done jointly with SOHO. Ulysses will have a second perihelion passage over both poles early in the next century and joint observations with the other spacecrafts will most certainly enhance our knowledge of the solar polar regions.

TRACE (the Transition Region and Coronal Explorer) is a one of NASA's "faster, cheaper, better" missions, scheduled for launch in early 1998. It will operate in a polar orbit with 8 month intervals of continuous Sun viewing. For more information see *http://www.space.lockheed.com/TRACE/welcome.html*. The TRACE mission operations team intends to merge its daily meeting with that of SOHO, use the same planning tools and procedures as SOHO,

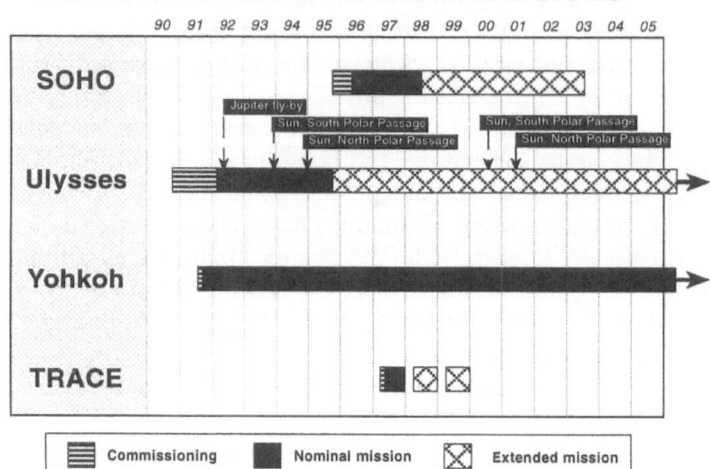

Fig. 2. Timelines for four major coronal and heliospheric science missions

and in general use the SOHO UV quicklook data as a reference for target and program selection for their high resolution, high cadence, UV and EUV partial field-of-view studies.

Finally ACE, the Advanced Composition Explorer, a solar wind studies mission, is also located near the first Lagrangian point. ACE science data will complement and enhance the data from SOHO's solar wind analyzers, CELIAS, COSTEP, and ERNE.

New GBO's such as the new French-Italian vector magnetograph THEMIS at the Canary Islands, and the Chromospheric Helium Imaging Photometer (CHIP) operating at the Mauna Loa Solar Observatory, as part of the HAO Advanced Coronal Observing System, will further expand the range of potential science objectives for joint studies.

References

St.Cyr, O.C., Sánchez–Duarte L. , Martens, P.C.H. Gurman, J.B., and Larduinat, E. (1995): SOHO Ground Segment, Science Operations, and Data Products. Solar Phys. **162**, 39–59

SOHO: an example of Project Management

Fabrizio Felici

SOHO Project Manager
ESA - European Space Agency-Noordwijk
The Netherlands

1 Introduction

The Solar and Heliospheric Observatory, SOHO, is the most comprehensive space mission ever devoted to the study of the Sun and its nearby cosmic environment known as the heliosphere. From the vantage point of a halo orbit around the first Lagrangian point, L1, SOHO's twelve scientific instruments observe and measure structures.

The SOHO mission is an international collaboration between ESA, European national authorities and NASA. ESA took the lead in the collaboration between the two agencies by procuring the spacecraft (including integration of the twelve instruments and environmental testing of the satellite) through European Industry. The instruments were built under the leadership of Principal Investigators - nine of them funded by European national authorities, and three by NASA. Further support was given by CO-investigators and Associated Scientists with European and US national funding. NASA provided the SOHO launch aboard an Atlas-2AS vehicle and it also takes care of mission operations as well as communications with the satellite via the Deep Space Network. Overall responsibility for the mission remains with ESA.

2 The mission's beginnings

Both SOHO and Cluster were proposed - initially in competition - in November 1982. Given their common aim, to be pursued in different contexts and by different methods, both missions were subsequently included as a Cornerstone called the Solar Terrestrial Science Programme (STSP), in ESA's Horizon 2000 Programme. This long- term scientific programme was approved by ESA Council, meeting at Ministerial Level, in early 1985. ESA's Science Programme Committee (SPC) then selected STSP in November 1985 as the first Cornerstone of the ESA Horizon 2000 programme to be implemented.

SOHO's complex multi-disciplinary payload, which included novel experiments in the still relatively young solar discipline of helioseismology, was being conceptually developed in parallel. This parallel development provided the correct foundation and interfaces for the industrial studies and the further contractual steps that the Agency would need to take vis-à-vis industry.

The eventual outcome of all of the foregoing activities was an ESA STSP, which however exceeded the European budget allocated to it at the time of selection. ESA therefore initiated a dialogue with NASA on the possibility of their contributing in terms of the scientific payload, the onboard equipment or other mission elements (launcher, operations, etc.). Certain cooperative elements were identified as early as 1986 and potential US industrial suppliers were visited by NASA and ESA representatives. This activity led to an inter-agency Memorandum of Understanding (MOU) defining the framework and principles of the proposed cooperation between ESA and NASA.

This agreement with NASA, together with the descoping and rationalization of several aspects of the mission and its payload, brought SOHO and Cluster within the allowable financial envelope. This in turn cleared the way for the industrial tendering phase. In the meantime, the overall STSP payload had been selected in March 1988, also after extensive discussion and descoping under the aegis of two scientific committees led by Profs. H. Balsiger and D. Southwood. All of the elements were then in place to initiate a joint Invitation-to-Tender (ITT) to European industry.

3   The mission takes shape

The submission of the industrial Phase-B proposal was followed by an extensive evaluation by senior ESA management of all cost and schedule, management and contract, product assurance, technical and test-facility aspects. Phase-B then began in December 1989.

The selection of the industrial teams at lower levels was pursued through a series of proposal and evaluation cycles, with the close involvement of the two ESA Project Teams and their Prime Contractors. The result was a good matching of the final industrial organization with the mandatory geographical return.

4   The building of the satellite: The industrial interface

A novel feature of the SOHO programme has been the multiplicity of its interfaces across several widely geographically separated organizations (Fig. 1). The first, and financially most relevant interface for the SOHO Project Team has been that with the industrial grouping led by Matra Marconi Space (France), which has developed, built and tested the SOHO spacecraft during the main development phase (Phase C/D). A *cost-plus* contract with a target cost and a *neutral zone* were defined at the start of Phase C/D and updated as the project evolved, with a series of Change Review Boards. SOHO's test campaigns were conducted using the *coordinated European test facilities*, requiring interfacing with the IABG, Ottobrunn (D) and Interspace, Toulouse

(F) facilities for the structural-model and flight- model test campaigns, respectively.

SOHO Mission Constituents

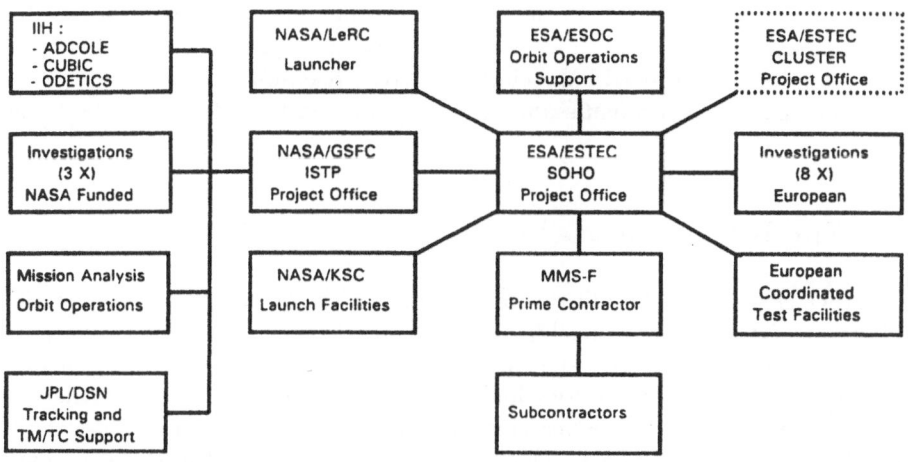

Fig. 1. SOHO mission constituents

The most demanding interfacing effort, however, was undoubtedly that between ESA and the Prime Contractor in order to cope with the large and complex payload that was finally selected for SOHO. The high number of separate experiment units -32 in all- unavoidably brought delays in the definition of exact interfaces between the instruments and the spacecraft bus. A formal control over variations and changes to the Experiment Interface Document was established at the outset. Each Experiment Team request was evaluated by the relevant Project Team experts and, if found valid, passed to the Prime Contractor for comment before a final ESA decision on implementation.

One area in particular where some difficult decisions have to be made at certain stages was the management of the experiment mass budgets. Thanks to the early introduction of formal control procedures, the original payload allocation of 610 kg needed to be increased by just 40 kg by the time that the final flight models of the experiments were delivered.

The primary preoccupation of the ESA Project Team throughout these efforts was to provide for the maximum of scientific results from the mission whilst keeping costs and schedule under strict control. A good example in this respect is the work done on minimizing the effects on short-term spacecraft pointing of moving parts of the spacecraft (reaction wheels, tape recorder) or experiments (mechanism actuation), the so-called *jitter*. This was addressed by a Pointing Review Board, periodically attended by experimenters, industry experts and the ESA Project Team; during these meetings the results of analyses (initially) were discussed and further steps proposed. This open approach led to improved balancing of SOHO's reaction wheels, to a series of tests on specific experiments and, finally, to an end-to-end test on the flight model in January 1995.

5   The NASA interface

Another element of the SOHO programme which has called for managerial effort has been the relationship with NASA. As in other cooperative programmes between the two agencies, no exchange of funds was foreseen and decisions affecting both parties had to be taken by consensus. Nevertheless, SOHO is the first such cooperative programme in which overall mission responsibility rests with ESA. This in turn has meant that the Project Team has had to interface with three separate NASA centers:
- Lewis Research Center (LeRC), for the interfaces to the launch vehicle, supplied by General Dynamics (now Lockheed-Martin) - Kennedy Space Center (KSC) for the pre-launch ground operations, and - Goddard Spaceflight Center (GSFC), which has supplied the three US-led experiments and hardware items, and provides operational support ranging from flight dynamics to the interfaces with the Deep Space Network, and the feeding of information to the Flight Operations Team (FOT). This NASA-led team of contractors operates the spacecraft from GSFC, in close liaison with the Experimenter Teams, who operate their experiments in so-called *quasi-real- time* command mode.

The multiple GSFC interfaces absorbed a considerable amount of time, particularly at management level, for the SOHO Project Team. The three US experiments are in fact among the largest and most complex of the whole

SOHO complement and their progress was therefore closely followed in the Quarterly ESA/NASA Management Meetings that have taken place during Phase C/D.

Another novel interfacing aspect has stemmed from the fact that ESA is in charge of a mission for which the operations support is provided by a team and a facility not under its full direct control. This has required specific and continuing attention, particularly in terms of the transfer of knowledge of the spacecraft systems from the Prime Contractor and ESA to the Flight Operations Team, constituted by NASA contractors from Allied Signal. Mission and Science Operations Working Groups, meeting several times per year, have been the typical forum in which these exchanges have taken place.

6 Spacecraft design drivers

The design and manufacturing of the SOHO spacecraft (Fig. 2) has been driven by several main requirements:
1) the high pointing stability at short, medium and long term
2) the cleanliness requirement, especially driven by the UV instruments
3) the autonomy requirement to be able to detect anomalous conditions, compensate them if possible, otherwise reconfigure itself to a safe configuration and be able to survive at least 48 hours out of ground control.

7 Pointing

The various requirements on pointing stability have influenced deeply the design of the AOCS and in particular the type of sensors and of the thermal control system. The cost constraints have led to a choice of an aluminium structure system which was conditional to an excellent thermal stability in case of anomalous behaviours. In order to minimize distorsions induced by variable loads, the large instruments and some of their electronic boxes have been thermally isolated from the rest of the spacecraft, and their stability left to internal control systems. The rest of the spacecraft thermal control has made use of finely tunable active elements (heaters) distributed in all important areas to compensate ageing of thermal coatings, uncertainties of experiments real dissipation vs time and effects of possible switch-off of specific elements of the payload (substitution heaters). The in-orbit performance measurements have confirmed that all pointing requirements, and in particular the requirement of two monthly intervals between thermal adjustments, have been met. Another element affecting pointing which has led to exhaustive analysis and on ground testing was the medium to high frequency structure borne vibrations (jitter) produced by mechanisms running continuously (reaction wheels) or intermittently (tape recorder, gyros, many experiments

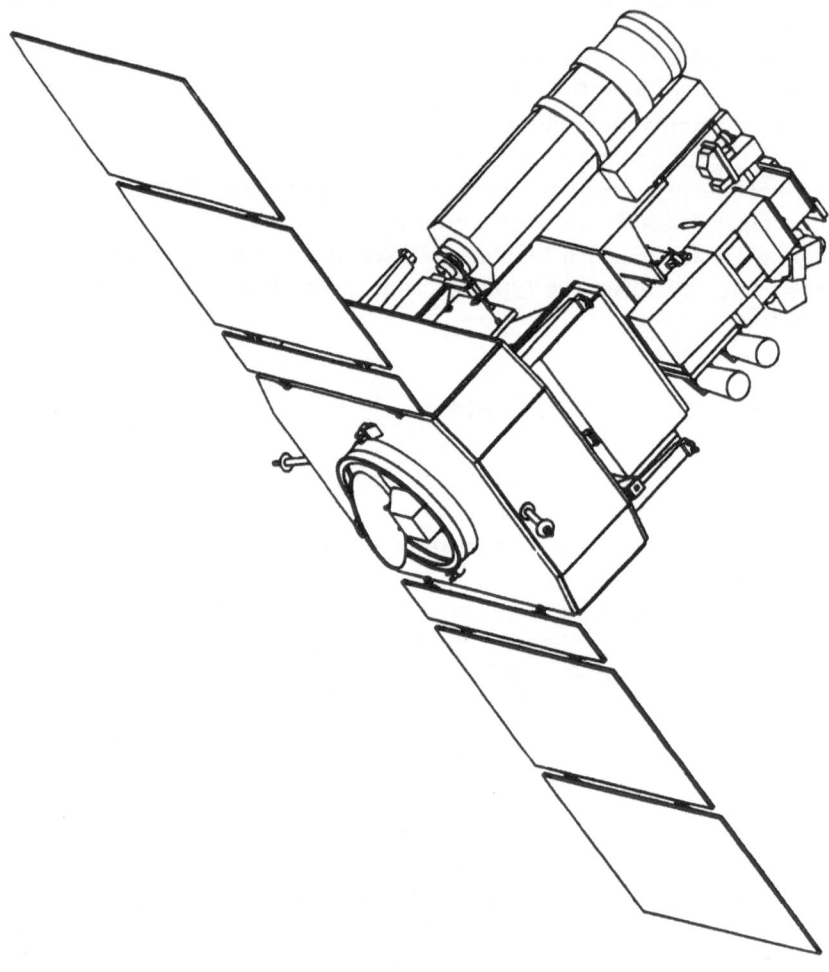

Fig. 2. The SOHO spacecraft

internal mechanisms). Certain specific measures have been implemented (extra balancing of reaction wheels, tape recorder and certain elements of the payload, in particular the UltraViolet Coronagraph Spectrometer -UVCS- revolving body), after exhaustive discussions between Scientific Teams and ESA/Prime Contractor on the best balanced approach from points of view of the scientific requirements, feasibility and cost/schedule. The excellent thermal stability shown in orbit by SOHO and the minimal jitter (measured in-orbit mainly by one internal compensating mechanism of the Michelson

Doppler Interferometer - MDI) have confirmed the soundness of the approach followed. No constraint is today imposed on the cycling of any instrument because of these requirements as was potentially envisaged on the basis of the pre-launch analysis.

8   Cleanliness

The cleanliness requirements have been fulfilled by methodical and continuous attention to the implementation of *normal* measures. A continuous nitrogen purging system has been built in the spacecraft flight model and an agreement with the launcher authority has allowed such a purge to continue in interrupted until lift-off. All payload module elements and experiments have been *baked out* before integration in the flight model to control the materials selected and to limit outgassing in orbit. Particle contamination has been controlled during integration using clean rooms up to class 100, canopies and laminar airflow tent during environmental testing and by periodic cleaning of surfaces. The lack of measurable degradation of the SOHO observation after several months of operation in orbit is a tribute to all engineers and scientists who have worked on the spacecraft for the two years of the flight model integration and testing programme.

9   Autonomy

The autonomy built in the spacecraft is linked to its intermittent daily contact with the ground stations, and the need to insure that the sun pointing attitude is kept regardless of malfunctioning and anomalous event that may affect the spacecraft or the payload. This autonomy goes from the automatic switch over of compensation heater for each instrument installed to keep pointing stable in case of switch-off of one or more experiment sensors, to the safing of the spacecraft with automatic switch over to redundant chains in case of detection of the onset of overall pointing out of limits.

10   The Payload

The SOHO mission's three principal scientific objectives are:
1) study of the solar interior, using helioseismology techniques and the measurement of the solar irradiance,
2) study of the heating mechanisms of the solar corona, and
3) investigation of the solar wind and its acceleration processes. The spacecraft's scientific payload consists of a set of state-of-the-art instruments (32 unit overall for 655 kg total mass, see Table 1., for the major parameters) developed and furnished by twelve international Principal Investigator (PI) consortia (see Table 2), involving 39 institutes from fifteen countries: Belgium,

Table 1. Major SOHO Parameters

Spacecraft Total Mass	1866 Kg
Payload Mass	655 Kg
Power	1400 W
S/C dimensions	H 3.89m x W 2.5m x L 9.52 m
(Solar Arrays deployed)	
Pointing Requirements (boresight)	short term 1"/15min
	medium term 10"/6 months
	long term 8'
Onboard Propellant system	252 kg Mono propellant-Hydrazine,
	Helium Pressurant,
	Blow-Down system 22-10 Bar.
Mission Operational Orbit	6 month Orbit around L1 Earth/Sun
	Libration Point at 1,5 Million Km.
Mission Operational Phase	29 months
Telemetry Rates	1.19 Kbps only HouseKeeping (HK)
	54.6 Kbps=5% HK+95% Science
	214.25 Kbps=1.1.% HK+20% Science+79% MDI
	or TR dump or idle.
RF Output	10 W
Telecommand Rate	2 Kbps

Denmark, Finland, France, Germany, Ireland, Italy, Japan, Netherlands, Norway, Russia, Spain, Switzerland, the United Kingdom and the United States. Nine of the consortia have been led by European PIs and three by US PIs.

Table 2. SOHO's Principal Investigators

Investigation	Principal Investigators
GOLF	A. Gabriel, IAS, Orsay (F)
VIRGO	C. Fröhlich, PMOD, Davos (CH)
MDI/SOI	P. Sherrer,Standford Univ (USA)
SUMER	K. Wilhelm, MPAe Lindau (D)
CDS	R. Harrison, RAL, Chilton (GB)
EIT	J.-P. Delaboudinière, IAS, Orsay (F)
UVCS	J. Kohl, SAO, Cambridge (USA)
LASCO	G. Brueckner, NRL, Wash.(USA)
SWAN	J.-L. Bertaux, SA, Verrières le B. (F)
CELIAS	P. Bochsler, Univ. Bern (CH)
COSTEP	H. Kunov, Univ. Kiel (D)
ERNE	J. Torsti, Univ. Turku (SF)

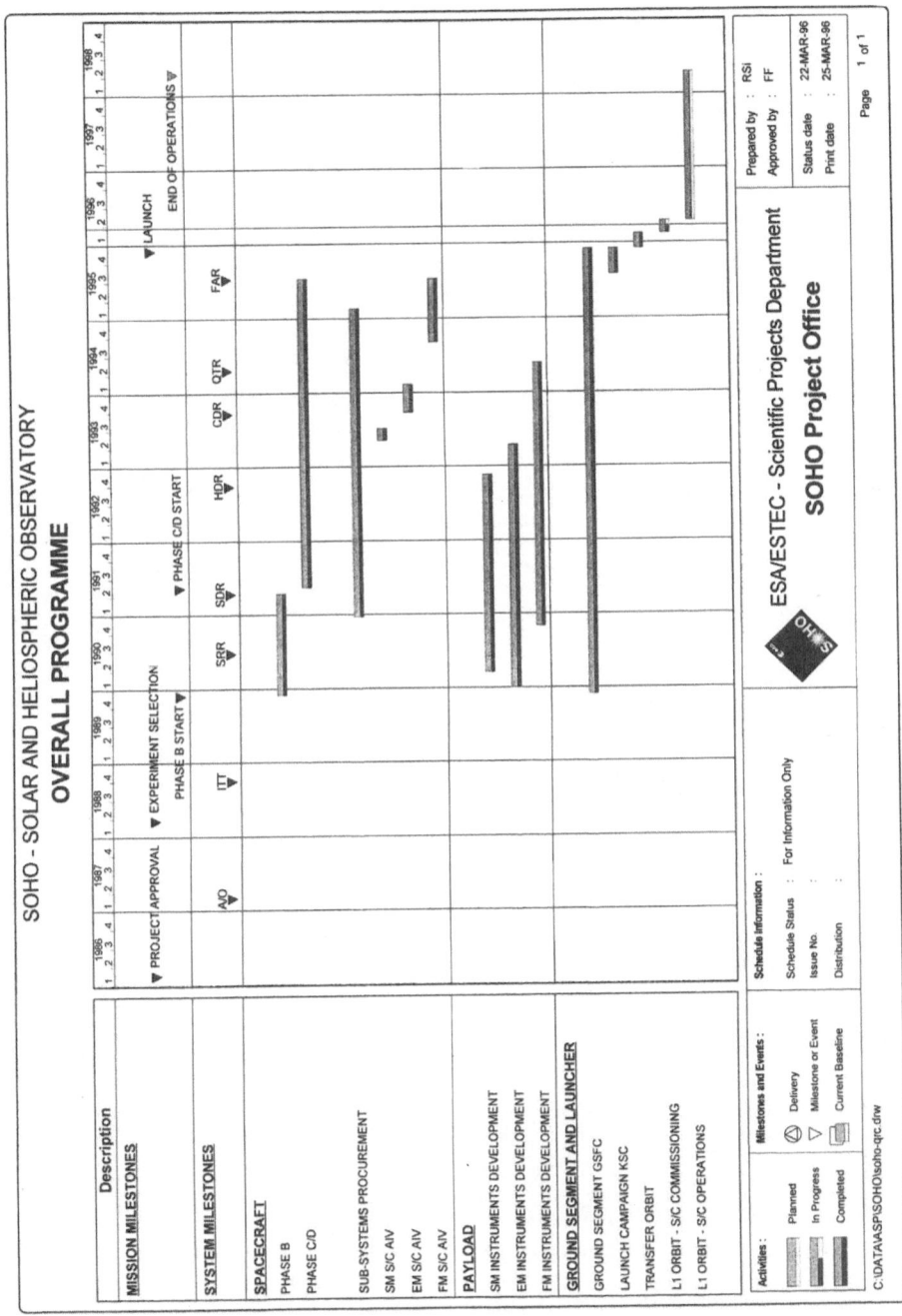

Future ESA Projects

Roger-M. Bonnet

European Space Agency, 8-10 rue Mario Nikis 75015 Paris, France

1   Introduction

The European Space Agency (ESA) is a truly international space organisation federating 14 Member States in Europe. In many respects ESA could
indeed be identified as a European NASA. However, a more focussed image
shows striking differences with the famous American agency.
ESA is a much smaller agency than NASA: in 1997 its budget was about 2.5
billion ECU[1] roughly 3 billion US$ or a fraction 1/4 or 1/5 that of NASA.
Its scientific budget is 350 million ECUs and that of NASA for the same
disciplines, 2.1 billion US$. The overall staff of ESA totals 1800 employees,
as compared with the many tens of thousands of NASA employees. Furthermore, ESA is not the only space agency in Europe. The four big Member
States (France, Germany, Italy, Great Britain) also possess their own space
agencies. The French space agency CNES, in particular, is bigger than ESA
in staff members.
ESA encompasses two types of programmes: the mandatory programme which
includes both the Science Programme and the Technology Research Programme, and optional programmes which govern the running of the following activities: Earth Observation, Telecommunications, the Ariane launcher,
Manned Space Programmes including the European contributions to the International Space Station, and Microgravity (Figure 1). Apart from its Head
Office which is located in Paris, ESA possesses several establishments (Figure
2) (ESTEC, ESOC, ESRIN, EAC), as well as a launch base in Kourou, French
Guiana, and offices in Washington, Moscow and Brussels, and several science
control centres in Europe and in the United States at the Jet Propulsion Laboratory, the Space Telescope Science Institute and the Goddard Space Flight
Center where SOHO is operated. ESTEC (1075 employees), the technical
centre, is located in Holland at Noordwijk near Amsterdam airport. ESOC
(about 270 employees), the Centre for Satellite Tracking and Operations, is
located at Darmstadt near Frankfurt in Germany.

2   The Space Science Programme: Horizons 2000

The Science Programme encompasses space astronomy, solar system exploration, solar physics, plasma physics and fundamental physics. Earth sciences

---

[1] The ECU in 1997 is equivalent to he EURO and is worth 1.29 US$.

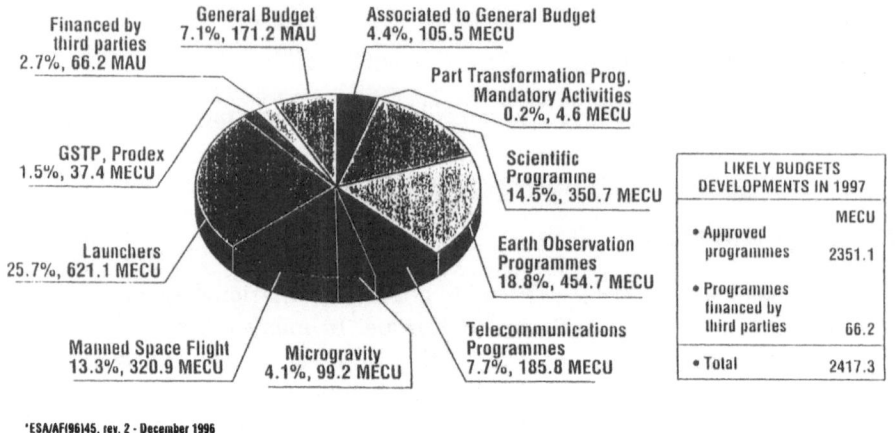

Financed by
third parties
2.7%, 66.2 MAU

General Budget
7.1%, 171.2 MAU

Associated to General Budget
4.4%, 105.5 MECU

Part Transformation Prog.
Mandatory Activities
0.2%, 4.6 MECU

GSTP, Prodex
1.5%, 37.4 MECU

Scientific
Programme
14.5%, 350.7 MECU

Launchers
25.7%, 621.1 MECU

Earth Observation
Programmes
18.8%, 454.7 MECU

Manned Space Flight
13.3%, 320.9 MECU

Microgravity
4.1%, 99.2 MECU

Telecommunications
Programmes
7.7%, 185.8 MECU

LIKELY BUDGETS DEVELOPMENTS IN 1997	
	MECU
• Approved programmes	2351.1
• Programmes financed by third parties	66.2
• Total	2417.3

'ESA/AF(96)45, rev. 2 - December 1996

MECU : Million of ECU (1 ECU = 0.84 British Pounds, 1.29 US Dollars, 1.76 Canadian Dollars, Rates applicable to 1997 budgets)

Fig. 1. The 1997 ESA Budgets for approved programmes + programmes financed by third parties - Total 2417,3 MAU

are not included and are part of the optional activities. The budget of the Science Programme is fixed every 3 years within a so-called Level of Resources which covers a period of five years. The voting of the Level of Resources requires unanimity of all the Member States. Inside the Level of Resources the budget is voted every year by a two-thirds majority. The five year continuity of the Level of Resources gives the programme a stability which is unique and which ensures the elaboration of a secure medium term plan. Since it is mandatory, and as such the very nucleus (or backbone) of all the Agency's activities, it also allows the setting up of a long-term plan.

The Science Programme is a legacy of ESRO, the organisation for Space Research in Europe, which merged into ESA in 1973-1974. For the last 30 years it has demonstrated the benefits of a truly international cooperation, as well as Europe's originality in space science. Today it provides the links which bind together the various scientific as well as industrial communities in the Member States. The future of European space science is built and reposes on this unique asset.

ESA's long-term plan in space science is called Horizons 2000 (Figure 3). Its first set of missions, the original Horizon 2000 programme, was established in 1984 and included missions to cover a 20-year period. Ten years later, in 1994, a new set of missions was defined and formed the Horizon 2000 Plus extension covering the period 2005-2016. Horizons 2000 is based on a set of cornerstones and medium to small missions. The cornerstones are identified a priori, hence 20 years before their latest launch date. This length of time is necessary to develop the key technologies required to implement the mis-

Fig. 2. ESA world locations.

sions. This is a very long time indeed which is the consequence of the very low financial yearly level of the Technology Programme at ESA (90 MECU for all ESA activities, as compared with the 500 MUS$ of NASA's space science budget alone). This long time is certainly a disadvantage which needs to be corrected in the implementation of future missions. Besides the cornerstones which ensure the stability of the programme, the medium and small missions ensure mobility and flexibility. All these elements permit a certain balance between the scientific disciplines reflecting the strength and areas of excellence of the European space science community. Cornerstones are allocated a budget which represents 2 years of the annual budget (700 MECU) and, for medium missions, a maximum of one year.

The four cornerstones (CS) of Horizons 2000 are:

- CS.1: Solar Terrestrial Science Programme (SOHO-Cluster to be launched in 2000).
- CS.2: The X-Ray Multimirror Mission (XMM to be launched in 1999).
- CS.3: The Comet Mission (Rosetta to be launched in 2003).
- CS.4: The Far Infrared Space Telescope (FIRST to be launched in 2005-2006).

In addition, Horizon 2000 Plus identifies three new cornerstones which could be launched as of 2009:

- a mission to Mercury
- an astrometric or infrared interferometer
- a gravitational wave laser antenna (LISA).

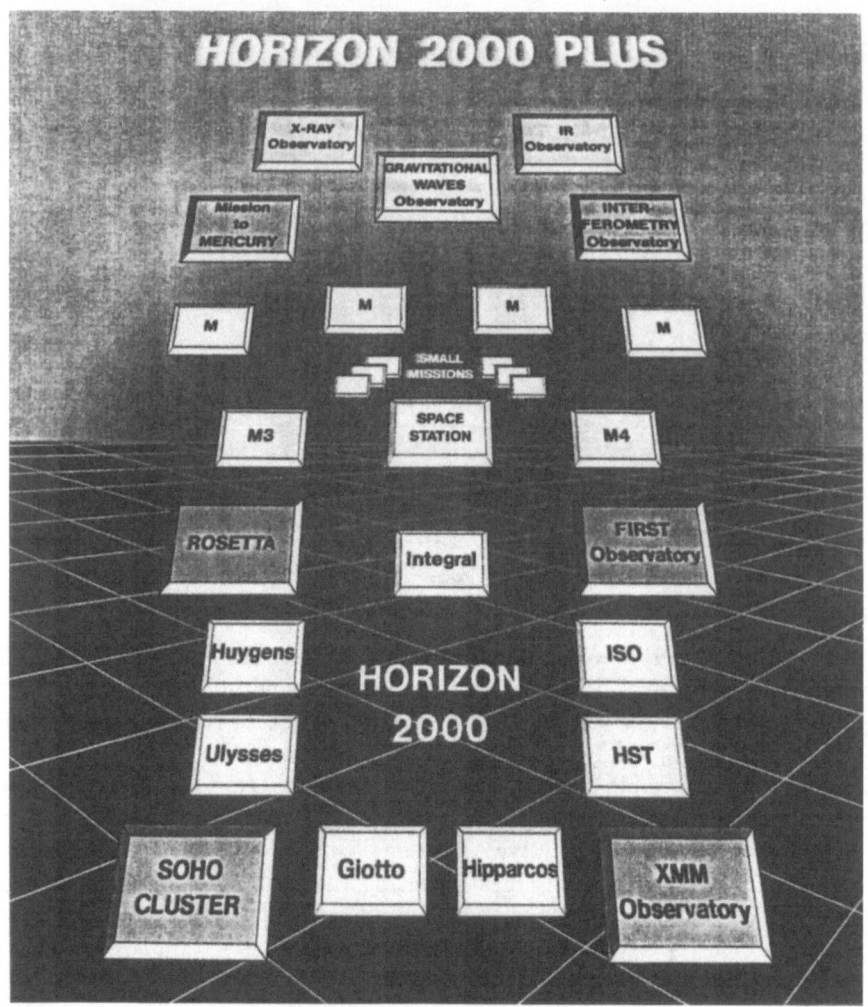

Fig. 3. Horizon 2000 Plus

The success of the Horizons 2000 programme is not only scientific: it placed Europe in a leading position in several areas of space science, and it also managed to convince the Member States to grant a yearly 5% increase in real terms for a period of 10 years. Unfortunately, in 1995, the ESA Council (meeting at ministerial level), even though it unanimously endorsed the content of the Horizon 2000 Plus extension, decided not to compensate the budget for inflation below 3%, resulting in a net loss of 3% in purchasing power until at least 1998 and possibly beyond.

3   The Astronomy Programme

With the past and future missions in Horizons 2000, ESA has acquired or is gaining indisputably dominant positions in high resolution astronomy (Hubble Space Telescope and Next Generation Space Telescope), in particular in astrometry (Hipparcos, GAIA), in high energy astrophysics (COS-B, EXOSAT, XMM and Integral), and in infrared astronomy (ISO, Planck Surveyor, FIRST, space interferometry).

3.1   Astrometry

ESA's astrometry mission, Hipparcos, was launched in 1989 and, although in an incorrect orbit, following the malfunction of its apogee boost motor, operated flawlessly for four uninterrupted years until August 1993. Its recently published catalogue gives positions, parallaxes (distances) and proper motions for 118 000 stars with an accuracy of 10-3 arc seconds, twice what was expected, and up to two orders of magnitude better than anything achieved from the ground. These measurements constitute an absolute reference for astrophysicists and cosmologists world-wide, renewing one of the oldest disciplines of astronomy and placing it under clear European leadership.

Based on this success, Horizons 2000 Plus has proposed a cornerstone mission in astrometry, called GAIA, based on the use of interferometry techniques and aiming at a resolution of a few micro arcseconds, 100 times better than Hipparcos, for several millions of stars in our galaxy and in several nearby galaxies (Figure 4). GAIA could fly as early as 2009 using an Ariane 5 rocket.

3.2   High Energy Astrophysics

Clearly ESA has been a pioneer in the field of gamma-ray astronomy with COS-B, launched in 1975, establishing the first map of gamma-ray emissions in our galaxy, and EXOSAT, a high resolution X-ray satellite for imaging and spectroscopy, launched in 1983. When it is launched in 1999, XMM, the second cornerstone, will be the largest X-ray telescope in 1 KeV. XMM will provide medium to high resolution spectroscopy and imaging with a few arc seconds resolution for the weakest objects in the X-ray sky between 0.1 KeV and 10 KeV.

The satellite used for XMM provides an excellent standard bus which will be used for the other main high energy mission, INTEGRAL, a high angular gamma-ray observatory, operating between 15 KeV and 10 MeV for high resolution spectroscopy and imagery. With its full high sensitivity, INTEGRAL will place Europe in a leading role in gamma-ray astronomy (Figure 5). The mission is of the size and ambition of a cornerstone but the re-use of the XMM bus allows its budget to remain in the category of a medium mission. INTEGRAL will be launched in 2001 by a Russian Proton rocket in the framework of a cooperative programme with Russia.

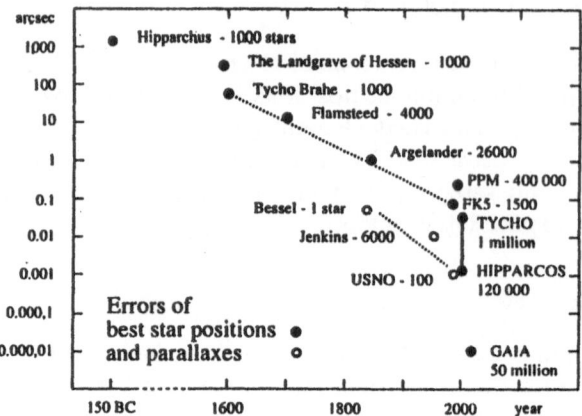

Fig. 4. Errors of star positions in the most accurate catalogues. Tycho Brahe achieved a jump in accuracy through the first "big science" in history. After four centuries with more gradual improvement another much larger jump in accuracy is obtained by the HIPPARCOS and TYCHO catalogues containing a total of one million stars. GAIA will improve these measurements by 2 orders of magnitude.

3.3   Infrared Astronomy

Until recently, this area of science was entirely dominated by the US, which launched the first two infrared explorer class missions : IRAS, a survey mission in 1983, and COBE, in 1991, to study the cosmic background radiation. The situation changed in 1995 with the very successful launch by an Ariane 4 rocket of ISO, ESA's Infrared Space Observatory, entirely developed in Europe (NASA contributes a second ground station to cover the entire 24 h orbit). With its 60 cm telescope and set of modern sensitive detectors, ISO has a sensitivity several orders of magnitude higher than IRAS. ISO also extends the spectral coverage of IRAS (2 to 80 mm) to 200 microns. This part of the spectrum is ideal to study the cold interstellar medium and the formation of stars and planetary systems. The four focal plane instruments (2 spectrometers, 1 photometer and an imaging camera) are used by a large number of astronomers throughout the world, in particular in the US which benefits from 20% of the observing time. Among its many exclusive discoveries, ISO has detected water in nearly all regions of space and a large number of molecules never observed before.

ISO also places Europe in a position of leadership. This position is reinforced by the fourth cornerstone of Horizons 2000, the Far Infrared Space Telescope, FIRST, now under pre- development at the technology level. With its 3.5 m telescope, and cryogenically cooled set of cameras, spectrometers and photometers, FIRST will allow the observation of the last unexplored region of the electromagnetic spectrum between 200 and 600 microns. This

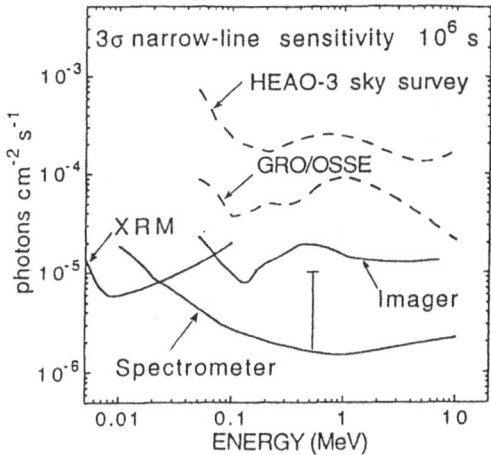

Fig. 5. Integral spectrometer, Imager and X-ray monitor sensitivity as compared with previous instruments.

region is particularly suited for the observations of highly red-shifted objects and would therefore allow the study of galaxy formation as close in time to the Big Bang as possible.

FIRST will be launched in 2006 or 2007 by an Ariane 5 and placed in an orbit around the second Lagrangian point L2 at 1.5 million km from the Earth.

The third medium size mission of Horizons 2000 will also be an infrared mission: Planck -Surveyor was selected in 1996 and is a second or third generation mission after COBE. It aims at observing the spatial distribution of the cosmic background radiation with an angular resolution of 10 arcmin nearly two orders of magnitude higher than COBE. This will allow the observation of brightness fluctuations which are specific to different models of the early Universe, and to discriminate between those (Figure 6). Planck will also be placed in orbit around L2. The coincidence with the orbit of FIRST has induced a study of the possible combination of both missions on the same spacecraft. Such an option, which is actively studied at this moment, would obviously lead to large savings in Horizons 2000. The choice of the final option will be made in 1998. This is why the launch of Planck cannot be given precisely at this time and could be as early as 2004 in a stand alone mission or in 2006 if launched with FIRST.

Last but not least, Horizons 2000 Plus has identified an infrared interferometry cornerstone, opening the way to very high resolution infrared astronomy. This mission is presently being studied and would be in competition with GAIA for a launch slot which could be as early as 2009-2010. High resolution infrared astronomy will open the very exciting area of terrestrial planets detection around other stars.

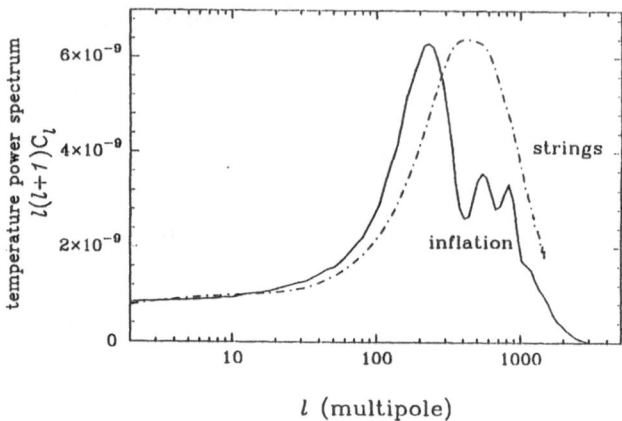

Fig. 6. The Power Spectrum of the Cosmic Microwave Background in a String-seeded model.

3.4   High Resolution Astronomy

With Hipparcos, GAIA, the infrared interferometer, ESA has played and will play a major role in high resolution astronomy. It should be recalled that Europe, through ESA, already plays a substantial role in the utilisation of the Hubble Space Telescope, a joint project with NASA. Discussions are on-going between the two agencies to continue to jointly operate and exploit Hubble until at least 2005. Beyond 2005, ESA and NASA are also discussing the joint development of the successor to Hubble, the so-called Next Generation Space Telescope (NGST), which NASA envisages to launch around 2007.

4   Planetary Exploration and Solar Physics

Planetary exploration has certainly been victim of the decisions taken back in the early 1970s by ESA and ESRO advisers not to develop missions in this field then largely dominated by the Americans and the Soviets. The Europeans could participate as guests in the US and Soviet missions, but never in a position of leadership. It was only in 1980, with the decisions to launch a spacecraft to Halley's comet and with Horizons 2000, that this unfortunate trend could be corrected. Now, ESA also occupies leading positions in this exciting area of space science.

4.1   Cometary Exploration

With the launch of the Giotto mission in 1985, and its successful flyby of Halley's comet, in 1986, at a distance of only 600 km, Europe with ESA regained an undisputable position of leadership. Giotto obtained the first high resolution images of a comet nucleus ever, and made the first in situ structural and chemical analysis of the dust and gas emission of the comet at very close distance to the nucleus. The observations of Giotto confirmed unambiguously the common origin of comets and interstellar clouds in the vicinity of the Solar System. In spite of the serious damage suffered by Giotto in crossing through the coma of Halley, the spacecraft, after 6 years of hibernation, could be reactivated and observe another comet, Grigg-Skjellerup in July 1992 at only 200 km distance.

This leading position led ESA naturally to include a cornerstone in Horizons 2000 dedicated to cometary exploration. Rosetta, presently under development in industry; will be launched in 2003 and will reach comet Wirtanen 8 years later. The spacecraft will orbit the nucleus and study its activation while the comet approaches perihelion, measuring its dust and molecular composition. It will also drop a small lander on the nucleus allowing the analysis of cometary material in situ.

4.2   Planetary Exploration

The first and very crucial step for ESA in planetary exploration, will be the landing on the surface of Titan in 2004 of the Huygens probe. The probe will be carried on board the US mission Cassini which will study the rings and satellite of Saturn as well as the giant planet itself (Cassini was launched on October 15, 1997 on board a Titan IV rocket). Titan is the largest of Saturn's moons. It is surrounded by a thick atmosphere of nitrogen with traces of ethane and methane. Ultraviolet radiation from the Sun, as well as a large lightning activity and bombardment by highly energetic electrons induces the formation of organic matter in the upper atmosphere. This prebiotic material has been falling like rain on the surface of Titan since the formation of the Saturnian system, 4,5 billions years ago. This material has accumulated on the surface of Titan and one of Huygens's tasks will be to study its thickness and composition as well as the atmosphere itself, and to take images of the lands of what is called an "Earth in the refrigerator" (its surface temperature is -180° C).

After its 7 year journey to Saturn through the interplanetary medium, Cassini will drop Huygens into Titan's atmosphere. It will land slowly after a 2 hour descent, thanks to a set of three parachutes (Figure 7). It will communicate with the Earth through Cassini, in contact with the probe for no more than 30 minutes. The probe is powered by its own batteries whose autonomy will not limit the communications time. When it lands on the surface of Titan, Huygens will have accomplished the most remote landing ever achieved by mankind.

Horizons 2000 has selected two other planets of particular interest for European scientists: Mars and Mercury. Mars is an obvious target and has been considered as a priority by the Survey Committee which framed Horizons 2000. However, no specific missions were identified for the exploration of the red planet. This was left to the normal competitive selection process, in the framework of medium missions. A first attempt failed when the Planck Surveyor was selected against Intermarsnet, a mission to land a network of seismic and atmospheric stations. Intermarsnet which was supposed to be a cooperative venture with NASA fell victim to the American agency's plans which were incompatible with a launch date of 2005 for the mission. That eliminated Mars from the competition at that time.

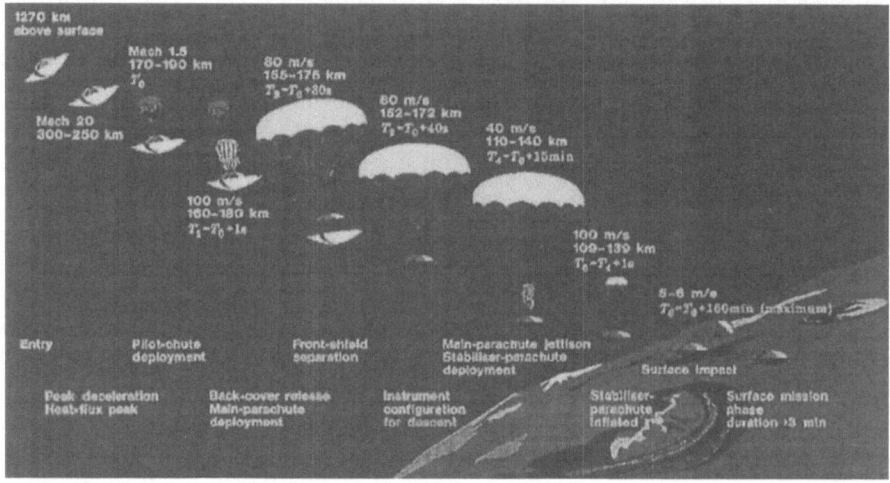

Fig. 7. Entry and descent scenarios of Huygens in the atmosphere of Titan.

However, the American approach to the exploration of Mars and the strong priority of Horizon 2000 for that planet led ESA's Space Science Advisory Committee to propose a small Mars mission for launch in 2003. The mission, called Mars Express, would be compatible with a budget of 150 MECU and would carry an orbiter equipped with sets of instruments, mostly spares from the lost Russian Mars-96 mission, together with a sounding radar to probe the underground water reserves and the permafrost, looking for possible places where to search for life or fossil forms of it. In addition, the orbiter may carry enough weight to land 150 kgs of instruments on the surface, including robots, seismic or even biological stations. The final decision to start Mars Express will be taken in 1998, providing ESA's science budget stops decreasing. Mercury on the contrary has been selected as a target for a cornerstone mission. The least known planet in the Solar System (apart from Pluto, impossible to reach by a European mission because of the unavailability in Europe of radioactive power generators, RTG's) would be a scientifically challenging target for both the planetary community and the plasma physicists who are intrigued by the presence of a magnetosphere around a planet which has no intrinsic magnetic field ! The Mercury mission could be launched with an Ariane 5 in 2009.

4.3   Plasma Physics and Solar Physics

Plasma Physics has always been a very strong component of the space science community in Europe. Several missions have been undertaken since the early years of ESRO including GEOS-1 and GEOS-2 and ISEE-2 which investigated the Earth magnetosphere and ionosphere. Europe's role is now culminating with the Solar Terrestrial Science Programme (STSP), the first cornerstone of Horizons 2000 made up of 2 key missions, Cluster and, of course, SOHO.

Cluster (Figure 8) is a fleet of four identical satellites to study the time and space fluctuations of the Earth's magnetosphere in three dimensions, as they are influenced by the solar wind and coronal mass ejections, or for solar activity in general. The 4 Clusters were launched by the first Ariane 5 on 4 June 1996. Unfortunately, the spectacular failure of the rocket put a premature end to the mission. After several months of difficult negotiations, a replacement mission could be decided. Cluster II, which is exactly identical to its defunct predecessor, will be launched in 2000 by two consecutive flights of Soyuz, a launcher commercialised by the French-Russian Starsem company. When Cluster II flies in 2000, SOHO will hopefully still be in operation, close to solar maximum. Hence, the main objectives of STSP, with Cluster, SOHO and Ulysses, the solar polar mission, flying at the same time, will be fulfilled.Solar Physics is also an area of scientific priority in Horizons 2000. SOHO, the most successful solar physics mission ever in the ESA programme, together with Ulysses (both joint ventures with NASA) clearly place Europe in a forefront position, as has been amply demonstrated in the course of this summer school. It will not be easy to define the next solar physics mission after SOHO and several candidates are being envisaged at this moment. Will it be a solar probe, to study the Sun and its corona in situ ? Will it be a stereoscopic mission to study solar structures in three dimensions from 2 or more spacecraft placed at wide solar angles on the Earths orbit ? Will it be a super high resolution imaging mission ? All three options are presently being investigated. The earliest time for launch is 2007-2008. Figure 9 illustrates the trajectory of a possible solar probe studied by ESA named Vulcan.

Fig. 8. The four Cluster satellites in the clear room at Kourou waiting for their launch on the first Ariane 5 launcher in 1996.

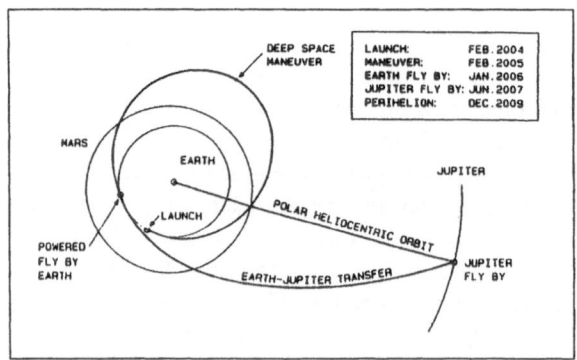

Fig. 9. Vulcan's ballistic mission profile.

5   Fundamental Physics

In the process of formulating Horizons 2000 Plus, a new and very active part of the science community manifested itself for the first time, providing a large number of high quality mission concepts in the field of fundamental physics, including general relativity and gravitational waves astronomy. As a result, Horizons 2000 now includes two missions in these disciplines.

The first one, called MiniStep, a joint venture with NASA , will test the Equivalence Principle at the level of 10-17 , with a small satellite (hence its name). The mission is foreseen for launch in 2004. By far, the most visionary mission of Horizons 2000 is LISA, a laser interferometer to directly detect gravitational waves of low frequency, away from all earth gravity perturbations. LISA (Figure 10) will be based on a set of 3 satellites orbiting at 20° from the Earth and following it, forming a triangle of 5 millions km side. Such an ambitious mission would be able to detect gravitational waves from massive black holes and neutrons stars as well as the (hypothetical) background of cosmic gravitational wave radiation which, as a remnant of the Big Bang, should pervade the entire Universe as does the microwave background. Since gravitational waves are decoupled from matter, as opposed to photons which are trapped in the too dense Universe until 300 000 years after the Big Bang, LISA should allow the observations of phenomena which occurred right at the moment of the Big Bang.

This mission is very demanding technologically as well as from the point of view of tracking 3 satellites so far away from the Earth and measuring their distances with an accuracy better than the size of a hydrogen atom ! Studies will soon start on LISA and its launch could be as early as 2009-2010.

6   Conclusions: The Challenge of Implementing Horizons 2000 !

So far, we have dealt only with the ESA projects as framed within the Horizons 2000 Programme. Of course, a myriad of national activities also exists within each Member State of ESA. In principle, these national programmes should complement Horizons 2000 rather than competing with it ! European coordination is part of the ESA Convention and the existence of a reference programme like Horizons 2000 has certainly greatly contributed to ensuring that the Member States and ESA do not duplicate their efforts. The added value of the European programme is clearly its continuity.

Continuity is not assured, even at medium term range, at national level. Hence, a fortiori, Europe's only long term plan in space science is the ESA plan ! However, there should be much more rationalisation between the national and the ESA activities so that their addition would result in a programme of critical mass when compared with the equivalent NASA science programme. How to avoid further division or fragmentation of space science activities in Europe is a genuine challenge, especially when comparing, as said in the introduction, the relatively modest European financial effort, with its US equivalent.

This is even more true in the wake of the recent decisions taken by European Ministers in 1995 in Toulouse who, although they gave their full support to Horizons 2000, did not agree to allocate the required financial level, essentially flat in purchasing power until 2002. This shortage of money, together with the need to replace Cluster (see section 4.3), have forced an in depth reanalysis of the implementation of Horizons 2000 and of the working methods traditionally used by ESA in the management of its space science missions.

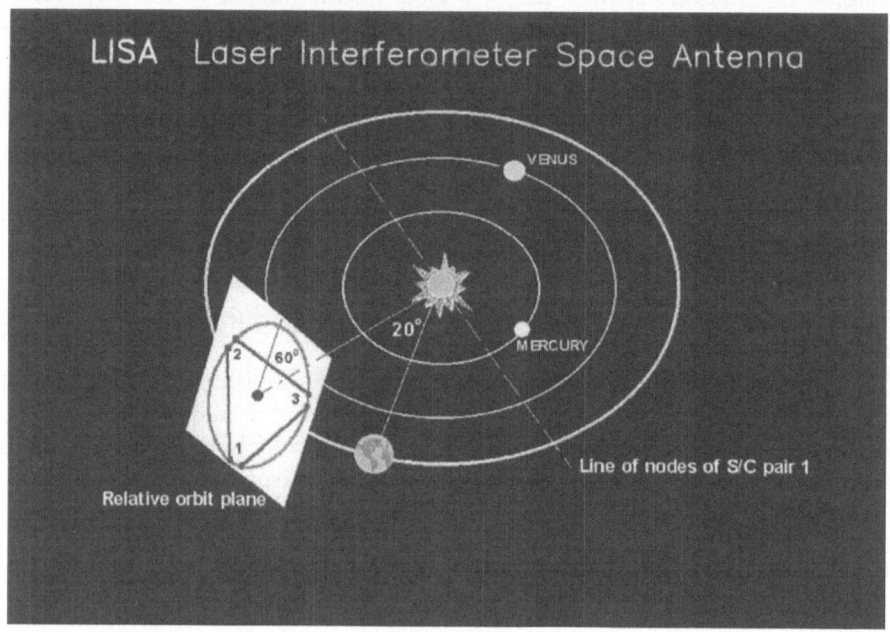

Fig. 10. LISA. The Laser Interferometer Space Antenna to detect the low frequency (1-10-4Hz) gravitational waves.

This is obviously necessary when looking at NASA's approach followed in the past three or four years, leading to develop missions much faster (three or four years instead of 10) and at a fraction of their previously foreseen cost (less than 200 MUS$ instead of 700). NASA is apparently facing this challenge with some success, as is well illustrated by the two recent successful Mars missions : the Mars Pathfinder and the Mars Global Surveyor (replacing Mars Observer lost in 1993 in the vicinity of Mars).

The American approach is largely based on an intensive and well focussed technological preparation. Not only do NASA managers and scientists benefit from the tremendous technological effort of the US Department of Defence, in particular in the Star Wars programme, but also, from within NASA's space science budget where, from a budget of more than 2 billions US$ (in 1997) some 500 MUS$ are assigned to advanced technologies and in orbit demonstrations. This tremendous effort, to be compared to an equivalent total of about 90 M$ for the whole of ESA, shows the imbalance and the strong disadvantage of Europe when compared to the largest space agency in the world. Under these circumstances can ESA, and Europe, be competitive ?

This has only been possible for ESA in being much more selective in its priorities, which means that the balance of the programme can only be secured across main disciplines but certainly not in each subdiscipline. The complementarity with national programmes may help in this respect : ESA programming the reference mission in each discipline, and national programmes undertaking more focussed missions in each subdiscipline.

It is with all the above considerations in mind that Horizons 2000 has been re-analysed in the course of 1997. Its scientific content has in no way been put into question but the management and the implementation of its missions have been reviewed in depth.

First of all, savings were identified from all ongoing projects and a 10% reduction was imposed on all science expenditures, thereby opening the possibility of minimizing the impacts of introducing Cluster II and a small Mars mission into the programme. In parallel, completely new -and sometime revolutionary- management methods, have been identified which could lead to additonal savings. Developing and demonstrating in flight key technologies may also induce savings, in particular for cornerstone missions, as well as increasing the total number of missions.

The new scenario shown in Figure 11 does re-introduce the balance in the programme and open the possibility for ESA to participate in the exploration of Mars as early as 2003 (corresponding to the best launch opportunity between now and 2010) and in the NGST as early as 2007, at the level of 150 MECU and 175 MECU respectively. This assumes however that the budget will stop decreasing after 1998 and will remain constant for 5 years onwards with a slight increase afterwards. It also assumes that, through a common management structure, and possibly a merging of Planck Surveyor and FIRST (section 3.5), savings equivalent to the cost of a medium mission (350 MECU) can be achieved. Furthermore, the concept of Small Missions for Advanced Research in Technology (SMART) has been introduced. Four such missions, costing no more than 50 MECU to the science budget can be programmed between 2001 and 2009. For example, on the assumption that a test of Solar Electric Propulsion is successful on SMART-1, it would be possible to launch the Mercury cornerstone practically at any time and not only in January 2009.

The basic assumption of this scenario is that the decline of the budget of the Science Programme will stop in 1998 and a constant budget in real terms will be agreed until 2002. More pessimistic assumptions will affect ESA's pro-

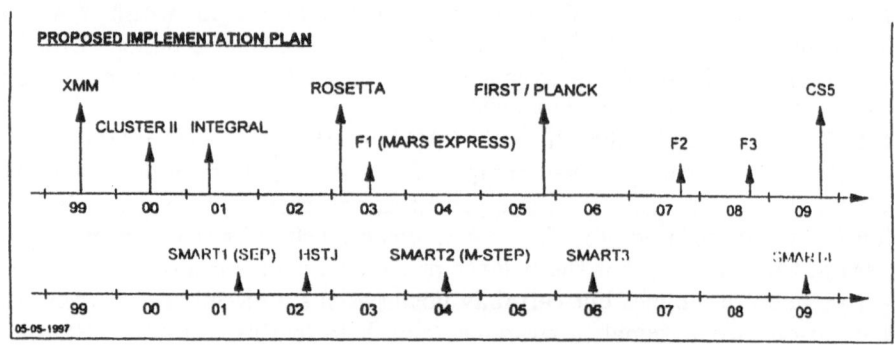

Fig. 11. New Horizons 2000 Implementation scenario.

posed involvement in the exploration of Mars and most probably the schedule of FIRST-Planck. This would be a shame, and the science community of Europe would have to resign from any attempt to maintain its role. It would no longer be a leader, but a follower for a long long time !

References : Details of all missions and of the Horizons 2000 content can be found on:
http ://www.estec.esa.nl/spdwww/

Lecture Notes in Physics

For information about Vols. 1–469
please contact your bookseller or Springer-Verlag

New Series m: Monographs